Fuzzy Impulsive Dynamic Equations on Time Scales

This book presents the main tools for investigations of the existence and uniqueness, as well as the existence, of multiple solutions for initial—and boundary—value problems for fuzzy impulsive dynamic equations on time scales.

Time-scale theory is relatively new. The basic theory attempts to unify both approaches of dynamic modeling: difference and differential equations. Similar ideas have been used before and go back to the introduction of the Riemann-Stieltjes integral, which unifies sums and integrals. Many results in differential equations easily carry over to the corresponding results for difference equations, while other results seem to be totally different in nature.

For these reasons, the theory of dynamic equations is an active area of research. The time scale calculus can be applied to any field in which dynamic processes are described by discrete or continuous time models. The calculus of time scales has various applications involving non-continuous domains such as certain insect populations, phytoremediation of metals, wound healing, maximization problems in economics, and traffic problems.

Fuzzy Impulsive Dynamic Equations on Time Scales is intended for researchers and students in engineering and science. The eight chapters in the book are organized in a way that is pedagogically accessible. Each chapter concludes with a section on practical problems to develop further understanding.

Advances in Applied Mathematics
Series Editor: Daniel Zwillinger

Quantum Computation
Helmut Bez and Tony Croft

Computational Mathematics
An Introduction to Numerical Analysis and Scientific Computing with Python
Dimitrios Mitsotakis

Delay Ordinary and Partial Differential Equations
Andrei D. Polyanin
Vsevolod G. Sorkin
Alexi I. Zhurov

Clean Numerical Simulation
Shijun Liao

Multiplicative Partial Differential Equations
Svetlin Georgiev and Khaled Zennir

Engineering Statistics
A Matrix-Vector Approach with MATLAB®
Lester W. Schmerr Jr.

General Quantum Numerical Analysis
Svetlin Georgiev and Khaled Zennir

An Introduction to Partial Differential Equations with MATLAB®
Matthew P. Coleman and Vladislav Bukshtynov

Handbook of Exact Solutions to Mathematical Equations
Andrei D. Polyanin

Introducing Game Theory and its Applications, Second Edition
Elliott Mendelson and Dan Zwillinger

Modeling Operations Research and Business Analytics
William P. Fox and Robert E. Burks

Decision Analysis through Modeling and Game Theory
William P. Fox

Advances in High-Order Predictive Modelling
Dan Gabriel Cacuci

Introduction to Financial Mathematics, Second Edition
Kevin J. Hastings

General Quantum Variational Calculus
Svetlin G. Georgiev and Khaled Zennir

Advanced Engineering Mathematics with MATLAB®, Sixth Edition
Dean G. Duffy

Exact Methods for Non-linear PDEs
Andrei D. Polyanin

Fuzzy Impulsive Dynamic Equations on Time Scales
Svetlin G. Georgiev and Khaled Zennir

https://www.routledge.com/Advances-in-Applied-Mathematics/book-series/CRCADVAPPMTH?pd=published,forthcoming&pg=1&pp=12&so=pub&view=list

Fuzzy Impulsive Dynamic Equations on Time Scales

Svetlin G. Georgiev and Khaled Zennir

CRC Press
Taylor & Francis Group
Boca Raton London New York

CRC Press is an imprint of the
Taylor & Francis Group, an **informa** business
A CHAPMAN & HALL BOOK

First edition published 2026
by CRC Press
2385 NW Executive Center Drive, Suite 320, Boca Raton FL 33431

and by CRC Press
4 Park Square, Milton Park, Abingdon, Oxon, OX14 4RN

CRC Press is an imprint of Taylor & Francis Group, LLC

ISBN: 978-1-041-10060-7 (hbk)
ISBN: 978-1-041-10497-1 (pbk)
ISBN: 978-1-003-65536-7 (ebk)

DOI: 10.1201/9781003655367

Typeset in Nimbus Roman
by KnowledgeWorks Global Ltd.

Publisher's note: This book has been prepared from camera-ready copy provided by the authors.

Contents

Preface

Time-scale theory was first initiated by Stefan Hilger in 1988 in his PhD thesis to unify both approaches of dynamic modeling: difference and differential equations. Similar ideas have been used before and go back to the introduction of the Riemann-Stieltjes integral, which unifies sums and integrals. Many results in differential equations easily carry over to the corresponding results for difference equations, while other results seem to be totally different in nature.

For these reasons, the theory of dynamic equations is an active area of research. The time scale calculus can be applied to any field in which dynamic processes are described by discrete- or continuous-time models. So, the calculus of time scales has various applications involving non-continuous domains such as certain bug populations, phytoremediation of metals, wound healing, maximization problems in economics, and traffic problems.

In *Fuzzy Impulsive Dynamic Equations on Time Scales*, the main tools for investigations of the existence and uniqueness of the solutions as well as the existence of multiple solutions for initial- and boundary-value problems for fuzzy impulsive dynamic equations on time scales are studied. This book is devoted to the qualitative theory of fuzzy impulsive dynamic equations on time scales. The book is intended for undergraduate students and undergraduate students in engineering and science courses. There are eight chapters in this book. The chapters in the book are organized in a way that is pedagogically accessible. Each chapter concludes with a section on practical problems.

In Chapter 1, a brief history of fuzzy logic is given. Some areas of mathematics based on fuzzy logic are discussed. Fuzzy logic applications are considered in engineering, physics, chemistry, biology, economics, and computer science. Chapter 2 deals with fuzzy functions on time scales. We focus on a new class derivative and a new class integral of such fuzzy functions. The corresponding fundamental properties of the introduced derivative and integral are studied and discussed. Chapter 3 is devoted to the existence and stability of the solutions of first-order fuzzy impulsive dynamic equations. In Chapter 4, some boundary value problems for first-order fuzzy impulsive dynamic equations are investigated. The existence of solutions of second-order fuzzy impulsive dynamic equations is considered in Chapter 5. Chapter 6 deals with some boundary value problems for second-order fuzzy impulsive dynamic equations. Chapter 7 is devoted to fuzzy impulsive dynamic equations on time scales. They are given criteria for oscillations of first-order, second-order, and higher-order fuzzy impulsive dynamic equations. In Chapter 8, linear fuzzy impulsive dynamical systems are introduced. Some criteria for the stability of the solutions of homogeneous and non-homogeneous linear fuzzy impulsive dynamical systems are deduced.

This book is addressed to a wide audience of specialists, such as mathematicians, physicists, engineers, and biologists. It can be used as a textbook at the graduate level and as a reference book for several disciplines.

August 2025

Svetlin G. Georgiev (Paris)
Khaled Zennir (Riyadh)

CHAPTER 1

An Introduction to Fuzzy Logic

In this chapter, a brief history of fuzzy logic is given. Some areas of mathematics based on fuzzy logic are discussed. Fuzzy logic applications are considered in engineering, physics, chemistry, biology, economics, and computer science.

1.1 HISTORY OF FUZZY LOGIC

Lukasiewicz and Lesniewski founded in the twenties of the XXth century a school of logic in Warsaw that became one of the most important teams in the world. Lukasiewicz introduced the idea of a uniform distribution of truth values in the interval $[0,1]$. If the values of n are considered, then

$$0, \quad \frac{1}{n-1}, \quad \frac{2}{n-1}, \quad \ldots, \quad \frac{n-2}{n-1}, \quad 1,$$

are possible truth values. If there is an infinite number of truth values, then we should take $\mathbb{Q} \cap [0,1]$. The negation is defined as follows:

$$n(x) = 1 - x, \quad x \in [0,1],$$

and

$$x \oplus y = \min\{1, x+y\}, \quad x, y \in [0,1],$$

is also fixed.

DOI: 10.1201/9781003655367-1

In the line of Lukasiewicz studies, in [44], fuzzy sets were introduced in 1965 in his work "Fuzzy Sets". Because of this work, Zadeh is acknowledged as the founder of fuzzy logic. The ideas on fuzzy sets were soon applied to different areas such as artificial intelligence, natural language, decision making, expert systems, neural networks, control theory, and so on.

Intuitively, a classical set is any collection of definite and distinct objects that are conceived as a whole. Objects that are included in a set are usually called its members. Any classical set must satisfy two requirements. First, members of each set must be distinguishable from each other. Second, any given object is either a member of the set or not. Any classical set has a sharp boundary that separates objects that are members of the set from those that are not its members. Fuzzy sets differ from classical sets by rejecting the second requirement. As a consequence, the boundaries of the fuzzy sets are not necessarily sharp.

It is well known that all objects of concern are members of a set called a universe of discourse, a universe, or a universal set. From now on, we denote by U a non-empty universe, either finite or infinite.

Definition 1.1.1 *A fuzzy set or type-1 fuzzy set A is a mapping $A : U \to [0,1]$. The value $A(x)$, $x \in U$, is referred to as the degree of membership of the element x to the fuzzy set A.*

An equivalent definition for a fuzzy set is given as follows.

Definition 1.1.2 *A fuzzy set A is defined by*

$$A = \{(x, \mu_A(x)) : \quad x \in U\},$$

with $\mu_A : U \to [0,1]$, explicitly providing the membership degree of each $x \in U$.

Note that Definition 1.1.2 attempts to identify the fuzzy set with the mapping graph A given in Definition 1.1.1. It is observed that each standard fuzzy set is associated with a special family of classical sets. These classical sets are now known as level-cuts. We denote these sets for any given fuzzy set A by A^α, where $\alpha \in [0,1]$.

Definition 1.1.3 *Given a standard fuzzy set A, the full family of its level cuts is defined as follows:*

$$A^\alpha = \{x \in U : A(x) \geq \alpha\},$$

for any $\alpha \in [0,1]$.

These sets are used to define the concepts of convexity and boundedness for fuzzy sets defined on the n-dimensional Euclidean space $\mathbb{R}^n, n \geq 1$.

Definition 1.1.4 *A fuzzy set A defined on* $\mathbb{R}^n$, $n \geq 1$, *is said to be convex (bounded) if and only if its level cuts are convex (bounded) in the classical sense for any* $\alpha \in [0,1]$.

One particular principle is introduced for extending various concepts of classical mathematics to their counterparts in mathematics based on fuzzy sets.

Definition 1.1.5 *Any concept obtained by this principle is called a cut-worthy concept, i.e., a concept extended from classical mathematics via the level cut representation of fuzzy sets.*

Some operations on fuzzy sets are discussed.

Definition 1.1.6 *For each standard fuzzy set A, its complement* $\overline{A}$ *is defined by the formula*

$$\overline{A}(x) = 1 - A(x) \quad \textit{for any} \quad x \in U.$$

Furthermore, he defines the intersection and union of standard fuzzy sets.

Definition 1.1.7 *For any standard fuzzy sets A and B, we define their union as follows*

$$(A \cup B)(x) = \max\{A(x), B(x)\} \quad \textit{for any} \quad x \in U.$$

Definition 1.1.8 *For any standard fuzzy sets A and B, their intersection is defined as follows*

$$(A \cap B)(x) = \min\{A(x), B(x)\} \quad \textit{for any} \quad x \in U.$$

These three operations are now referred to in the literature as standard operations on fuzzy sets. The meaning of a containment of one fuzzy set in another is defined in the following way.

Definition 1.1.9 *We say that a fuzzy set A is contained in a fuzzy set B and we will denote it by* $A \subset B$ *if and only if*

$$A(x) \leq B(x) \quad \textit{for any} \quad x \in U.$$

Moreover, various properties of the standard operations on fuzzy sets are derived. In 1965, another work [45] was published in which the role of fuzzy sets in system theory is discussed, the discrete-time fuzzy dynamic systems are defined, and some basic issues involved in the applications of systems under fuzzy constraints are also discussed. The notion of fuzzy classes of systems, such as the class of systems that are "approximately equivalent" to a given system, the class of systems that are "approximately linear", and the class of systems that are "adaptive" is introduced.

Another area in which fuzzy sets and relations appeared is the axiomatic theory of many-valued sets. It started in [39], where the set theories within the finitely and infinitely valued Lukasiewicz logics and considered standard as well as finitely valued fuzzy sets are explored. Cumulative hierarchies of fuzzy sets within finitely valued Lukasiewicz logics are examined. However, the publication [44] in 1965 represents the turning point that marks the emergence of fuzzy logic.

Several researchers considered this new idea so interesting and potentially important that they opted to work on its further development. Joseph A. Goguen stands out as an important contribution to fuzzy logic. In [20], an important generalization of standard fuzzy sets is developed to the so-called *L* fuzzy sets by replacing the unit interval of membership grades with a more general algebraic structure of a complete residuated lattice. Basic ideas for developing logic for reasoning with inexact concepts are developed in which fuzzy sets play the role of inexact predicates. Later, contributions have been made in several other areas, such as theoretical computer science, artificial intelligence, and systems science. The applicability of fuzzy sets to pattern recognition is investigated in the Ph.D. thesis [7] and the first formulation of fuzzy topological spaces was developed in [8].

Some well-known researchers also contributed to the development of fuzzy logic at this early stage. Richard Bellman and Robert Kalaba joined Zadeh and developed the role of fuzzy sets in abstraction, generalization, and pattern classification.

These details of the very early history of fuzzy logic are listed here to show how quickly some of the fundamental ideas were developed in various ways. In the 1970s, an impressive growth of the literature on fuzzy logic appeared, and many new ideas were introduced. This literature involved a series of books on fuzzy set theory.

In the meantime, fuzzy logic continued to be a subject of many investigations throughout the 1970s, 1980s, and present times.

1.2 MATHEMATICS BASED ON FUZZY LOGIC

Classical mathematics is based on the theory of sets and classical logic. In other words, classical mathematics studies notions that may be defined in terms of the notion of set and that mathematical reasoning is governed by the laws of classical logic. Classical mathematics admits the principle of bivalence: Membership in sets, as well as the validity of mathematical statements about sets, and thus about all mathematical objects, are two-valued. Statements about mathematical objects admit degrees of truth. Since these include statements about membership, the objects include fuzzy sets. Classical logic is not appropriate for reasoning with statements that admit degrees of truth. It needs to be replaced with a new one. Fuzzy logic is a candidate in this respect. The admission of truth degrees leads to an important mathematics which is called mathematics based on fuzzy logic. By admitting degrees of truth and admitting fuzzy sets, one does not exclude ordinary sets and other objects of classical mathematics from consideration. Thus, notions and statements of mathematics based on fuzzy logic may concern not only fuzzy sets but also ordinary sets and classical structures. Statements about fuzzy sets and fuzzy relations may both be of high value. Fuzzy logic-based mathematics extends classical mathematics.

1.3 THE ROLE OF FUZZY LOGIC

The role of fuzzy logic in mathematics based on fuzzy logic is similar to the role of classical logic for classical mathematics. One may take a given fuzzy logic and develop theories within this logic. In classical mathematics, the intuitive way of working with sets is accompanied by mainly informal usage of classical logic. Therefore, mathematicians express their definitions, theorems, and proofs in ordinary language. For instance, when we express the transitivity of a given relation R, we write

$$\begin{aligned} &\text{for any} \quad u,v,w \in U, \quad \text{if} \quad (u,v) \in R \quad \text{and} \quad (v,w) \in R, \\ &\text{then} \quad (u,w) \in R. \end{aligned} \tag{1.1}$$

For such a claim, one need not resort to the formal apparatus of classical logic. In this case, its informal use is sufficient. When R is a fuzzy relation with truth degrees in L, things become more complex. The propositions $(u,v) \in R$ and $(v,w) \in R$ are then each a matter of degree, and the above informal approach cannot be used. One needs to resort to using the apparatus of fuzzy logic more explicitly. Fuzzy logic offers the following general approach. First, we take a language of appropriate fuzzy logic and express (1.1) by a formula of this language. If we take the predicate logic $BL\forall$, its language contains a single binary relation symbol r, and then (1.1) takes the following form

$$\phi_{Tra} : \forall u \forall v \forall w((r(u,v) \otimes r(v,w)) \to r(u,w)).$$

Now, we take a structure **L** of truth degrees and an **L**-structure

$$\mathbf{M} = (M, r^M),$$

where $U = M$ and $R = r^m$. If

$$\mathbf{L} = (L, \vee, \wedge, \otimes \rightarrow 0, 1),$$

then the fact that (1.1) is satisfied by the transitivity relation R is understood to mean that ϕ_{Tra} is true in **M**, that is

$$||\phi_{Tra}||_{\mathbf{M}} = 1.$$

Using the principles of prediction fuzzy logic, the fact (1.1) can be rewritten as follows:

$$\text{for any} \quad u, v, w \in U \quad \text{we have} \quad R(u,v) \otimes R(v,w) \leq R(u,w). \quad (1.2)$$

The fact (1.2) is the meaning of (1.1) for fuzzy relations. In other words, a fuzzy relation R is transitive if and only if it satisfies (1.2), and thus we obtain the definition of transitivity of a fuzzy relation. For the standard Gödel and Lukasiewicz algebra, the condition

$$R(u,v) \otimes R(v,w) \leq R(u,w),$$

reads

$$\min(R(u,v), R(v,w)) \leq R(u,w),$$

and

$$\max(0, R(u,v) + R(v,w) - 1) \leq R(u,w),$$

respectively, which express the Zadeh notion [46] and the Ruspini notion [36] of transitivity.

This approach offers a way to fuzzify classical concepts, that is, write the definitions of a classical concept using a fuzzy logical formula and interpret it according to the principles of fuzzy logic. Reliance on fuzzy logic makes it natural to consider the fuzzy properties of fuzzy relations and to consider theorems involving such properties.

1.4 SELECTED AREAS OF MATHEMATICS BASED ON FUZZY LOGIC

We represent here some branches of mathematics based on fuzzy logic and their developments.

1.4.1 Sets and Relations

By introducing the notion of a fuzzy set, Zadeh [44] set the stage for developing the theory of fuzzy sets and fuzzy relations. The seminal works of Goguen [20] are of great importance for this topic. Goguen assumes that the set L of truth degrees must form a structure

$$\mathbf{L} = (L, \wedge, \otimes \to 0, 1),$$

that must be a common residuated algebra and $\mathbf{L}$ can be regarded as a parameter. Thus, a general theory of $\mathbf{L}$-sets and $\mathbf{L}$-relations may be developed. A given condition may have different meanings for different $\mathbf{L}$s. For example, the meaning of transitivity of a fuzzy relation R expressed by

$$R(u,v) \otimes R(v,w) \leq R(u,w),$$

takes the form

$$\min(R(u,v), R(v,w)) \leq R(u,w).$$

If $\mathbf{L}$ is the Gödel algebra and it takes the form

$$\max(0, R(u,v) + R(v,w) - 1) \leq R(u,w).$$

If $\mathbf{L}$ is the Lukasiewicz algebra. In this view, the characteristic functions of the classical sets and relations are all $\mathbf{L}$-sets and $\mathbf{L}$-relations for the particular case of $\mathbf{L}$ being the two elements of Boolean algebra $L = \{0, 1\}$. Morphisms $\mathbf{L}_1 \to \mathbf{L}_2$ are investigated. This view was adopted by [19], where general fuzzy rational systems with possibly different $\mathbf{L}$s are considered. For example, a morphism consists of two mappings: $f : U_1 \to U_2$ of universes and $g : L_1 \to L_2$ of sets of truth degrees. These ideas are reflected in the distinction of the fuzzy topology of fixed-basis and variable-basis, where "basis" refers to L.

Noting that the set L^U of all L-sets in U is equipped with operations and relations inherited from those in L. For example, the truth function $\otimes$ of the conjunction of L induces the corresponding operation $\otimes$ which is defined as follows

$$(A \otimes B)(u) = A(u) \otimes B(u),$$

for any $A, B \in L^U$ and for any $u \in U$. Here, $A \otimes B$ is also called the $\otimes$ intersection of A and B and is often denoted by $A \cap_{\otimes} B$. The partial order $\leq$ on L extends to L^U in the following manner

$$A \leq B \quad \text{if and only if} \quad A(u) \leq B(u) \quad \text{for any} \quad u \in U,$$

in which case one commonly uses $A \subseteq B$ and calls $\subseteq$ the inclusion of fuzzy sets. Thus, we obtained the structure

$$\mathbf{L}^U = (L^U, \vee, \wedge, \otimes \rightarrow 0, 1),$$

of all fuzzy sets $A : U \rightarrow L$ is the direct product of **L**s indexed by U and its apparatus obeys any law valid in **L**. If **L** is a residuated lattice, then so is $\mathbf{L}^U$.

In addition, further operations, relations, and fuzzy relations were defined for fuzzy sets. Various characteristics for fuzzy sets were defined, such as the height or measure of fuzzy sets. The height of a fuzzy set A is defined by

$$h(A) = \bigvee_{u \in U} A(u).$$

The simple average of the fuzzy sets A and B is defined as follows

$$[h(A,B)](u) = \frac{A(u) + B(u)}{2}.$$

Note that the language in which the former notions are defined is a language of a class of structures which contains the two-element Boolean algebra 2, and thus the notions make sense for classical sets. They are in fact generalizations of classical notions to fuzzy sets. These notions may be degenerate for classical sets. For instance, the

α cut of a classical set A is the whole universe for $\alpha = 0$ and A itself for $\alpha = 1$, which are the only options since there is no other degree of truth in the classical case. These notions can be regarded as new notions that appear in the theory of fuzzy sets.

The idea of considering fuzzy sets as random sets has appeared in several reviews. More precisely, if $(\Omega, \mathscr{F}, P)$ is a probability space, a random set S in a universe U is a measurable set-valued function $S : \Omega \to 2^U$. Then, the random set S induces a fuzzy set A_S by placing

$$A_S(u) = P\{\omega \in \Omega : u \in S(\omega)\}.$$

This relationship between fuzzy sets and random sets is not one-to-one and has been studied in several works.

All of these investigations have led to many subsequent contributions which have improved and developed them further.

1.4.2 Algebra

The research on fuzzy sub-algebras and other structures began with "Fuzzy Groups" by [35]. It represents a significant direction in the application of fuzzy sets to algebraic concepts. The notion of fuzzy sub-algebra derives from the concept that a fuzzy set is closed with respect to an operation. Rosenfeld introduced the notion of a fuzzy sub-groupoid and a fuzzy sub-group. He establishes a number of results for these notions.

Rosenfeld calls a fuzzy sub-groupoid of a classical groupoid $(A, \cdot)$ any fuzzy set $A : U \to [0, 1]$ which satisfies the following inequality

$$\min(A(u), A(v)) \leq A(u \cdot v).$$

If U is a group, then A is its fuzzy sub-group if it is its fuzzy sub-groupoid such that

$$A(u) \leq A(u^{-1}), \tag{1.3}$$

for any $u \in U$. The Rosenfeld idea generalizes the concept of sub-groupoid and sub-group.

Consider a first-order language with a binary function symbol $\circ$ and a unary relation symbol r. By classical logic, we have that a subset A of M is a classical sub-group of $(M,\cdot)$ if and only if the following formula

$$\forall x \forall y (r(x) \otimes r(y) \to r(x \otimes y)),$$

is true in the structure

$$\mathbf{M} = \left(M, r^{\mathbf{M}}, \circ^{\mathbf{M}}\right),$$

where the interpretations $r^{\mathbf{M}}$ and $\circ^{\mathbf{M}}$ of symbols r and $\circ$ equal A and $\cdot$, respectively. Then, the notion of fuzzy sub-groupoid is obtained by taking the same formula and interpreting it in first-order fuzzy logic with the set $L = [0,1]$ of truth degrees and min as the truth function $\otimes$.

In [24], the Rosenfeld concept is developed and fuzzy subspaces of vector spaces are examined. They defined a fuzzy subspace of a vector space U in a field K as a fuzzy set $A : U \to [0,1]$ such that

$$A + A \subseteq A,$$

and

$$k \cdot A \subseteq A,$$

for each $k \in K$. Here, the operations $+$ and $\cdot$ are based on the Zadeh extensional principle. Katsaras and Liu proved that this definition is equivalent to the conditions

$$(A(u), A(v)) \leq A(u+v),$$

and

$$A(u) \leq A(ku),$$

for any $u, v \in U$ and for any $k \in K$. They also utilize the then-recent notion of fuzzy topology and define fuzzy topological vector spaces as vector spaces U with a fuzzy topology τ on U, such that vector addition and scalar multiplication are continuous with respect to product topologies based on τ and a topology in the field of scalars, and develop fuzzy subspaces of fuzzy topological vector spaces.

Independently, the searchers in [1], defined fuzzy topologies on algebras and fuzzy substructures of such fuzzy topological algebras. They generalized the Rosenfeld notion of fuzzy sub-group as follows

$$A(u) \otimes A(v) \leq A(U \circ v),$$

in which $\otimes$ is a t-norm, and strengthened (1.3) by the condition

$$A(u) = A(u^{-1}),$$

and added the condition

$$A(e) = 1.$$

In [9], the author adds observations regarding the α-cuts, i.e. a fuzzy set A is a fuzzy sub-group if and only if A^{α} is an ordinary sub-group for each

$$\alpha \leq A(u).$$

In [12], a relationship is found between fuzzy sub-groups and fuzzy equivalence relations, both with respect to a t-norm.

A different approach was introduced by [4]. Their idea is to look at the concept of algebra from the point of view of fuzzy logic and to develop universal algebra. For a type (F, σ) and a complete residuated lattice **L**, an algebra with fuzzy equality, shortly **L**-algebra is a structure

$$\mathbf{M}(M, \approx^{\mathbf{M}}, F^{\mathbf{M}}),$$

such that $(M, F^{\mathbf{M}})$ is an ordinary algebra of type (F, σ), $R \approx^{\mathbf{M}}$ is a fuzzy equality on M, and each operator $f^{\mathbf{M}} \in F^{\mathbf{M}}$ is compatible with $\approx^{\mathbf{M}}$ in that

$$\left(m_1 \approx^{\mathbf{M}} n_1, \ldots, m_k \approx^{\mathbf{M}} n_k\right) \leq f^{\mathbf{M}}(m_1, \ldots, m_k) \approx^{\mathbf{M}} f^{\mathbf{M}}(n_1, \ldots, n_k),$$

for any k-ary operation $f^{\mathbf{M}}$ and any $m_j, n_j \in M$, $j \in \{1, \ldots, k\}$. The degrees $m \approx^{\mathbf{M}} n$ are interpreted as degrees of similarity of m and n. Algebras with fuzzy equalities are structures of a predicate fuzzy logic with a language containing $\approx$ as a single relation symbol, $f \in F$ of function symbols, and in which the structure **L** of truth degrees may be any complete residuated lattice. The compatibility condition means that the formula described as

$$\text{if} \quad x_j \quad \text{and} \quad y_j \quad \text{are pairwise similar for} \quad j \in \{1, \ldots, k\},$$

then

$$f(x_1, \ldots, x_k) \quad \text{and} \quad f(y_1, \ldots, y_k) \quad \text{are similar,}$$

is true.

For $L = \{0, 1\}$, algebras with fuzzy equalities may be identified with classical algebras.

The idea of algebras with fuzzy equalities is closely related to that of metric algebras. A metric algebra is a triplet

$$\mathbf{M} = \left(M, \rho^{\mathbf{M}}, F^{\mathbf{M}} \right),$$

consisting of an arbitrary algebra $(M, F^{\mathbf{M}})$ and a metric $\rho^{\mathbf{M}}$ on M. In the particular case of $\otimes$ being a continuous Archimedean t-norm, every algebra with fuzzy equality may be transformed into a metric algebra.

Many results on algebras with fuzzy equalities in the spirit of classical results in universal algebra are expounded in the book by [4]. Among them are those regarding basic structural notions such as subalgebras, morphisms, congruences, factorization, direct unions and limits, reduced products, free algebras, varieties, semi-varieties, and quasi-varieties.

In addition to fuzzy sub-algebras and algebras with fuzzy equalities, several other approaches have appeared in the literature. For example, the Demirci vague algebras [10], vague algebras based on the Demirci fuzzy functions, which are conceived as certain fuzzy relations on sets equipped with fuzzy equalities with which they are required to be compatible. This means that the Demirci vague algebras are more general than algebras with fuzzy equalities.

Closure operators and related structures, such as closure systems, interior operators and systems, and Galois connections are fundamental for many areas of mathematics, including set theory, algebra, topology, and geometry. In the literature on fuzzy sets, fuzzy closure operators appeared for the first time in the work by [29]. The Lowen conditions generalize the Kuratowski axioms for topological closure. Later, fuzzy closure operators and related structures appeared in a number of studies.

1.4.3 Topology

The concepts and results on classical topology from the fuzzy logic point of view started in [8]. Fuzzy topology represents one of the earliest areas of mathematics based on fuzzy logic.

A classical topology in a set X is a family τ of its subsets that contains the empty set, contains X, and is closed with respect to arbitrary unions and with respect to finite intersections. The sets in τ are called open, and their complements are called closed.

The first research devoted to fuzzy topology is [21]. The aims were to generalize for fuzzy topology some basic results of classical topology. Several of the subsequent contributions assume that the standard interval $[0,1]$ of truth degrees is replaced by a complete lattice L and the standard negation is replaced by an order-reversing involutive function. In [20], a complete residuated lattice L with a possibly non-commutative truth function $\otimes$ of conjunction was used. Thus, Goguen defines the notion of an L-topological space as a pair (X,τ), when $\tau \in L^X$ satisfies the above conditions except that the last one is replaced by the following condition

$$A_1,\ldots,A_k \in \tau \quad \text{implies} \quad A_1 \otimes \ldots \otimes A_k \in \tau.$$

Various fuzzifications of classical topological concepts evolved as further aspects and connections were taken into account. The notion of a compact fuzzy topological space was presented by [8]. This generalization of the classical compactness is subject to a number of problems.

The early research introduced the notion of a fuzzy unit interval due to [23]. Hutton needed a suitable fuzzy topology that generalized the classical topology in $[0,1]$. The notion was later extended to the notion of a fuzzy real line by [13], which generalizes the usual topology of reals.

Further examinations of the notion of fuzzy topology include the associated interior and closure operators. The notion of a point in fuzzy topology and the related concept of separation have been proven for the particularly peculiar and have been approached in a number of ways. Related are examinations of the local structure of fuzzy topological spaces, and in particular the concept of neighborhood. Examinations concern the notion of convergence to which two approaches have been developed. One is based on the concept of a net and the other on that of filter. Filter theory led to further examinations of neighborhoods and spaces whose structures can be recovered from neighborhood systems.

The notions of metric and metrizability have proved to be of special interest. Various connections to probabilistic metric spaces and probabilistic topologies are also worth mentioning.

Category theory became a working language in fuzzy topological spaces. It was necessary to compare the various notions of fuzzy topology, and thus to compare the respective categories of fuzzy topological spaces.

Two related views which have been developed are known as fixed-basis fuzzy topology and variable-basis fuzzy topology.

In many contributions, fuzzy topology was shown to be capable of contributing to solutions of mathematical problems for which classical topology is not. For example, in [30], fuzzy topologies in classical metric spaces were shown to provide new insights into these spaces relevant to the approximation theory.

1.4.4 Quantities

A large part of classical mathematics deals with exact quantities. Such quantities are represented by the concept of numbers. Fuzzy sets on the real line are natural models for such quantities. This is the basic reason why these fuzzy sets have played a special role in the development of fuzzy logic and mathematics based on fuzzy logic.

A fuzzy quantity is understood as a fuzzy set A in $\mathbb{R}^n$ that has some additional, intuitively appealing, or technically convenient properties.

In general, a fuzzy quantity $A : \mathbb{R}^n \to [0,1]$ is usually required to perform the test.

1. be normal in that $A(u) = 1$ for some $u \in \mathbb{R}^n$.

2. be upper semi-continuous at each point.

3. have a bounded support set, i.e., the set
$$\{u \in \mathbb{R}^n : A(u) > 0\},$$
is bounded.

Note that upper semi-continuity can be phrased in terms of cuts: A is upper semi-continuous if and only if for each $\alpha \in [0,1]$, the α-cut A^α is a closed set.

Considerable interest has been shown in convex fuzzy quantities, that is, fuzzy quantities A for which
$$A(\lambda u + (1-\lambda)v) \geq \min(A(u), A(v)),$$
for any $u, v \in \mathbb{R}^n$ and for any $\lambda \in [0,1]$. The convexity of A means that each α-cut A^α is a convex subset of $\mathbb{R}^n$. Observe that the convexity of fuzzy quantities in $\mathbb{R}$ means that the α-cuts A^α of A are intervals, open or closed.

Fuzzy intervals are simply fuzzy normal quantities with bounded supports whose α-cuts are closed intervals.

Fuzzy sets are studied in $\mathbb{R}$ the basic arithmetic operations $+$, $-$, $\cdot$, and $/$, obtained by the extension principle. They observed several properties such as preservation of normality by all operations and convexity by $+$, $-$, and $\cdot$, but not $/$. They also proved that the operations $+$ and $\cdot$ are associative, commutative, have singletons $\{1/0\}$ and $\{1/1\}$, denoted commonly by 0 and 1, respectively, as their neutral elements, but also that there are no inverse elements with respect to the operations $+$ and $\cdot$ and that the distributive law
$$A \cdot (B + C) = A \cdot B + A \cdot C,$$

holds for positive fuzzy numbers, i.e., for fuzzy numbers A for which

$$A(u) = 0 \quad \text{for any} \quad u < 0,$$

but not in general. Mizimuto and Tanaka defined the operations $\prod$ and $\sqcup$ in fuzzy sets in $\mathbb{R}$ derived from the operations of minimum and maximum in $\mathbb{R}$ by the extension principle.

A more substantial departure is inspired by problems with solving of equations such as

$$A + X = C,$$

for which the difference $C - A$ based on the extension principle need not be a solution. Such problems appeared in several works. In [37], it is observed that the problem of solving equations such as

$$A + X = B,$$

is a particular case of solving fuzzy relational equations.

Numerous later studies have been devoted to various practically motivated questions, including numerical characteristics of fuzzy intervals such as: the mean value, measures of specificity and fuzziness or defuzzifications, and those based on defining fuzzy ordering-like relations.

1.4.5 Mathematical Analysis

Mathematical analysis is one of the areas in which generalizations based on replacing real numbers by fuzzy quantities have been developed in a number of researches. The research and methods in this area follow those in the set-valued analysis by [2] and the use of the extension principle and the level cut representations of fuzzy sets. Fundamental to mathematical analysis are properties of the topology induced by a given metric on $\mathbb{R}^n$ such as: openness and closedness, compactness, or completeness, and the same holds true for analysis with fuzzy quantities.

Denote by $\mathscr{F}(\mathbb{R}^n)$ the set of all fuzzy intervals. Note that several results may be obtained without the convexity assumption involved. Any α-cuts A^α, $\alpha \in (0,1]$, as well as the closure $\overline{\text{supp} A}$ of the support are compact subsets of $\mathbb{R}^n$. For two fuzzy quantities A and B, define the supremum metric d_∞ and the parameterized metric d_p, $1 \le p < \infty$, as follows

$$d_\infty(A,B) = \sup_{\alpha\in[0,1]} d_H(A^\alpha, B^\alpha),$$

and

$$d_p(A,B) = \left(\int_0^1 (d_H(A^\alpha, B^\alpha))^p \, d\alpha\right)^{\frac{1}{p}},$$

respectively. Here, d_H denotes the Hausdorff metric.

The metric d_∞ was introduced and also appears in fuzzy random variables. The d_p metrics appear again by [26] in the context of random variables.

It turned out that functions assigning fuzzy quantities to reals are mathematically interesting and feasible generalizations of classical real functions, and such functions appear in the area of fuzzy random variables.

By fuzzy-set-valued mappings, we understand functions of the form $F : K \to \mathscr{F}(\mathbb{R}^n)$, where K denotes a set in $\mathbb{R}^k$. Definitions of some properties are obtained as particular cases of classical notions. As an example, we will give a definition for a continuous function at a point.

Definition 1.4.1 *A function $F : K \to \mathscr{F}(\mathbb{R}^n)$ is said to be a continuous function at $t_0 \in K$ if for any $\varepsilon > 0$ there is a $\delta = \delta(\varepsilon) > 0$ such that for any $t \in K$ for which*

$$\|t - t_0\| < \delta,$$

one has

$$d_\infty(F(t), F(t_0)) < \varepsilon.$$

There are several notions of the differentiability of functions valued by fuzzy sets, including the de Blasi differentiability [11], the Hukuhara differentiability [34], and the Bobylev differentiability [5]. We will represent the first one.

Definition 1.4.2 *A function $F : K \to \mathscr{F}(\mathbb{R}^n)$ is called de Blasi differentiable at $t_0 \in K$ if there exists an upper semicontinuous and homogeneous mapping $D^F_{t_0} : \mathbb{R}^k \to \mathscr{F}(\mathbb{R}^n)$, called the de Blasi differential of F at t_0, satisfying*

$$\lim_{\|\Delta t\| \to 0} \frac{d_\infty\left(F(t_0 + \Delta t), F(t_0) + D^F_{t_0}(\Delta t)\right)}{\Delta t} = 0,$$

where homogeneity of $D^F_{t_0}$ means

$$D^F_{t_0}(\lambda t) = \lambda D^F_{t_0}(t),$$

for any $\lambda > 0$ and for any $t \in \mathbb{R}^k$.

Note that the de Blasi differential is unique if it exists. It is linear in that

$$D^{\lambda F + \beta G}_{t_0} = \lambda D^F_{t_0} + \beta D^G_{t_0},$$

and its existence at t_0 implies the continuity of F at t_0.

The integration of fuzzy-set-valued functions has been examined for the case $k = 1$ appears in considerations regarding fuzzy random variables. For simplicity, consider functions of the form $F : [0,1] \to \mathscr{F}(\mathbb{R}^n)$.

Definition 1.4.3 *A function $F : [0,1] \to \mathscr{F}(\mathbb{R}^n)$ is said to be integrable bounded if there exists an integrable function $f : [0,1] \to \mathbb{R}$ such that*

$$\|F(t)\| \leq f(t), \quad t \in [0,1].$$

Here $\|\cdot\|$ is the norm induced by the metric d_∞.

It follows that any α-cut F^α of an integrable bounded F is integrable and bounded as a set-valued function.

Definition 1.4.4 *If $F : [0,1] \to \mathscr{F}(\mathbb{R}^n)$ is strongly measurable and integrable bounded, then the integrable boundedness of any F^α ensures that F^α is Aumann integrable in $[0,1]$ in the sense of set value analysis. If then there exists a fuzzy set $A \in \mathscr{F}(\mathbb{R}^n)$ such that*

$$A^\alpha = \int_0^1 F^\alpha(t)dt,$$

for any $\alpha \in [0,1]$*, where*

$$\int_0^1 F^{\alpha}(t)dt,$$

is the Aumann integral of F^{α}*, one calls* F *integrable over* $[0,1]$*, denotes*

$$A = \int_0^1 F(t)dt,$$

and calls it the integral of F *over* $R[0,1]$*.*

A continuous function is always integrable. The integral has many properties analogous to those of the Aumann integral, such as

$$\begin{aligned}\int_0^1 (\beta F(t) + \gamma G(t))dt &= \beta \int_0^1 F(t)dt + \gamma \int_0^1 G(t)dt,\\ \int_0^1 (\beta F(t))dt &= \beta \int_0^1 F(t)dt,\end{aligned}$$

or the integral inequality

$$d_{\infty}\left(\int_0^1 F(t)dt, \int_0^1 G(t)dt\right) \le \int_0^1 d_{\infty}(F,G)dt.$$

A visible direction in the mathematical analysis of functions that involve fuzzy quantities is represented by fuzzy differential equations. One version of the problems considered is given by the initial value problem

$$\begin{aligned}\frac{dx}{dt}(t) &= f(x(t)), \quad t_1 < t < t_2,\\ x(0) &= 0,\end{aligned} \tag{1.4}$$

where $t_1 < 0 < t_2$, $f : \mathscr{F}(\mathbb{R}) \to \mathscr{F}(\mathbb{R})$ is a given function, $x_0 \in \mathscr{F}(\mathbb{R})$ is a fuzzy interval representing the initial value. The goal is to find a fuzzy-set-valued function $x : (t_1, t_2) \to \mathscr{F}(\mathbb{R})$ satisfying (1.4) in which usually the Hukuhara derivative and the metric d_∞ are used.

Note that several other formulations of the initial value problems as well as other problems known from classical differential equations have been proposed.

1.4.6 Probability and Statistics

In one early work [47], a generalization of the notion of event is proposed and fuzzy events are conceived as particular fuzzy sets $A : \Omega \to [0,1]$ with $A(\omega)$ interpreted as the degree to which A applies the outcome $\omega \in \Omega$. To measure probabilities of fuzzy events, Zadeh assumed that there is a given probability space $(\Omega, \mathscr{B}, P)$ with $\Omega = \mathbb{R}^n$, $\mathscr{B}$ the σ-field of Borel sets in $\mathbb{R}^n$, and P a probability measure on $\mathscr{B}$. He defined fuzzy events as Borel-measurable fuzzy sets $A \in [0,1]^\Omega$, defined the probability of a fuzzy event by the Lebesgue-Stieltjes integral as follows

$$m(A) = \int_\Omega A dP, \tag{1.5}$$

and presented some observations regarding the closeness of fuzzy events under fuzzy-set operations, independence, conditional probability, and entropy. The Zadeh formula generalizes the classical formula

$$\int_\Omega \kappa_A dP,$$

where κ_A is the characteristic function of a classical event A.

In [25], an attempt to generalize notions of classical measure theory such as that of σ-algebra, measure, and probability is started from the point of view of fuzzy sets.

As fundamental notion is that of tribe on Ω with respect to a given t-norm $\otimes$, or a $\otimes$-tribe, which is a subset $\mathscr{G}$ of $[0,1]^\Omega$ satisfying

$$0 \in \mathscr{G},$$

$$\text{for each} \quad A \in \mathscr{G} \quad \text{we have} \quad \neg A \in \mathscr{G}, \tag{1.6}$$

$$\text{for each sequence} \quad \{A_n\}_{n\in\mathbb{N}} \in \mathscr{G} \quad \text{we have} \quad \otimes_{n=1}^{\infty} A_n \in \mathscr{G}, \tag{1.7}$$

where with $\neg$ is denoted the standard negation,

$$\neg a = 1 - a,$$

and the operations on fuzzy sets are based on $\neg$ and $\otimes$, and defined componentwise, i.e.

$$(\neg A)(\omega) = \neg(A(\omega)).$$

Note that the infinite $\otimes$-based intersection $\otimes_{n=1}^{\infty} A_n$ always exists and if condition (1.7) is replaced by requiring that

$$A, B \in \mathscr{G} \quad \text{imply} \quad A \otimes B \in \mathscr{G},$$

we obtain the notion of clan. Due to (1.6) and the duality of t-norms and their t-conorms, $\otimes$ may be replaced by its dual t-norm $\oplus$. Since the operations based on t-norms, t-conorms, and the standard negation generalize the classical intersection, disjunction, and complement, it follows that the notion of tribe generalizes σ-algebras. Hence, modulo identification of sets with crisp fuzzy sets, the collection of subsets of Ω is a σ-algebra if and only if it is a tribe with respect to an arbitrary t-norm.

If $\mathscr{A}$ is an ordinary σ-algebra in Ω and $\otimes$ is Borel-measurable, then the set $\mathscr{A}^{\vee}$ of all measurable fuzzy sets in $[0,1]^{\Omega}$ is a tribe with respect to $\otimes$, which is called a generated tribe or a full tribe. The Zadeh system of fuzzy events is a tribe-generated system. In contrast, the collection $\mathscr{F}^{\vee}$ of all crisp fuzzy sets of $\mathscr{G}$ forms a tribe that forms a σ algebra. The properties of $\otimes$-tribes depend on those of the t-norm $\otimes$.

For a t-norm $\otimes$ and its t-conorm $\oplus$, a $\otimes$-measure on a $\otimes$-tribe is a function $m : \mathscr{G} \to [-\infty, \infty]$ assuming at most one of the values $-\infty$ and ∞ which satisfies

$$m(\emptyset) = 0,$$

$$m(A \otimes B) + m(A \oplus B) = m(A) + m(B),$$

$$\lim_{n\to\infty} m(A_n) = m(C) \quad \text{whenever} \quad A_n \nearrow C,$$

for any $A, B, A_n \in \mathscr{G}$, $n \in \mathbb{N}$, and for some $C \in \mathscr{G}$. A measure m is called a probability $\otimes$-measure if it assumes only values in $[0,1]$ and $m(\Omega) = 1$. If $\mathscr{G}$ consists of crisp fuzzy sets, the $\otimes$-measures are the ordinary measures modulo identification of crisp fuzzy sets with ordinary sets.

Conversely, if P is a probability measure in a σ-algebra $\mathscr{A}$ and $\mathscr{G}$ its generated tribe, then

$$m_p(A) = \int_{\Omega} A dP,$$

defines a probability measure $\otimes$ in $\mathscr{G}$, from which the ordinary probability measure P can be recovered. Many other properties of tribes and measures, representations, decompositions, and other properties of $\otimes$-measures have been obtained.

The concept of random variables has been studied from the point of view of fuzzy sets. For a given probability space $(\Omega, \mathscr{B}, P)$, a random variable is a mapping $X : \Omega \to \mathbb{R}$ that is measurable with respect to the σ-algebra $\mathscr{B}$ and the Borel σ-algebra in $\mathbb{R}$. Instead of real numbers, a fuzzy random variable assigns fuzzy sets of real numbers to the experimental outcomes.

Numerous problems involving the notion of fuzzy random variables have been explored, including integration and differentiation of probabilistic settings, weak and strong laws of large numbers, limit theorems, martingales, decision models, regression models, and so on.

1.4.7 Geometry

Classical geometric concepts such as the point, line, triangle, or incidence are clear-cut. There are several possible approaches for fuzzifying geometrical concepts.

In a series of works, Rosenfeld has presented an approach to geometrical notions. Central to this approach are various intuitive geometric notions, from topological properties of connectedness, adjacency, or separation, to metric notions such as area, perimeter, diameter, and distance, as well as various further notions such as those regarding convexity. For example, two points u and v are connected in a fuzzy set A if there exists a path π in $\mathbb{R}^n$ from u to v such that

$$A(u) \vee A(v) \leq \bigvee_{w \in \pi} A(w).$$

The degree

$$s_A(\pi) = \bigvee_{w \in \pi} A(w),$$

is called the strength of π with respect to A. The degree of connectness of u and v with respect to A is determined as follows

$$c_A(u,v) = \bigwedge_{\pi} s_A(\pi).$$

Metric notions here are defined using the available classical notions. For example, the area of $A \in [-0,1]^{\mathbb{R}^2}$ is defined as an integral of A in $\mathbb{R}^2$. Rosenfeld also fuzzified the classical notions of triangles, rectangles, and other shapes. The level cuts A^α of fuzzy rectangles are then just classical rectangles.

Various approaches have been initiated in [6]. The authors define a fuzzy point at $(u,v) \in \mathbb{R}^2$ as a fuzzy quantity P on $\mathbb{R}^2$, also denoted by $P(u,v)$, whose core contains only (u,v). For two fuzzy points $P_1(u_1,v_1)$ and $P_2(u_2,v_2)$, they also define fuzzy distance as a fuzzy set $D(P_1,P_2)$ in $\mathbb{R}$ defined by its α-cuts as follows

$$D(P_1,P_2)^\alpha = \{d(u,v) : u \in P_1(u_1,v_1)^\alpha, v \in P_2(u_2,v_2)^\alpha\},$$

and prove that $D(P_1,P_2)$ is a fuzzy number. They define a line as a set of pairs (X,Y) of fuzzy numbers satisfying

$$AX + BY = C,$$

where A, B, and C are given fuzzy numbers with operations defined by the extension principle, as well as a possibly less restrictive equation

$$Y = AX + b.$$

They introduce four notions of a line as fuzzy sets L_{12}, L_{21}, L_3, L_4 in $\mathbb{R}^2$ whose α-levels are given as follows:

$$L_{11}^{\alpha} = \{(x,y) \in \mathbb{R}^2 : ax + by = c, \quad a \in A^{\alpha}, b \in B^{\alpha}, c \in C^{\alpha}\},$$

$$L_{12}^{\alpha} = \{(x,y) \in \mathbb{R}^2 : y = ax + b, \quad a \in A^{\alpha}, b \in B^{\alpha}\},$$

$$L_{3}^{\alpha} = \{(x,y) \in \mathbb{R}^2 : y - v = a(x-u), \quad (u,v) \in \mathbb{R}^2, a \in A^{\alpha}\},$$

$$L_{4}^{\alpha} = \left\{(x,y) \in \mathbb{R}^2 : \frac{y - v_1}{x - u_1} = \frac{v_2 - v_1}{u_2 - u_1}, \right.$$
$$\left. (u_1, v_1) \in \mathbb{R}^2, \quad (u_2, v_2) \in \mathbb{R}^2 \right\}.$$

The notions of parallel lines and intersection of lines, circle and polygon, circumference, area, and perimeter, which they define as fuzzy numbers, are examined.

1.4.8 Time Scale Calculus

Time scale calculus is a unification of the theory of difference equations with that of differential equations, unifying integral and differential calculus with the calculus of finite differences, offering a formalism for studying hybrid systems. It has applications in any field that requires simultaneous modeling of discrete and continuous data. It gives a new definition of a derivative such that if one differentiates a function is defined on the real numbers, then the definition is equivalent to standard differentiation, but if one uses a function defined on the integers, then it is equivalent to the forward difference operator.

Time scale calculus was introduced in 1988 by Stefan Hilger [22]. Similar ideas have been used before and go back at least to the introduction of the Riemann–Stieltjes integral, which unifies sums and integrals.

A time scale is an arbitrary non-empty closed subset of the real numbers. We will denote a time scale by the symbol $\mathbb{T}$. For example, the set $\mathbb{T} = [0,1]$, where $[0,1]$ is the real-valued interval, is a time scale. The interval $(0,1)$ is not a time scale. The basic operators in time scale calculus are the forward jump operator and the backward jump operator, whose definitions are as follows.

Definition 1.4.5 *1. The forward jump operator $\sigma : \mathbb{T} \to \mathbb{T}$ is defined by*

$$\sigma(t) = \inf\{s \in \mathbb{T} : s > t\} \quad \text{for} \quad t \in \mathbb{T}.$$

2. *The backward jump operator $\rho : \mathbb{T} \to \mathbb{T}$ is defined by*

$$\rho(t) = \sup\{s \in \mathbb{T} : s < t\} \quad \text{for} \quad t \in \mathbb{T}.$$

3. *The graininess function $\mu : \mathbb{T} \to [0,\infty)$ is defined by*

$$\mu(t) = \sigma(t) - t \quad \text{for} \quad t \in \mathbb{T}.$$

We set

$$\inf = \sup \mathbb{T}, \quad \sup = \inf \mathbb{T}.$$

We note that $\sigma(t) \geq t$ for any $t \in \mathbb{T}$ and that $\rho(t) \leq t$ for any $t \in \mathbb{T}$.

Let $\mathbb{T}$ be a time scale with forward jump operator and backward jump operator σ and ρ, respectively.

Definition 1.4.6 *We define the set*

$$\mathbb{T}^\kappa = \begin{cases} \mathbb{T} \backslash (\rho(\sup \mathbb{T}), \sup \mathbb{T}] & \text{if} \quad \sup \mathbb{T} < \infty \\ \mathbb{T} \quad \text{otherwise.} \end{cases}$$

We have the following classification of the points on the time scale $\mathbb{T}$.

Definition 1.4.7 *For $t \in \mathbb{T}$, we have the following cases.*

1. *If $\sigma(t) > t$, then we say that t is right-scattered.*
2. *If $t < \sup \mathbb{T}$ and $\sigma(t) = t$, then we say that t is right-dense.*

3. *If $\rho(t) < t$, then we say that t is left-scattered.*

4. *If $t > \inf \mathbb{T}$ and $\rho(t) = t$, then we say that t is left-dense.*

5. *If t is left-scattered and right-scattered at the same time, then we say that t is isolated.*

6. *If t is left-dense and right-dense at the same time, then we say that t is dense.*

The Hilger derivative or delta derivative of a function f at a point $t \in \mathbb{T}$ can be defined as follows.

Definition 1.4.8 *Assume that $f : \mathbb{T} \to \mathbb{R}$ is a function and let $t \in \mathbb{T}^{\kappa}$. We define $f^{\Delta}(t)$ to be the number, provided that it exists, such that, for any $\varepsilon > 0$, there is a neighborhood U of t, $U = (t - \delta, t + \delta) \cap \mathbb{T}$ for some $\delta > 0$, such that*

$$|f(\sigma(t)) - f(s) - f^{\Delta}(t)(\sigma(t) - s)| \leq \varepsilon \, |\sigma(t) - s| \quad \text{for} \quad \text{all} \quad s \in U.$$

We call $f^{\Delta}(t)$ the delta or Hilger derivative of f at t. We say that f is delta or Hilger differentiable, shortly differentiable, in $\mathbb{T}^{\kappa}$ if $f^{\Delta}(t)$ exists for all $t \in \mathbb{T}^{\kappa}$. The function $f^{\Delta} : \mathbb{T} \to \mathbb{R}$ is said to be the delta derivative or Hilger derivative, shortly derivative, of f in $\mathbb{T}^{\kappa}$.

If $\mathbb{T} = \mathbb{R}$, then the delta derivative coincides with the classical derivative. Note that the delta derivative is well defined. The Hilger derivative has the following important properties.

Theorem 1.4.9 *Assume that $f : \mathbb{T} \to \mathbb{R}$ is a function and let $t \in \mathbb{T}^{\kappa}$. Then we have the following.*

1. *If f is differentiable at t, then f is continuous at t.*

2. *If f is continuous at t and t is right-scattered, then f is differentiable at t with*

$$f^{\Delta}(t) = \frac{f(\sigma(t)) - f(t)}{\mu(t)}.$$

3. *If t is right-dense, then f is differentiable at t iff the limit*

$$\lim_{s \to t} \frac{f(t) - f(s)}{t - s}$$

exists as a finite number. In this case,

$$f^{\Delta}(t) = \lim_{s \to t} \frac{f(t) - f(s)}{t - s}.$$

4. *If f is differentiable at t, then*

$$f(\sigma(t)) = f(t) + \mu(t) f^{\Delta}(t).$$

Definition 1.4.10 1. *A function $f : \mathbb{T} \to \mathbb{R}$ is called regulated provided its right-sided limits exist (finite) at all right-dense points in $\mathbb{T}$ and its left-sided limits exist (finite) at all left-dense points in $\mathbb{T}$.*

2. *A continuous function $f : \mathbb{T} \to \mathbb{R}$ is called pre-differentiable with the region of differentiation D provided*

   (a) *$D \subset \mathbb{T}^{\kappa}$,*

   (b) *$\mathbb{T}^{\kappa} \backslash D$ is countable and contains no right-scattered elements of $\mathbb{T}$,*

   (c) *f is differentiable at each $t \in D$.*

Theorem 1.4.11 *Let $t_0 \in \mathbb{T}$, $x_0 \in \mathbb{R}$, $f : \mathbb{T}^{\kappa} \to \mathbb{R}$ be a given regulated map. Then there exists exactly one pre-differentiable function F satisfying*

$$F^{\Delta}(t) = f(t) \quad \text{for all} \quad t \in D, \quad F(t_0) = x_0.$$

Definition 1.4.12 1. *Assume $f : \mathbb{T} \to \mathbb{R}$ is a regulated function. Any function F in Theorem 1.4.11 is called a pre-antiderivative of f. We define the indefinite integral of a regulated function f by*

$$\int f(t) \Delta t = F(t) + c,$$

where c is an arbitrary constant and F is a pre-antiderivative of f. We define the Cauchy integral by

$$\int_{\tau}^{s} f(t)\Delta t = F(s) - F(\tau) \quad \text{for all} \quad \tau, \mathrm{s} \in \mathbb{T}.$$

2. *A function $F : \mathbb{T} \to \mathbb{R}$ is called an anti-derivative of $f : \mathbb{T} \to \mathbb{R}$ provided*

$$F^{\Delta}(t) = f(t) \quad \text{holds for all} \quad \mathrm{t} \in \mathbb{T}^{\kappa}.$$

Definition 1.4.13 *A function $f : \mathbb{T} \to \mathbb{R}$ is called rd-continuous provided that it is continuous at the right-dense points in $\mathbb{T}$ and its left-sided limits exist (finite) at the left-dense points in $\mathbb{T}$. The set of rd-continuous functions $f : \mathbb{T} \to \mathbb{R}$ is denoted by $\mathrm{C}_{\mathrm{rd}}(\mathbb{T})$. The set of functions $f : \mathbb{T} \to \mathbb{R}$ that are differentiable and whose derivatives are rd-continuous is denoted by $\mathrm{C}^1_{\mathrm{rd}}(\mathbb{T})$.*

Note that if f is rd-continuous, then f is regulated. The defined integral has the following properties.

Theorem 1.4.14 *If $a, b, c \in \mathbb{T}$, $\alpha \in \mathbb{R}$ and $f, g \in \mathrm{C}_{\mathrm{rd}}(\mathbb{T})$, then*

(i) $\displaystyle\int_a^b (f(t) + g(t))\Delta t = \int_a^b f(t)\Delta t + \int_a^b g(t)\Delta t,$

(ii) $\displaystyle\int_a^b (\alpha f)(t)\Delta t = \alpha \int_a^b f(t)\Delta t,$

(iii) $\displaystyle\int_a^b f(t)\Delta t = -\int_b^a f(t)\Delta t,$

(iv) $\displaystyle\int_a^b f(t)\Delta t = \int_a^c f(t)\Delta t + \int_c^b f(t)\Delta t,$

(v) $\displaystyle\int_a^b f(\sigma(t))g^{\Delta}(t)\Delta t = (fg)(b) - (fg)(a) - \int_a^b f^{\Delta}(t)g(t)\Delta t,$

(vi) $\displaystyle\int_a^b f(t)g^{\Delta}(t)\Delta t = (fg)(b) - (fg)(a) - \int_a^b f^{\Delta}(t)g(\sigma(t))\Delta t,$

(vii) $\int_a^a f(t)\Delta t = 0,$

(viii) if $|f(t)| \leq g(t)$ *on* $[a,b)$, *then*

$$\left|\int_a^b f(t)\Delta t\right| \leq \int_a^b g(t)\Delta t,$$

(ix) if $f(t) \geq 0$ *for all* $a \leq t < b$, *then* $\int_a^b f(t)\Delta t \geq 0.$

Many results concerning differential equations carry over quite easily to the corresponding results for difference equations, while other results seem to be completely different from their continuous counterparts. The study of dynamic equations on time scales reveals such discrepancies and helps avoid proving results twice, once for differential equations and once again for difference equations. The general idea is to prove a result for a dynamic equation where the domain of the unknown function is a so-called time scale (also known as a time-set), which may be an arbitrary closed subset of the reals. In this way, the results apply not only to the set of real numbers or a set of integers but to more general time scales such as a Cantor set.

The three most popular examples of calculus on time scales are differential calculus, difference calculus, and quantum calculus. Dynamic equations on a time scale have potential for applications, such as in population dynamics. For example, they can model insect populations that evolve continuously while in season, die out in winter while their eggs are incubating or dormant, and then hatch in a new season, giving rise to a non-overlapping population.

Functional dynamic equations on time scales, fractional dynamic calculus and fractional dynamic equations on time scales, boundary value problems for dynamic equations on time scales, integral equations on time scales, impulsive equations on time scales, and projector analysis of dynamic equations on time scales were explored by [15, 16]. The dynamic geometry on time scales and its basic concepts were investigated by [17].

Recently, in [31], the Hukuhara differentiability of functions with interval values and differential equations with intervals is introduced on time scales. In [42], the following definitions for the Hukuhara delta derivative of the first type and the second type on time scales are introduced.

Definition 1.4.15 *Assume that $f : \mathbb{T} \to F(\mathbb{R})$ is a fuzzy function and $t \in \mathbb{T}^{\kappa}$. Then f is said to be first type right fuzzy delta differentiable at t, shortly right δ_H-differentiable at t, if there exists an element $\delta_H^+ f(t) \in F(\mathbb{R})$ with the property that, for any given $\varepsilon > 0$, there exists a neighborhood $U_{\mathbb{T}}$ of t, i.e., $U_{\mathbb{T}} = (t-\delta, t+\delta)\bigcap \mathbb{T}$ for some $\delta >$ 0, such that for all $t+h \in U_{\mathbb{T}}$ the H-difference $f(t+h) \ominus_H f(\sigma(t))$ exists and*

$$D\big(f(t+h) \ominus_H f(\sigma(t)), \delta_H^+ f(t)(h-\mu(t))\big) \leq \varepsilon |h-\mu(t)|$$

with $0 \leq h < \delta$. In this case, $\delta_H^+ f(t)$ is said to be the derivative of the fuzzy delta of the first type right of f at t, shortly right of δ_H-derivative of f at t.

Definition 1.4.16 *Assume that $f : \mathbb{T} \to F(\mathbb{R})$ is a fuzzy function and $t \in \mathbb{T}^{\kappa}$. Then f is said to be first type left fuzzy delta differentiable at t, shortly left δ_H-differentiable at t, if there exists an element $\delta_H^- f(t) \in F(\mathbb{R})$ with the property that, for any given $\varepsilon > 0$, there exists a neighborhood $U_{\mathbb{T}}$ of t, i.e., $U_{\mathbb{T}} = (t-\delta, t+\delta)\bigcap \mathbb{T}$ for some $\delta > 0$, such that for all $t-h \in U_{\mathbb{T}}$ the H-difference $f(\sigma(t)) \ominus_H f(t-h)$ exists and*

$$D\big(f(\sigma(t)) \ominus_H f(t-h), \delta_H^- f(t)(h+\mu(t))\big) \leq \varepsilon (h+\mu(t))$$

with $0 \leq h < \delta$.

Definition 1.4.17 *Let $f : \mathbb{T} \to F(\mathbb{R})$ be a fuzzy function and $t \in \mathbb{T}^{\kappa}$. Then f is said to be first type fuzzy delta differentiable at t, shortly δ_H-differentiable at t, if f is both first type left and right fuzzy delta differentiable at $t \in \mathbb{T}^{\kappa}$ and $\delta_H^- f(t) = \delta_H^+ f(t)$, and we will denote it by $\delta_H f(t)$. We call $\delta_H f(t)$ the first type fuzzy delta derivative of f at* *t, shortly δ_H-derivative of f at t. We say that f is first type fuzzy delta differentiable at t, shortly δ_H-differentiable at t, if its δ_H-derivative exists at t. We say that f is first type fuzzy delta differentiable on $\mathbb{T}^\kappa$, shortly δ_H-differentiable on $\mathbb{T}^\kappa$, if its δ_H-derivative exists at each $t \in \mathbb{T}^\kappa$. The fuzzy function $\delta_H f : \mathbb{T}^\kappa \to F(\mathbb{R})$ is then called the first-type fuzzy delta derivative, shortly δ_H-derivative of f in $\mathbb{T}^\kappa$.*

Definition 1.4.18 *Assume that $f : \mathbb{T} \to F(\mathbb{R})$ is a fuzzy function and $t \in \mathbb{T}^\kappa$. Then f is said to be second type right fuzzy delta differentiable at t, shortly right δ_{SH}-differentiable at t, if there exists an element $\delta^+_{SH} f(t) \in F(\mathbb{R})$ with the property that for any given $\varepsilon > 0$ there exists a neighborhood $U_\mathbb{T}$ of t, i.e., $U_\mathbb{T} = (t-\delta, t+\delta)\bigcap \mathbb{T}$ for some $\delta > 0$ such that for all $t+h \in U_\mathbb{T}$ the H-difference $f(\sigma(t)) \ominus_H f(t+h)$ exists and*

$$D\big(f(\sigma(t)) \ominus_H f(t+h), \delta^+_{SH} f(t)(-(h-\mu(t)))\big) \leq \varepsilon|h-\mu(t)|$$

with $0 \leq h < \delta$. In this case, $\delta^+_{SH} f(t)$ is said to be a second type right fuzzy delta derivative of f at t, shortly right δ_{SH}-derivative of f at t.

Definition 1.4.19 *Assume that $f : \mathbb{T} \to F(\mathbb{R})$ is a fuzzy function and $t \in \mathbb{T}^\kappa$. Then f is said to be second type left fuzzy delta differentiable at t, shortly left δ_{SH}-differentiable at t, if there exists an element $\delta^-_{SH} f(t) \in F(\mathbb{R})$ with the property that for any given $\varepsilon > 0$ there exists a neighborhood $U_\mathbb{T}$ of t, i.e., $U_\mathbb{T} = (t-\delta, t+\delta)\bigcap \mathbb{T}$ for some $\delta > 0$ such that for all $t-h \in U_\mathbb{T}$ the H-difference $f(t-h) \ominus_H f(\sigma(t))$ exists and*

$$D\big(f(t-h) \ominus_H f(\sigma(t)), \delta^-_{SH} f(t)(-(h+\mu(t)))\big) \leq \varepsilon(h+\mu(t))$$

with $0 \leq h < \delta$.

Definition 1.4.20 *Let $f : \mathbb{T} \to F(\mathbb{R})$ be a fuzzy function and $t \in \mathbb{T}^\kappa$. Then f is said to be the second type fuzzy delta differentiable at t, shortly δ_{SH}-differentiable at t, if f is both second type left and right* *fuzzy delta differentiable at* $t \in \mathbb{T}^{\kappa}$ *and* $\delta_{SH}^{-} f(t) = \delta_{SH}^{+} f(t)$, *and we will denote it by* $\delta_{SH} f(t)$. *We call* $\delta_{SH} f(t)$ *the second type fuzzy delta derivative of* f *at* t, *shortly the* δ_{SH}*-derivative of* f *at* t. *We say that* f *is second type fuzzy delta differentiable at* t, *shortly* δ_{SH}*-differentiable at* t, *if its* δ_{SH}*-derivative exists at* t. *We say that* f *is a fuzzy delta of the second type differentiable in* $\mathbb{T}^{\kappa}$, *shortly* δ_{SH}*-differentiable in* $\mathbb{T}^{\kappa}$, *if its* δ_{SH}*-derivative exists at each* $t \in \mathbb{T}^{\kappa}$. *The fuzzy function* $\delta_{SH} f : \mathbb{T}^{\kappa} \to F(\mathbb{R})$ *is then called the second type fuzzy delta derivative, shortly* δ_{SH}*-derivative of* f *in* $\mathbb{T}^{\kappa}$.

Some of the properties of the Hukuhara delta derivative for fuzzy functions on time scales and presented sufficient conditions for the existence and uniqueness of solutions to fuzzy dynamic equations on time scales are investigated. In [40], the authors studied the generalized differentiability and integrability of fuzzy functions on time scales. Some results on the calculus of fuzzy functions on time scales are presented, and the concept of Henstock-Kurzweil integrals for fuzzy functions is used. Fuzzy dynamic equations on time scales under the second Hukuhara delta derivative were studied in [41]. They presented some properties of the second Hukuhara delta derivative for fuzzy functions and established the existence and uniqueness criteria for fuzzy dynamic equations on time scales, using Banach contraction mapping. The use of fuzzy dynamic equations is a natural way to model dynamical systems under possible uncertainty. The generalized differentiability of fuzzy number-valued functions was introduced and studied in [3]. Fuzzy differential equations under generalized derivative were investigated in [33]. Difference equations appear naturally as discrete analogs and as numerical solutions of differential and delay differential equations that model various diverse phenomena in biology, ecology, physiology, physics, engineering, and economics. Fuzzy difference equations are difference equations with fuzzy parameters and fuzzy initial values, and the solution is a sequence of fuzzy numbers. Fuzzy difference equations are important for the analysis of real-world phenomena. Lakshmikantham and Vatsala presented the basic theory of fuzzy difference equations in [28].

Recently, there has been an increasing investigation regarding the existence of solutions for fuzzy difference equations. First-order linear fuzzy dynamic equations are one of the simplest fuzzy dynamic equations which may appear in many applications. Bede et al. in [3] studied first-order linear fuzzy differential equations under the generalized differentiability concept and presented solutions to this problem in some cases. Later, solutions to this problem are presented in general cases. In [18] the basic classes of fuzzy dynamic equations on time scales are investigated.

In this book, some classes of fuzzy impulsive dynamic equations on arbitrary time scales are investigated. Integral representations of the solutions of the first-order and second-order fuzzy impulsive dynamic equations are deduced. Criteria for the existence and uniqueness of solutions for first-order and second-order fuzzy impulsive dynamic equations are given. Some boundary value problems are investigated for first- and second-order fuzzy impulsive dynamic equations on arbitrary time scales. Oscillations of a fuzzy function are defined. The solutions of the first-order and second-order basic-class fuzzy impulsive dynamic equations on arbitrary time scales for oscillations are investigated.

1.5 EASTERN VS. WESTERN PERSPECTIVE

The well-known American professor Bart Kosko highlighted the differences between Eastern and Western philosophies regarding the concept of truth; in summary, in opposition to Aristotle's Buddha. In fact, Kosko said that Western philosophy, Aristotle's successor, has accepted uncritically the bivalent system as the system that is useful, but overly simplifying complex reality. Put simply: What has won in simplicity is lost in accuracy. By contrast, Eastern philosophies—Buddha, Lao Tse, Confucius, etc.—always have accepted the strict unity of opposites, of what they call (as we know) the yin and yang.

On the other hand, if it is true that Aristotle was the great introducer of absolute bivalence, we should not ignore that he did not completely overlook aspects that could be fuzzy propositions, as when he commented that: "In any case, what is said according to these (qualities) supports arguably the most and least", or when we talk about that we can come to knowledge, but without the certainty of it. If Aristotle did not study this concept, it may have been lacking the necessary mathematical knowledge for development. It was not until the appearance of a calculus increasingly systematic and operational, combinatorial, and probability theory, or the new theory, now known as "Crisp", or Classic, Sets, initiated by Cantor, as well as modern statistics and matrix calculations. As mentioned earlier, Aristotle did not have the mathematical apparatus for developing fuzzy logic. The gestation of this construct starts with Newton and Leibniz, who developed calculus in the seventeenth century. Despite the obsession with accuracy in mathematics, they have made a perfectly applicable sentence by Einstein: To the extent that mathematics refers to reality, this is not true. And to the extent they are true, do not refer to reality. But what Einstein does not say is undoubtedly that the deductive apparatus developed by mathematics facilitates the understanding of reality. The explanation given by Aristotle movement is replaced by the most innovative in Newton, but thanks to a sharp appoggiatura in calculus, without which it would not have been possible. However, the infinitesimal calculus in depth was only used for the study of physics in the following centuries, experiencing spectacular growth with Euler, Laplace, Lagrange, Fourier, and so on. Today permeates all sciences, both social and human and natural. This Calculus Mathematics introduced the question of degree: to what degree is modified by seamlessly changing B, where A is a dependent variable B?

Classical Aristotelian logic has therefore been shown, and for a long time, to be quite effective in the science so-called "hard", such as mathematics or physics. But it is insufficient when the predicates contain imprecision, uncertainty, or vagueness; on the other hand, is how the brain actually works and human reasoning, and in general, it is how systems behave around us. Fuzzy logic has also helped the software interpret judgments of this kind.

1.6 APPLICATIONS OF FUZZY LOGIC

In fuzzy logic, the statements are not absolutely true or absolutely false. One thing can be true for 5% (technically, its "degree of truth" is 0.05). And variables (or categories) are no longer numbers but have no linguistically accurate names or borders (high or low, rich or poor, fat or thin, healthy or sick, old or young, hot or cold, or conversely, "is normal" and with what degree this occurs, etc.), operators that modify them are of the "lot", "pretty", "too", "almost", or "not too much." These are called fuzzy modifiers ("hedges" or even "fuzzy modifiers"). This is attributed to a value to each of these features, diffuse, and that value is accentuated or diluted according, for example, to the various powers or roots of the initial values or with polynomial expressions that depend on these functions. They are called modifiers "Intensification" or "Expansion" (Dilation), respectively, with gradual effects and variation. The feature of working with border regions is not fully defined in this area of mathematics.

In 1973, with the basic theory of Zadeh fuzzy controllers, other researchers began to apply fuzzy logic to various mechanical and industrial processes, improving on the existing hitherto. Established several research groups at universities fuzzy Japan. Thus, Professors Terano and Shibata in Tokyo, together with Professors Tanaka and Asai in Osaka, made significant contributions to both the development of fuzzy logic theory and its applications.

In 1980, Professor Ebrahim Mandani in the United Kingdom [32] designed the first fuzzy controller for a steam engine, which would be applied to control a cement plant in Denmark and did - of course - so with great success.

In 1987 Hitachi used a fuzzy controller for the Sendai train control, which used an innovative system created by man. Since then, the controller has been doing its job very efficiently. It was also during this year of 1987 that the company Omron developed the first commercial fuzzy controllers. So, the year 1987 is considered the "fuzzy boom" due to the large number of products based on fuzzy logic to be traded. In 1993, fuzzy logic was applied to control chemical injection water treatment plants for the first time in Japan. It was right there in the country Nippon and South Korea, where fuzzy logic has reached a higher level, creating close partnerships between government, universities and industries [38].

Parallel to the study of the applications of fuzzy logic, Professors Takagi and Sugeno developed the first approach to construct fuzzy rules (Fuzzy Rules), from training data or training (Fuzzy Learning). The applications of fuzzy logic in everyday life since then have grown rapidly. In fact, it is already part of it. For example, some brands of washing machines that use fuzzy logic are Electrolux, AEG and Miele, and use these computational methods to moderate the program if washing clothes "is not very dirty", which is a vague concept. The technique is also widespread in ABS, braking cars, auto-focus cameras, or elevator control, spam filters, also called "spam", and now ubiquitous video games. Manufacturers do not want to give much publicity to the fuzziness implicit in these developments for an obvious reason. To say that your cars' brake controlled by fuzzy logic does not belong to the class of messages that can sell more cars. To build a fuzzy system, an engineer might begin with a set of fuzzy rules from an expert. An engineer might define the degrees of membership in various fuzzy input and output sets with sets of curves.

The Fuzzy Rules, or rules of a fuzzy system, define a set of overlapping patches that relate a full range of inputs to a full range of outputs. In that sense, the fuzzy system approximates some mathematical function or equation of cause and effect. A very important result says that fuzzy systems can approximate any continuous mathematical function. Bart Kosko [27] proved this uniform convergence theorem by showing that enough small fuzzy patches can sufficiently cover the graph of any function or input/output relation. The theorem also shows that we can pick in advance the maximum error of the approximation and be sure that there exists a finite number of fuzzy rules that achieve it. And recent advances in Neural Networks, or Neural Nets (in terms of programs that learn from experience), and Genetic Algorithms (as programs evolve over time) are certainly a fitting complement to fuzzy logic. Another key reason for increased research in this field would be the interest in Neural Networks and Fuzzy Systems resemblance. The relations between the two techniques were wanted, thus obtaining Neuro-Fuzzy Systems, which use neural network-based learning methods to identify and optimize their parameters. Then, as we say, genetic algorithms appeared, which together with neural networks and "Fuzzy Systems" are powerful and therefore of great interest for future research, both current and mathematics, and for the most new, it is already here, rapidly taking shape [38].

Neural networks are collections of "neurons" and "synapses" that change their values in response to input from surrounding neurons and synapses. The neural network acts like a computer because it maps inputs to outputs. The neurons and synapses may be silicon components, or equations in software, that simulate their behavior.

Supervised learning, reached through supervised networks, tunes the rules of a fuzzy system as if they were synapses. The user provides the first set of rules, which the neural net refines by running through hundreds of thousands of inputs, slightly varying the fuzzy sets each time to see how well the system performs. The network tends to keep the changes that improve performance and ignore the others.

Unsupervised learning, reached through unsupervised neural networks, blindly clusters data into groups (by a procedure called clustering), the members of which resemble one another. There may be no right or wrong response or way to organize the data. It is in advance. Fuzzy modeling is often used to transform the knowledge of an expert into a mathematical model. The emphasis is on constructing a fuzzy expert system that replaces the human expert. Also, as a tool that can assist human observers in the difficult task of transforming their observations into a mathematical model. In many fields of science, human observers have provided linguistic descriptions and explanations of various systems. However, to study these phenomena, one needs to construct a suitable mathematical model, a process that usually requires a very subtle mathematical understanding. Fuzzy modeling is a much more direct and natural approach to transform the linguistic description into such a model. A fuzzy model represents the real system in a form that corresponds closely to the way humans perceive it. Thus, the model is easily understandable and each parameter has a readily perceivable meaning. The model can be easily altered to incorporate new phenomena, and if its behavior is different from expected, it is usually easy to find which rule should be modified and how. Furthermore, the mathematical procedures used in fuzzy modeling have been tested many times, and their techniques are relatively well documented. The possible "bad reputation" or the bitter controversy over fuzzy logic, apart from prejudice, will be possibly wrong with the chosen name, as noted at the beginning. What is fuzzy logic is not in itself, which has a precise mathematical definition, but the world over which it applies, including our perception of its boundaries and categories [43].

CHAPTER 2

Fuzzy Calculus on Time Scales

In this chapter, we introduced new classes of delta derivatives called the first-type fuzzy delta derivative and the second-type fuzzy delta derivative. They are proven to have their existence and uniqueness. Some of their basic properties are deduced, such as fuzzy delta differentiation of a sum of two fuzzy functions and fuzzy delta differentiation of a multiplication of a fuzzy function with a constant. They are defined α-levels of fuzzy functions and they are given formulae for their fuzzy delta derivatives. They are defined as first-type fuzzy delta integration and second-type fuzzy delta integration of fuzzy functions, and some of their basic properties are deduced.

Suppose that $\mathbb{T}$ is a time scale with the forward jump operator and the delta differentiation operator σ and Δ, respectively. With $F(\mathbb{R})$ we denote the space of the real fuzzy numbers, and with $D(\cdot,\cdot)$ we will denote the Hausdorff distance between the real fuzzy numbers. For more details on fuzzy numbers and the Hausdorff distance between the real fuzzy numbers, we refer the reader to the appendix of this book.

2.1 FIRST TYPE FUZZY DELTA DIFFERENTIATION

Definition 2.1.1 *Assume that $f : \mathbb{T} \to F(\mathbb{R})$ is a fuzzy function and $t \in \mathbb{T}^\kappa$. Then f is said to be the first type right fuzzy delta*

DOI: 10.1201/9781003655367-2

differentiable at t*, shortly right* δ_H*-differentiable at* t*, if there exists an element* $\delta_H^+ f(t) \in F(\mathbb{R})$ *with the property that, for any given* $\varepsilon > 0$*, there exists a neighborhood* $U_{\mathbb{T}}$ *of* t*, i.e.,* $U_{\mathbb{T}} = (t-\delta, t+\delta) \bigcap \mathbb{T}$ *for some* $\delta > 0$*, such that for all* $t+h \in U_{\mathbb{T}}$ *the H-difference* $f(t+h) \ominus_H f(\sigma(t))$ *exists and*

$$D\big(f(t+h) \ominus_H f(\sigma(t)), \delta_H^+ f(t)(h-\mu(t))\big) \leq \varepsilon |h-\mu(t)|,$$

with $0 \leq h < \delta$*. In this case,* $\delta_H^+ f(t)$ *is said to be the derivative on the right of the fuzzy delta of the first type of* f *in* t*, shortly right of the derivative* δ_H *of* f *in* t*.*

Definition 2.1.2 *Assume that* $f : \mathbb{T} \to F(\mathbb{R})$ *is a fuzzy function and* $t \in \mathbb{T}^\kappa$*. Then* f *is said to be first type left fuzzy delta differentiable at* t*, shortly left* δ_H*-differentiable at* t*, if there exists an element* $\delta_H^- f(t) \in F(\mathbb{R})$ *with the property that, for any given* $\varepsilon > 0$*, there exists a neighborhood* $U_{\mathbb{T}}$ *of* t*, i.e.,* $U_{\mathbb{T}} = (t-\delta, t+\delta) \bigcap \mathbb{T}$ *for some* $\delta > 0$*, such that for all* $t-h \in U_{\mathbb{T}}$ *the H-difference* $f(\sigma(t)) \ominus_H f(t-h)$ *exists and*

$$D\big(f(\sigma(t)) \ominus_H f(t-h), \delta_H^- f(t)(h+\mu(t))\big) \leq \varepsilon (h+\mu(t)),$$

with $0 \leq h < \delta$*.*

Definition 2.1.3 *Let* $f : \mathbb{T} \to F(\mathbb{R})$ *be a fuzzy function and* $t \in \mathbb{T}^\kappa$*. Then* f *is said to be first-type fuzzy delta differentiable at* t*, shortly* δ_H*-differentiable at* t*, if* f *is both first-type left and right fuzzy delta differentiable at* $t \in \mathbb{T}^\kappa$ *and* $\delta_H^- f(t) = \delta_H^+ f(t)$*, and we will denote it by* $\delta_H f(t)$*. We call* $\delta_H f(t)$ *the first-type fuzzy delta derivative of* f *at* t*, shortly* δ_H*-derivative of* f *at* t*. We say that* f *is first-type fuzzy delta differentiable at* t*, shortly* δ_H*-differentiable at* t*, if its* δ_H*-derivative exists at* t*. We say that* f *is fuzzy delta differentiable in the first type on* $\mathbb{T}^\kappa$*, soon* δ_H *differentiable on* $\mathbb{T}^\kappa$*, if its* δ_H *derivative exists in each* $t \in \mathbb{T}^\kappa$*. The fuzzy function* $\delta_H f : \mathbb{T}^\kappa \to F(\mathbb{R})$ *is then called the first-type fuzzy delta derivative, shortly* δ_H*-derivative of* f *in* $\mathbb{T}^\kappa$*.*

Theorem 2.1.4 *If the δ_H-derivative of f in $t \in \mathbb{T}^\kappa$ exists, then it is unique. Hence, the δ_H-derivative is well defined.*

Proof 2.1.5 *Assume that $\delta_H^1 f(t)$ and $\delta_H^2 f(t)$ are the δ_H-derivative of f at $t \in \mathbb{T}^\kappa$. Take $\varepsilon > 0$ arbitrarily. Then there exists a $\delta = \delta(\varepsilon) > 0$ such that for any $t+h \in (t-\delta, t+\delta)$ the H-difference $f(t+h) \ominus_H f(\sigma(t))$ exists and*

$$D\left(\delta_H^1 f(t)(h-\mu(t)), f(t+h) \ominus_H f(\sigma(t))\right) \leq \varepsilon|h-\mu(t)|,$$

$$D\left(\delta_H^2 f(t)(h-\mu(t)), f(t+h) \ominus_H f(\sigma(t))\right) \leq \varepsilon|h-\mu(t)|,$$

with $0 \leq h < \delta$. Then

$$\begin{aligned}
&D\left(\delta_H^1 f(t), \delta_H^2 f(t)\right)\\
&= \frac{1}{|h-\mu(t)|} D\left(\delta_H^1 f(t)(h-\mu(t)), \delta_H^2 f(t)(h-\mu(t))\right)\\
&= \frac{1}{|h-\mu(t)|} D\Big(\delta_H^1 f(t)(h-\mu(t)) + f(t+h) \ominus_H f(\sigma(t)),\\
&\qquad \delta_H^2 f(t)(h-\mu(t)) + f(t+h) \ominus_H f(\sigma(t))\Big)\\
&\leq \frac{1}{|h-\mu(t)|} D\left(\delta_H^1 f(t)(h-\mu(t)), f(t+h) \ominus_H f(\sigma(t))\right)\\
&\quad + \frac{1}{|h-\mu(t)|} D\left(f(t+h) \ominus_H f(\sigma(t)), \delta_H^2 f(t)(h-\mu(t))\right)\\
&\leq \varepsilon + \varepsilon\\
&= 2\varepsilon,
\end{aligned}$$

for all $h-\mu(t) \neq 0$. Since $\varepsilon > 0$ was arbitrarily chosen, we conclude that

$$D\left(\delta_H^1 f(t), \delta_H^2 f(t)\right) = 0.$$

Therefore, $\delta_H^1 f(t) = \delta_H^2 f(t)$. *This completes the proof.*

Theorem 2.1.6 *Assume that* $f : \mathbb{T} \to F(\mathbb{R})$ *is a continuous function at* $t_1 \in \mathbb{T}^\kappa$ *and* t_1 *is right-scattered. Then* f *is* δ_H*-differentiable at* t_1 *and*

$$\delta_H f(t_1) = \frac{f(\sigma(t_1)) \ominus_H f(t_1)}{\mu(t_1)}.$$

Proof 2.1.7 *We have*

$$\begin{aligned}\lim_{h\to 0+} \frac{f(t_1+h) \ominus_H f(\sigma(t_1))}{h-\mu(t_1)} &= \frac{f(t_1) \ominus_H f(\sigma(t_1))}{-\mu(t_1)} \\ &= \frac{f(\sigma(t_1)) \ominus_H f(t_1)}{\mu(t_1)},\end{aligned}$$

and

$$\lim_{h\to 0+} \frac{f(\sigma(t_1)) \ominus_H f(t_1-h)}{h+\mu(t_1)} = \frac{f(\sigma(t_1)) \ominus_H f(t_1)}{\mu(t_1)}.$$

Hence,

$$\begin{aligned}&\lim_{h\to 0+} D\left(\frac{f(t_1+h) \ominus_H f(\sigma(t_1))}{h-\mu(t_1)}, \frac{f(\sigma(t_1)) \ominus_H f(t_1)}{\mu(t_1)}\right) \\ &= D\left(\frac{f(\sigma(t_1)) \ominus_H f(t_1)}{\mu(t_1)}, \frac{f(\sigma(t_1)) \ominus_H f(t_1)}{\mu(t_1)}\right) \\ &= 0,\end{aligned}$$

and

$$\begin{aligned}&\lim_{h\to 0+} D\left(\frac{f(\sigma(t_1)) \ominus_H f(t_1-h)}{h+\mu(t_1)}, \frac{f(\sigma(t_1)) \ominus_H f(t_1)}{\mu(t_1)}\right) \\ &= D\left(\frac{f(\sigma(t_1)) \ominus_H f(t_1)}{\mu(t_1)}, \frac{f(\sigma(t_1)) \ominus_H f(t_1)}{\mu(t_1)}\right) \\ &= 0.\end{aligned}$$

This completes the proof.

Example 2.1.8 *Let* $\mathbb{T} = \mathbb{N}$. *Consider*

$$f(t) = \left(t, t+1, t^2+1\right), \quad t \in \mathbb{T}.$$

Here

$$\begin{aligned}
\sigma(t) &= t+1, \\
\mu(t) &= 1, \\
f(\sigma(t)) &= \left(\sigma(t), \sigma(t)+1, (\sigma(t))^2+1\right) \\
&= \left(t+1, t+1+1, (t+1)^2+1\right) \\
&= \left(t+1, t+2, t^2+2t+2\right), \\
f(\sigma(t)) \ominus_H f(t) &= (1, 1, 2t+1), \\
\frac{f(\sigma(t)) \ominus_H f(t)}{\mu(t)} &= (1, 1, 2t+1), \quad t \in \mathbb{T}.
\end{aligned}$$

Therefore

$$\delta_H f(t) = (1, 1, 2t+1), \quad t \in \mathbb{T}.$$

Example 2.1.9 *Let* $\mathbb{T} = 2^{\mathbb{N}_0}$. *Consider*

$$f(t) = \left(t^2 - t, t^2+1, t^3+1\right), \quad t \in \mathbb{T}.$$

We have

$$\begin{aligned}
\sigma(t) &= 2t, \\
\mu(t) &= t,
\end{aligned}$$

$$\begin{aligned} f(\sigma(t)) &= \left((\sigma(t))^2 - \sigma(t), (\sigma(t))^2 + 1, (\sigma(t))^3 + 1\right) \\ &= \left(4t^2 - 2t, 4t^2 + 1, 8t^3 + 1\right), \\ f(\sigma(t)) \ominus_H f(t) &= \left(3t^2 - t, 3t^2, 7t^3\right), \\ \frac{f(\sigma(t)) \ominus_H f(t)}{\mu(t)} &= \left(3t - 1, 3t, 7t^2\right), \quad t \in \mathbb{T}. \end{aligned}$$

Therefore

$$\delta_H f(t) = \left(3t - 1, 3t, 7t^2\right), \quad t \in \mathbb{T}.$$

Example 2.1.10 *Let* $\mathbb{T} = [0,1] \bigcup \{3,7,8\}$. *Consider*

$$f(t) = (-3t + 2, t, 3t + 5), \quad t \in \mathbb{T}^{\kappa}.$$

We will find $\delta_H f(1)$. *We have*

$$\begin{aligned} \sigma(1) &= 3, \\ \mu(1) &= \sigma(1) - 1 \\ &= 3 - 1 \\ &= 2, \\ f(1) &= (-1, 1, 8), \\ f(\sigma(1)) &= (-3\sigma(1) + 2, \sigma(1), 3\sigma(1) + 5) \\ &= (2 - 9, 3, 9 + 5) \\ &= (-7, 3, 14), \\ f(\sigma(1)) \ominus_H f(1) &= (-6, 2, 6), \\ \frac{f(\sigma(1)) \ominus_H f(1)}{\mu(1)} &= (-3, 1, 3). \end{aligned}$$

Consequently

$$\delta_H f(1) = (-3, 1, 3).$$

Exercise 2.1.11 *Let* $\mathbb{T} = \left[0, \frac{1}{2}\right] \bigcup [1,4] \bigcup \{11, 15\}$, *where* $\left[0, \frac{1}{2}\right]$ *and* $[1,4]$ *are the real-valued intervals. Let also,*

$$f(t) = \left(\frac{t+1}{t+2}, t^2 + \frac{t+1}{t+2}, t^3 + t^2 + t + 1 + \frac{t+1}{t+2}\right), \quad t \in \mathbb{T}.$$

Find

$$\delta_H f\left(\frac{1}{2}\right), \quad \delta_H f(4), \quad \delta_H f(11).$$

Theorem 2.1.12 *Assume that* $f : \mathbb{T} \to F(\mathbb{R})$ *is* δ_H*-differentiable at* $t \in \mathbb{T}^\kappa$. *Then* f *is continuous at* t.

Proof 2.1.13 *Take* $\varepsilon > 0$ *arbitrarily. Since* f *is* δ_H*-differentiable at* t, *there exists a* $\delta > 0$ *such that for any* $t-h, t+h \in (t-\delta, t+\delta)$, $0 \le h < \delta$, $0 \le h < \varepsilon$, *the H-differences* $f(t+h) \ominus_H f(\sigma(t))$ *and* $f(\sigma(t)) \ominus_H f(t-h)$ *exist and*

$$\begin{aligned} D(f(t+h) \ominus_H f(\sigma(t)), \delta_H f(t)(h-\mu(t))) &\le \varepsilon |h-\mu(t)|, \\ D(f(\sigma(t)) \ominus_H f(t-h), \delta_H f(t)(h+\mu(t))) &\le \varepsilon (h+\mu(t)). \end{aligned}$$

In particular, we have

$$D(f(\sigma(t)) \ominus_H f(t), \delta_H f(t)\mu(t)) \le \varepsilon \mu(t).$$

Hence, for $t-h, t+h \in (t-\delta, t+\delta)$, $0 \le h < \delta$, *we get*

$$\begin{aligned} D(f(t+h), f(t)) &= D(f(t+h) \ominus_H f(\sigma(t)), f(t) \ominus_H f(\sigma(t))) \\ &= D\Big(f(t+h) \ominus_H f(\sigma(t)) + \delta_H f(t)(h-\mu(t)), \\ &\qquad f(t) \ominus_H f(\sigma(t)) + \delta_H f(t)(h-\mu(t)) \Big) \\ &\le D(f(t+h) \ominus_H f(\sigma(t)), \delta_H f(t)(h-\mu(t))) \end{aligned}$$

$$
\begin{aligned}
&+D\left(\delta_H f(t)(h-\mu(t)), f(t)\ominus_H f(\sigma(t))\right)\\
\leq\ & \varepsilon(h-\mu(t))+D\left(\delta_H f(t)h,\widetilde{0}\right)\\
&+D\left(\delta_H f(t)(-\mu(t)), f(t)\ominus_H f(\sigma(t))\right)\\
=\ & \varepsilon(h-\mu(t))+hD(\delta_H f(t),\widetilde{0})\\
&+D\left(f(\sigma(t))\ominus_H f(t), \delta_H f(t)\mu(t)\right)\\
\leq\ & \varepsilon(h-\mu(t))+h\|\delta_H f(t)\|+\varepsilon\mu(t)\\
=\ & h\left(\varepsilon+\|\delta_H f(t)\|\right)\\
<\ & \varepsilon\left(\varepsilon+\|\delta_H f(t)\|\right),
\end{aligned}
$$

and

$$
\begin{aligned}
D(f(t),f(t-h)) &= D\left(f(t)\ominus_H f(\sigma(t)), f(t-h)\ominus_H f(\sigma(t))\right)\\
&= D\left(f(\sigma(t))\ominus_H f(t), f(\sigma(t))\ominus_H f(t-h)\right)\\
&= D\Big(f(\sigma(t))\ominus_H f(t)+\delta_H f(t)(h+\mu(t)),\\
&\qquad f(\sigma(t))\ominus_H f(t-h)+\delta_H f(t)(h+\mu(t))\Big)\\
&\leq D\left(\delta_H f(t)(h+\mu(t)), f(\sigma(t))\ominus_H f(t-h)\right)\\
&\quad +D\left(f(\sigma(t))\ominus_H f(t), \delta_H f(t)(h+\mu(t))\right)
\end{aligned}
$$

$$\leq \varepsilon(h+\mu(t))+D(f(\sigma(t))\ominus_H f(t),\delta_H f(t)\mu(t))$$

$$+D\left(\delta_H f(t)h,\widetilde{0}\right)$$

$$\leq \varepsilon(h+\mu(t))+\varepsilon\mu(t)+hD\left(\delta_H f(t),\widetilde{0}\right)$$

$$< \varepsilon(\varepsilon+\mu(t))+\varepsilon\mu(t)+\varepsilon\|\delta_H f(t)\|$$

$$= \varepsilon\left(\varepsilon+2\mu(t)+\|\delta_H f(t)\|\right).$$

This completes the proof.

Theorem 2.1.14 *Let $f:\mathbb{T}\to F(\mathbb{R})$ be a fuzzy function and let $t\in\mathbb{T}^\kappa$ be right-dense. Then f is δ_H-differentiable at t if and only if the limits*

$$\lim_{h\to 0+}\frac{f(t+h)\ominus_H f(t)}{h} \quad \text{and} \quad \lim_{h\to 0+}\frac{f(t)\ominus_H f(t-h)}{h}, \tag{2.1}$$

exist and satisfy the relations

$$\lim_{h\to 0+}\frac{f(t+h)\ominus_H f(t)}{h}=\lim_{h\to 0+}\frac{f(t)\ominus_H f(t-h)}{h}=\delta_H f(t).$$

Proof 2.1.15 *1. Assume that f is δ_H-differentiable at t. Take $\varepsilon>0$ arbitrarily. Since f is δ_H-differentiable at t, there exists a $\delta>0$ such that for $t-h,t+h\in(t-\delta,t+\delta)$, $0\leq h<\delta$, the H-differences $f(t+h)\ominus_H f(\sigma(t))$ and $f(\sigma(t))\ominus_H f(t-h)$ exist and*

$$D(f(t+h)\ominus_H f(\sigma(t)),\delta_H f(t)(h-\mu(t))) \leq \varepsilon|h-\mu(t)|,$$

$$D(f(\sigma(t))\ominus_H f(t-h),\delta_H f(t)(\mu(t)+h)) \leq \varepsilon(\mu(t)+h),$$

and using that $\sigma(t)=t$, we arrive at the following inequalities

$$\begin{aligned} D(f(t+h)\ominus_H f(t),\delta_H f(t)h) &\leq \varepsilon h,\\ D(f(t)\ominus_H f(t-h),\delta_H f(t)h) &\leq \varepsilon h. \end{aligned} \tag{2.2}$$

Thus,

$$D\left(\frac{f(t+h)\ominus_H f(t)}{h}, \delta_H f(t)\right) \leq \varepsilon,$$
$$D\left(\frac{f(t)\ominus_H f(t-h)}{h}, \delta_H f(t)\right) \leq \varepsilon, \tag{2.3}$$

for all $t-h, t+h \in (t-\delta, t+\delta)$, $0 < h < \delta$. *Because* $\varepsilon > 0$ *was arbitrarily chosen, we obtain the following relations*

$$\delta_H f(t) = \lim_{h\to 0+} \frac{f(t+h)\ominus_H f(t)}{h} = \lim_{h\to 0+} \frac{f(t)\ominus_H f(t-h)}{h}.$$

2. *Suppose that* t *is right-dense and the limits (2.1) exist. Then, for any* $\varepsilon > 0$, *there is a* $\delta > 0$ *such that for* $t-h, t+h \in (t-\delta, t+\delta)$, $0 < h < \delta$, *we have (2.3), whereupon we get (2.2). This completes the proof.*

Example 2.1.16 *Let* $\mathbb{T} = [0,1]\bigcup\left\{1+\frac{1}{2^k}\right\}_{k\in\mathbb{N}}$, *where* $[0,1]$ *is the real-valued interval. Let also,*

$$f(t) = \left(t, 3t+7, 7+3t+t^2\right), \quad t \in \mathbb{T}.$$

We will find $\delta_H f(1)$. *Note that* $\sigma(1) = 1$. *Consequently,* $t = 1$ *is a right-dense point. Take* $h > 0$ *arbitrarily. We have*

$$f(1) = (1, 10, 11),$$

$$\begin{aligned} f(1+h) &= \left(1+h, 3(1+h)+7, 7+3(1+h)+(1+h)^2\right) \\ &= \left(1+h, 10+3h, 10+3h+1+2h+h^2\right) \\ &= \left(1+h, 10+3h, 11+5h+h^2\right), \end{aligned}$$

$$f(1-h) = \left(1-h, 3(1-h)+7, 7+3(1-h)+(1-h)^2\right)$$

$$= \left(1-h, 10-3h, 10-3h+1-2h+h^2\right)$$

$$= \left(1-h, 10-3h, 11-5h+h^2\right),$$

$$f(1+h) \ominus_H f(1) = \left(h, 3h, 5h+h^2\right),$$

$$f(1) \ominus_H f(1-h) = \left(h, 3h, 5h-h^2\right),$$

$$\frac{f(1+h) \ominus_H f(1)}{h} = (1,3,5+h),$$

$$\frac{f(1) \ominus_H f(1-h)}{h} = (1,3,5-h),$$

$$\lim_{h\to 0+} \frac{f(1+h) \ominus_H f(1)}{h} = (1,3,5),$$

$$\lim_{h\to 0+} \frac{f(1) \ominus_H f(1-h)}{h} = (1,3,5).$$

Therefore

$$\delta_H f(1) = (1,3,5).$$

Exercise 2.1.17 *Let* $\mathbb{T} = \left\{\frac{1}{2^k}\right\}_{k\in\mathbb{N}} \bigcup \{0\}$ *and*

$$f(t) = \left(1+t^2, 1+3t+4t^2+\frac{t+4}{t+1}, 11+5t+7t^2+\frac{t+4}{t+1}\right), \quad t \in \mathbb{T}.$$

Find $\delta_H f(0)$.

Theorem 2.1.18 *Let* $f : \mathbb{T} \to F(\mathbb{R})$ *be* δ_H*-differentiable at* $t \in \mathbb{T}^\kappa$. *Then*

$$f(\sigma(t)) = f(t) + \mu(t) \cdot \delta_H f(t),$$

or

$$f(t) = f(\sigma(t)) + (-1) \cdot (\mu(t) \cdot \delta_H f(t)).$$

Proof 2.1.19 *1. Let $\sigma(t) = t$. Then $\mu(t) = 0$ and*

$$f(\sigma(t)) = f(t) + \mu(t) \cdot \delta_H f(t),$$

or

$$f(t) = f(\sigma(t)) + (-1)\mu(t) \cdot \delta_H f(t).$$

2. Let $\sigma(t) > t$. Then, by Theorem 2.1.6, we have

$$f(\sigma(t)) \ominus_H f(t) = \mu(t) \cdot \delta_H f(t).$$

Hence, using the definition of $\ominus_H$, we obtain

$$f(\sigma(t)) = f(t) + \mu(t) \cdot \delta_H f(t),$$

or

$$f(t) = f(\sigma(t)) + (-1) \cdot (\mu(t) \cdot \delta_H f(t)).$$

This completes the proof.

Example 2.1.20 *Let $\mathbb{T} = 2\mathbb{N}_0$ and*

$$f(t) = \left(\frac{t+1}{t+2}, t, t^2 + 5t + 1 \right), \quad t \in \mathbb{T}.$$

We have

$$\begin{aligned}
\sigma(t) &= t+2, \\
\mu(t) &= 2, \\
f(\sigma(t)) &= \left(\frac{\sigma(t)+1}{\sigma(t)+2}, \sigma(t), (\sigma(t))^2 + 5\sigma(t) + 1 \right) \\
&= \left(\frac{t+2+1}{t+2+2}, t+2, (t+2)^2 + 5(t+2) + 1 \right) \\
&= \left(\frac{t+3}{t+4}, t+2, t^2 + 4t + 4 + 5t + 11 \right) \\
&= \left(\frac{t+3}{t+4}, t+2, t^2 + 9t + 15 \right),
\end{aligned}$$

and

$$
\begin{aligned}
& f(\sigma(t)) \ominus_H f(t) \\
= & \left(\frac{t+3}{t+4} - \frac{t+1}{t+2}, t+2-t, t^2+9t+15-t^2-5t-1\right) \\
= & \left(\frac{(t+3)(t+2)-(t+1)(t+4)}{(t+2)(t+4)}, 2, 4t+14\right) \\
= & \left(\frac{t^2+5t+6-t^2-5t-4}{(t+2)(t+4)}, 2, 4t+14\right) \\
= & \left(\frac{2}{(t+2)(t+4)}, 2, 4t+14\right),
\end{aligned}
$$

and

$$
\begin{aligned}
\delta_H f(t) &= \frac{f(\sigma(t)) \ominus_H f(t)}{\mu(t)} \\
&= \left(\frac{1}{(t+2)(t+4)}, 1, 2t+7\right),
\end{aligned}
$$

and

$$
\begin{aligned}
f(t)+\mu(t)\cdot\delta_H f(t) &= \left(\frac{t+1}{t+2}, t, t^2+5t+1\right) \\
& + \left(\frac{2}{(t+2)(t+4)}, 2, 4t+14\right) \\
= & \left(\frac{t+1}{t+2} + \frac{2}{(t+2)(t+4)}, t+2, t^2+5t+1+4t+14\right) \\
= & \left(\frac{(t+1)(t+4)+2}{(t+2)(t+4)}, t+2, t^2+9t+15\right) \\
= & \left(\frac{t^2+5t+6}{(t+2)(t+4)}, t+2, t^2+9t+15\right)
\end{aligned}
$$

$$= \left(\frac{(t+2)(t+3)}{(t+2)(t+4)}, t+2, t^2+9t+15\right)$$

$$= \left(\frac{t+3}{t+4}, t+2, t^2+9t+15\right), \quad t \in \mathbb{T}.$$

Thus,

$$f(\sigma(t)) = f(t) + \mu(t) \cdot \delta_H f(t), \quad t \in \mathbb{T}.$$

Theorem 2.1.21 *Let $f, g : \mathbb{T} \to F(\mathbb{R})$ be δ_H-differentiable at $t \in \mathbb{T}^\kappa$. Then $f+g : \mathbb{T} \to F(\mathbb{R})$ is δ_H-differentiable at $t \in \mathbb{T}^\kappa$ and*

$$\delta_H(f+g)(t) = \delta_H f(t) + \delta_H g(t).$$

Proof 2.1.22 *Take $\varepsilon > 0$ arbitrarily. Since $f, g : \mathbb{T} \to F(\mathbb{R})$ are δ_H-differentiable at $t \in \mathbb{T}^\kappa$, there exists a $\delta > 0$ such that for $t-h, t+h \in (t-\delta, t+\delta)$, $0 \le h < \delta$, the H-differences $f(t+h) \ominus_H f(\sigma(t))$ and $f(\sigma(t)) \ominus_H f(t-h)$ exist and*

$$\begin{aligned} D(f(t+h) \ominus_H f(\sigma(t)), \delta_H f(t)(h-\mu(t))) &\le \frac{\varepsilon}{2}|h-\mu(t)|, \\ D(f(\sigma(t)) \ominus_H f(t-h), \delta_H f(t)(h+\mu(t))) &\le \frac{\varepsilon}{2}(h+\mu(t)), \\ D(g(t+h) \ominus_H g(\sigma(t)), \delta_H g(t)(h-\mu(t))) &\le \frac{\varepsilon}{2}|h-\mu(t)|, \\ D(g(\sigma(t)) \ominus_H g(t-h), \delta_H g(t)(h+\mu(t))) &\le \frac{\varepsilon}{2}(h+\mu(t)). \end{aligned}$$

Hence, for $t-h, t+h \in (t-\delta, t+\delta)$, $0 \le h < \delta$, we have

$$\begin{aligned} & D((f+g)(t+h) \ominus_H (f+g)(\sigma(t)), (\delta_H f(t) + \delta_H g(t))(h-\mu(t))) \\ &= D\Big((f(t+h) \ominus_H f(\sigma(t))) + (g(t+h) \ominus_H g(\sigma(t))), \\ &\quad \delta_H f(t)(h-\mu(t)) + \delta_H g(t)(h-\mu(t))\Big) \end{aligned}$$

$$\begin{aligned}
&\leq D(f(t+h)\ominus_H f(\sigma(t)), \delta_H f(t)(h-\mu(t))) \\
&\quad +D(g(t+h)\ominus_H g(\sigma(t)), \delta_H g(t)(h-\mu(t))) \\
&\leq \frac{\varepsilon}{2}|h-\mu(t)| + \frac{\varepsilon}{2}|h-\mu(t)| \\
&= \varepsilon|h-\mu(t)|,
\end{aligned}$$

and

$$\begin{aligned}
&D((f+g)(\sigma(t))\ominus_H (f+g)(t+h), (\delta_H f(t)+\delta_H g(t))\,(h+\mu(t))) \\
&= D\Big((f(\sigma(t))\ominus_H f(t-h)) + (g(\sigma(t))\ominus_H g(t-h)), \\
&\quad \delta_H f(t)(h+\mu(t)) + \delta_H g(t)(h+\mu(t))\Big) \\
&\leq D(f(\sigma(t))\ominus_H f(t-h), \delta_H f(t)(h+\mu(t))) \\
&\quad +D(g(\sigma(t))\ominus_H g(t-h), \delta_H g(t)(h+\mu(t))) \\
&\leq \frac{\varepsilon}{2}(h+\mu(t)) + \frac{\varepsilon}{2}(h+\mu(t)) \\
&= \varepsilon(h+\mu(t)).
\end{aligned}$$

This completes the proof.

Example 2.1.23 *Let* $\mathbb{T} = \left\{\frac{1}{3^k}\right\}_{k\in\mathbb{N}} \bigcup\{0\}$ *and*

$$\begin{aligned}
f(t) &= \left(\frac{t+1}{t+2}, 1, 2\right), \\
g(t) &= (t,3,4), \quad t\in\mathbb{N}.
\end{aligned}$$

Note that $\sigma(0) = 0$, i.e., $t = 0$ is a right-dense point. We have

$$\begin{aligned} f(0) &= \left(\frac{1}{2}, 1, 2\right), \\ g(0) &= (0, 3, 4). \end{aligned}$$

Take $h > 0$ arbitrarily. Then

$$\begin{aligned} f(h) &= \left(\frac{h+1}{h+2}, 1, 2\right), \\ f(-h) &= \left(\frac{1-h}{2-h}, 1, 2\right), \\ f(h) \ominus_H f(0) &= \left(\frac{h+1}{h+2} - \frac{1}{2}, 0, 0\right) \\ &= \left(\frac{2h+2-h-2}{2(h+2)}, 0, 0\right) \\ &= \left(\frac{h}{2(h+2)}, 0, 0\right), \\ f(0) \ominus_H f(-h) &= \left(\frac{1}{2} - \frac{1-h}{2-h}, 0, 0\right) \\ &= \left(\frac{2-h-2+2h}{2(2-h)}, 0, 0\right) \\ &= \left(\frac{h}{2(2-h)}, 0, 0\right), \\ \frac{f(h) \ominus_H f(0)}{h} &= \left(\frac{1}{2(h+2)}, 0, 0\right), \\ \frac{f(0) \ominus_H f(-h)}{h} &= \left(\frac{1}{2(2-h)}, 0, 0\right), \\ \lim_{h \to 0} \frac{f(h) \ominus_H f(0)}{h} &= \left(\frac{1}{4}, 0, 0\right), \end{aligned}$$

$$\lim_{h\to 0}\frac{f(0)\ominus_H f(-h)}{h} = \left(\frac{1}{4},0,0\right).$$

Therefore

$$\begin{aligned}\lim_{h\to 0}\frac{f(h)\ominus_H f(0)}{h} &= \lim_{h\to 0}\frac{f(0)\ominus_H f(-h)}{h}\\ &= \delta_H f(0)\\ &= \left(\frac{1}{4},0,0\right).\end{aligned}$$

Next,

$$\begin{aligned}g(h) &= (h,3,4),\\ g(-h) &= (-h,3,4),\\ g(h)\ominus_H g(0) &= (h,0,0),\\ g(0)\ominus_H g(-h) &= (h,0,0),\\ \frac{g(h)\ominus_H g(0)}{h} &= (1,0,0),\\ \frac{g(0)\ominus_H g(-h)}{h} &= (1,0,0),\\ \lim_{h\to 0}\frac{g(h)\ominus_H g(0)}{h} &= (1,0,0),\\ \lim_{h\to 0}\frac{g(0)\ominus_H g(-h)}{h} &= (1,0,0).\end{aligned}$$

Thus,

$$\lim_{h\to 0}\frac{g(h)\ominus_H g(0)}{h} = \lim_{h\to 0}\frac{g(0)\ominus_H g(-h)}{h}$$

$$
\begin{aligned}
&= \delta_H g(0) \\
&= (1,0,0).
\end{aligned}
$$

Then

$$
\begin{aligned}
\delta_H f(0) + \delta_H g(0) &= \left(\frac{1}{4} + 1, 0, 0\right) \\
&= \left(\frac{5}{4}, 0, 0\right).
\end{aligned} \tag{2.4}
$$

Also,

$$
\begin{aligned}
(f+g)(t) &= \left(\frac{t+1}{t+2} + t, 4, 6\right) \\
&= \left(\frac{t+1+t^2+2t}{t+2}, 4, 6\right) \\
&= \left(\frac{t^2+3t+1}{t+2}, 4, 6\right), \quad t \in \mathbb{T}, \\
(f+g)(0) &= \left(\frac{1}{2}, 4, 6\right), \\
(f+g)(h) &= \left(\frac{h^2+3h+1}{h+2}, 4, 6\right), \\
(f+g)(-h) &= \left(\frac{h^2-3h+1}{2-h}, 4, 6\right), \\
(f+g)(h) \ominus_H (f+g)(0) &= \left(\frac{h^2+3h+1}{h+2} - \frac{1}{2}, 0, 0\right) \\
&= \left(\frac{2h^2+6h+2-h-2}{2(h+2)}, 0, 0\right)
\end{aligned}
$$

$$
\begin{aligned}
&= \left(\frac{2h^2+5h}{2(h+2)},0,0\right),\\
\frac{(f+g)(h)\ominus_H(f+g)(0)}{h} &= \left(\frac{2h+5}{2(h+2)},0,0\right),\\
\lim_{h\to 0}\frac{(f+g)(h)\ominus_H(f+g)(0)}{h} &= \left(\frac{5}{4},0,0\right),\\
(f+g)(0)\ominus_H(f+g)(-h) &= \left(\frac{1}{2}-\frac{h^2-3h+1}{2-h},0,0\right)\\
&= \left(\frac{2-h-2h^2+6h-2}{2(2-h)},0,0\right)\\
&= \left(\frac{-2h^2+5h}{2(2-h)},0,0\right),\\
\frac{(f+g)(0)\ominus_H(f+g)(-h)}{h} &= \left(\frac{-2h+5}{2(2-h)},0,0\right),\\
\lim_{h\to 0}\frac{(f+g)(h)\ominus_H(f+g)(0)}{h} &= \left(\frac{5}{4},0,0\right),\\
\lim_{h\to 0}\frac{(f+g)(0)\ominus_H(f+g)(-h)}{h} &= \left(\frac{5}{4},0,0\right).
\end{aligned}
$$

Consequently

$$
\begin{aligned}
\lim_{h\to 0}\frac{(f+g)(h)\ominus_H(f+g)(0)}{h} &= \lim_{h\to 0}\frac{(f+g)(0)\ominus_H(f+g)(-h)}{h}\\
&= \delta_H(f+g)(0)\\
&= \left(\frac{5}{4},0,0\right).
\end{aligned}
$$

From here and from (2.4), *we arrive at*

$$
\delta_H f(0)+\delta_H g(0)=\delta_H(f+g)(0).
$$

Theorem 2.1.24 *Let $f : \mathbb{T} \to F(\mathbb{R})$ be δ_H-differentiable at $t \in \mathbb{T}^\kappa$. Then for any $\lambda \in \mathbb{R}$ the function $\lambda \cdot f : \mathbb{T} \to F(\mathbb{R})$ is δ_H-differentiable at $t \in \mathbb{T}^\kappa$ and*

$$\delta_H(\lambda \cdot f)(t) = \lambda \cdot \delta_H f(t).$$

Proof 2.1.25 *If $\lambda = 0$, then the assertion is evident. Let $\lambda \neq 0$. Take $\varepsilon > 0$ arbitrarily. Since $f : \mathbb{T} \to F(\mathbb{R})$ is δ_H-differentiable at $t \in \mathbb{T}^\kappa$, there exists a $\delta > 0$ such that for $t-h, t+h \in (t-\delta, t+\delta)$, $0 \leq h < \delta$, the H-differences $f(t+h) \ominus_H f(\sigma(t))$ and $f(\sigma(t)) \ominus_H f(t-h)$ exist and*

$$\begin{aligned}
D(f(t+h) \ominus_H f(\sigma(t)), \delta_H f(t)(h-\mu(t))) &\leq \frac{\varepsilon}{|\lambda|}|h-\mu(t)|, \\
D(f(\sigma(t)) \ominus_H f(t-h), \delta_H f(t)(h+\mu(t))) &\leq \frac{\varepsilon}{|\lambda|}(h+\mu(t)).
\end{aligned}$$

Hence, for $t-h, t+h \in (t-\delta, t+\delta)$, $0 \leq h < \delta$, we get

$$\begin{aligned}
&D((\lambda \cdot f)(t+h) \ominus_H (\lambda \cdot f)(\sigma(t)), \lambda \cdot \delta_H f(t)(h-\mu(t))) \\
&= D(\lambda \cdot (f(t+h) \ominus_H f(\sigma(t))), \lambda \cdot \delta_H f(t)(h-\mu(t))) \\
&= |\lambda| D(f(t+h) \ominus_H f(\sigma(t)), \delta_H f(t)(h-\mu(t))) \\
&\leq |\lambda| \frac{\varepsilon}{|\lambda|}|h-\mu(t)| \\
&= \varepsilon|h-\mu(t)|,
\end{aligned}$$

and

$$\begin{aligned}
&D((\lambda \cdot f)(\sigma(t)) \ominus_H (\lambda \cdot f)(t-h), \lambda \cdot \delta_H f(t)(h+\mu(t))) \\
&= D(\lambda \cdot (f(\sigma(t)) \ominus_H f(t-h)), \lambda \cdot \delta_H f(t)(h+\mu(t))) \\
&= |\lambda| D(f(\sigma(t)) \ominus_H f(t-h), \delta_H f(t)(h+\mu(t)))
\end{aligned}$$

$$\leq |\lambda|\frac{\varepsilon}{|\lambda|}(h+\mu(t))$$

$$= \varepsilon(h+\mu(t)).$$

This completes the proof.

Theorem 2.1.26 *Let $t \in \mathbb{T}^{\kappa}$, $f : \mathbb{T} \to F(\mathbb{R})$ and $f_\alpha(t) = [f(t)]^\alpha$, $\alpha \in [0,1]$. If f is δ_H-differentiable at t, then f_α is δ_H-differentiable at t and*

$$\delta_H[f(t)]^\alpha = \delta_H f_\alpha(t), \quad \alpha \in [0,1].$$

Proof 2.1.27 *Let f is δ_H-differentiable at t and let $\alpha \in [0,1]$ be arbitrarily chosen.*

1. *Let t be right-dense. Take $h > 0$ arbitrarily. Since f is δ_H-differentiable at t, there exist*

$$\lim_{h\to 0}\frac{f(t+h)\ominus_H f(t)}{h} \quad \text{and} \quad \lim_{h\to 0}\frac{f(t)\ominus_H f(t-h)}{h},$$

and

$$\begin{aligned}\delta_H f(t) &= \lim_{h\to 0}\frac{f(t+h)\ominus_H f(t)}{h}\\ &= \lim_{h\to 0}\frac{f(t)\ominus_H f(t-h)}{h}.\end{aligned}$$

Next,

$$\begin{aligned}[f(t+h)\ominus_H f(t)]^\alpha &= f_\alpha(t+h)\ominus_H f_\alpha(t),\\ [f(t)\ominus_H f(t-h)]^\alpha &= f_\alpha(t)\ominus_H f_\alpha(t-h),\\ \frac{[f(t+h)\ominus_H f(t)]^\alpha}{h} &= \frac{f_\alpha(t+h)\ominus_H f_\alpha(t)}{h},\\ \frac{[f(t)\ominus_H f(t-h)]^\alpha}{h} &= \frac{f_\alpha(t)\ominus_H f_\alpha(t-h)}{h}.\end{aligned}$$

Hence,

$$
\begin{aligned}
[\delta_H f(t)]^\alpha &= \lim_{h\to 0} \frac{[f(t+h)\ominus_H f(t)]^\alpha}{h} \\
&= \lim_{h\to 0} \frac{f\alpha(t+h)\ominus_H f\alpha(t)}{h} \\
&= \lim_{h\to 0} \frac{[f(t)\ominus_H f(t+h)]^\alpha}{h} \\
&= \lim_{h\to 0} \frac{f\alpha(t)\ominus_H f\alpha(t-h)}{h}.
\end{aligned}
$$

Thus, the limits

$$
\lim_{h\to 0} \frac{f\alpha(t+h)\ominus_H f\alpha(t)}{h} \quad \text{and} \quad \lim_{h\to 0} \frac{f\alpha(t)\ominus_H f\alpha(t-h)}{h},
$$

exist and

$$
\begin{aligned}
\lim_{h\to 0} \frac{f\alpha(t+h)\ominus_H f\alpha(t)}{h} &= \lim_{h\to 0} \frac{f\alpha(t)\ominus_H f\alpha(t-h)}{h} \\
&= \delta_H f\alpha(t) \\
&= [\delta_H f(t)]^\alpha.
\end{aligned}
$$

2. *Let t be right-scattered. Since f is δ_H-differentiable at t, we have*

$$
\delta_H f(t) = \frac{f(\sigma(t))\ominus_H f(t)}{\mu(t)}.
$$

From here,

$$
\begin{aligned}
[\delta_H f(t)]^\alpha &= \frac{f\alpha(\sigma(t))\ominus_H f\alpha(t)}{\mu(t)} \\
&= \delta_H f\alpha(t).
\end{aligned}
$$

This completes the proof.

Theorem 2.1.28 *Let $t \in \mathbb{T}^{\kappa}$, $f : \mathbb{T} \to F(\mathbb{R})$ be δ_H-differentiable at t. Let also,*

$$[f(t)]^{\alpha} = \left[\underline{f}^{\alpha}(t), \overline{f}^{\alpha}(t)\right], \quad \alpha \in [0,1].$$

Then $\underline{f}^{\alpha}$ and $\overline{f}^{\alpha}$ are Δ-differentiable at t and

$$[\delta_H f(t)]^{\alpha} = \left[\underline{f}^{\alpha\Delta}(t), \overline{f}^{\alpha\Delta}(t)\right], \quad \alpha \in [0,1].$$

Proof 2.1.29 *Let $\alpha \in [0,1]$ be arbitrarily chosen.*

1. *Suppose that t is right-dense. Then*

$$\begin{aligned} [\delta_H f(t)]^{\alpha} &= \left[\lim_{h \to 0} \frac{\underline{f}^{\alpha}(t+h) - \underline{f}^{\alpha}(t)}{h}, \lim_{h \to 0} \frac{\overline{f}^{\alpha}(t+h) - \overline{f}^{\alpha}(t)}{h}\right] \\ &= \left[\underline{f}^{\alpha\Delta}(t), \overline{f}^{\alpha\Delta}(t)\right]. \end{aligned}$$

2. *Suppose that t is right-scattered. Then*

$$\begin{aligned} [\delta_H f(t)]^{\alpha} &= \left[\frac{\underline{f}^{\alpha}(\sigma(t)) - \underline{f}^{\alpha}(t)}{\mu(t)}, \frac{\overline{f}^{\alpha}(\sigma(t)) - \overline{f}^{\alpha}(t)}{\mu(t)}\right] \\ &= \left[\underline{f}^{\alpha\Delta}(t), \overline{f}^{\alpha\Delta}(t)\right]. \end{aligned}$$

This completes the proof.

Example 2.1.30 *Let $\mathbb{T} = [-1,1] \bigcup 2^{\mathbb{N}}$, where $[-1,1]$ is the real-valued interval. Consider the function $f : \mathbb{T} \to F(\mathbb{R})$ whose α-levels are given by*

$$[f(t)]^{\alpha} = g(t) \cdot [-1,1], \quad t \in \mathbb{T}, \quad \alpha \in [0,1],$$

where

$$g(t) = \begin{cases} 1 + \sin t & \text{if} \quad t \neq 0 \\ 1 & \text{if} \quad t = 0. \end{cases}$$

We will find $[\delta_H f(0)]^\alpha$, $\alpha \in [0,1]$. Here $t = 0$ is a right-dense point. For $h > 0$ sufficiently small, we have

$$\begin{aligned}
f(h) &= g(h)\cdot[-1,1] \\
&= (1+\sin h)\cdot[-1,1] \\
&= [-1-\sin h, 1+\sin h], \\
f(0) &= g(0)\cdot[-1,1] \\
&= 1\cdot[-1,1] \\
&= [-1,1], \\
[f(h)\ominus_H f(0)]^\alpha &= \Big[\min\{-1-sinh+1, 1+\sin h-1\}, \\
&\quad \max\{-1-\sin h+1, \quad 1+\sin h-1\}\Big] \\
&= \Big[\min\{-\sin h, \sin h\}, \max\{-\sin h, \sin h\}\Big] \\
&= [-\sin h, \sin h], \\
\frac{1}{h}[f(h)\ominus_H f(0)]^\alpha &= \frac{1}{h}\cdot[-\sin h, \sin h] \\
&= \left[-\frac{\sin h}{h}, \frac{\sin h}{h}\right], \\
\lim_{h\to 0}\frac{1}{h}[f(h)\ominus_H f(0)]^\alpha &= \lim_{h\to 0}\left[-\frac{\sin h}{h}, \frac{\sin h}{h}\right] \\
&= [-1,1].
\end{aligned}$$

Example 2.1.31 *Let $\mathbb{T} = [-1,3]\bigcup\{5,6,9\}$, where $[-1,3]$ is the real-valued interval. Consider the function $f : \mathbb{T} \to F(\mathbb{R})$ whose α-levels are given by*

$$[f(t)]^\alpha = g(t)\cdot[-1,2], \quad t \in \mathbb{T}, \quad \alpha \in [0,1],$$

where

$$g(t) = \begin{cases} \dfrac{t(t+1)}{t+4} & \textit{if} \quad t \in [-1,3] \\ \\ t \quad \textit{if} \quad t \in \{5,6,9\}. \end{cases}$$

We will find

$$[\delta_H f(0)]^\alpha \quad \textit{and} \quad [\delta_H f(3)]^\alpha.$$

Note that 0 *is a right-dense point and* 3 *is a right-scattered point. We have, for* $h \in (0,3)$,

$$\begin{aligned}
g(h) &= \frac{h(h+1)}{h+4}, \\
f(h) &= g(h) \cdot [-1,2] \\
&= \frac{h(h+1)}{h+4} \cdot [-1,2] \\
&= \left[-\frac{h(h+1)}{h+4}, \frac{2h(h+1)}{h+4} \right], \\
g(0) &= 0, \\
f(0) &= g(0) \cdot [-1,2] \\
&= 0 \cdot [-1,2] \\
&= [0,0], \\
g(3) &= \frac{12}{7}, \\
f(3) &= g(3) \cdot [-1,2]
\end{aligned}$$

$$
\begin{aligned}
&= \frac{12}{7}\cdot[-1,2] \\
&= \left[-\frac{12}{7},\frac{24}{7}\right],
\end{aligned}
$$

$$
\begin{aligned}
\sigma(3) &= 5, \\
\mu(3) &= \sigma(3)-3 \\
&= 5-3 \\
&= 2, \\
g(\sigma(3)) &= g(5) \\
&= 5, \\
f(\sigma(3)) &= g(\sigma(3))\cdot[-1,2] \\
&= 5\cdot[-1,2] \\
&= [-5,10].
\end{aligned}
$$

Then

$$
\begin{aligned}
[f(h)\ominus_H f(0)]^\alpha &= \left[\min\left\{-\frac{h(h+1)}{h+4},\frac{2h(h+1)}{h+4}\right\},\right. \\
&\qquad \left.\max\left\{-\frac{h(h+1)}{h+4},\frac{2h(h+1)}{h+4}\right\}\right] \\
&= \left[-\frac{h(h+1)}{h+4},\frac{2h(h+1)}{h+4}\right],
\end{aligned}
$$

$$
\begin{aligned}
\frac{1}{h}\cdot[f(h)\ominus_H f(0)]^\alpha &= \frac{1}{h}\cdot\left[-\frac{h(h+1)}{h+4}, \frac{2h(h+1)}{h+4}\right] \\
&= \left[-\frac{h+1}{h+4}, \frac{2(h+1)}{h+4}\right], \\
[\delta_H f(0)]^\alpha &= \lim_{h\to 0+}\frac{1}{h}\cdot[f(h)\ominus_H f(0)]^\alpha \\
&= \lim_{h\to 0+}\left[-\frac{h+1}{h+4}, \frac{2(h+1)}{h+4}\right] \\
&= \left[-\frac{1}{4}, \frac{1}{2}\right], \\
[f(\sigma(3))\ominus_H f(3)]^\alpha &= \left[\min\left\{-5+\frac{12}{7}, 10-\frac{24}{7}\right\},\right. \\
&\qquad \left.\max\left\{-5+\frac{12}{7}, 10-\frac{24}{7}\right\}\right] \\
&= \left[\min\left\{-\frac{23}{7}, \frac{46}{7}\right\}, \max\left\{-\frac{23}{7}, \frac{46}{7}\right\}\right] \\
&= \left[-\frac{23}{7}, \frac{46}{7}\right], \\
[\delta_H f(3)]^\alpha &= \frac{1}{\sigma(3)}\cdot[f(\sigma(3))\ominus_H f(3)]^\alpha \\
&= \frac{1}{5}\cdot\left[-\frac{23}{7}, \frac{46}{7}\right] \\
&= \left[-\frac{23}{35}, \frac{46}{35}\right].
\end{aligned}
$$

Exercise 2.1.32 *Let* $\mathbb{T} = [-1,4]\bigcup\{5,11,21\}$, *where* $[-1,4]$ *is the real-valued interval. Consider the function* $f : \mathbb{T} \to F(\mathbb{R})$ *whose*

α-levels are given by

$$[f(t)]^\alpha = g(t)\cdot[-1,7],$$

where

$$g(t) = \begin{cases} \dfrac{t(t^2+t+1)}{2t^2+3t+4} & \text{if} \quad t \in [-1,4] \\ \\ t^2+1 \quad \text{if} \quad t \in \{5,11,21\}. \end{cases}$$

Find

$$[\delta_H f(0)]^\alpha, \quad [\delta_H f(4)]^\alpha, \quad [\delta_H f(5)]^\alpha.$$

Example 2.1.33 *Let $\mathbb{T} = 2^{\mathbb{N}_0}$ and $f : [1,8] \to F(\mathbb{R})$ be a fuzzy function whose α-levels are given by*

$$[f(t)]^\alpha = g(t)\cdot[1,8], \quad \alpha \in [0,1], \quad t \in [1,8],$$

where $g(t) = 1+t^2$, $t \in \mathbb{T}$. We have

$$\sigma(t) = 2t, \quad t \in \mathbb{T},$$

and

$$\begin{aligned} [f(t)]^\alpha &= g(t)\cdot[1,8] \\ &= [1+t^2, 8+8t^2], \\ [f(\sigma(t))]^\alpha &= [1+(\sigma(t))^2, 8+8(\sigma(t))^2] \\ &= [1+4t^2, 8+32t^2], \quad t \in [1,8], \quad \alpha \in [0,1]. \end{aligned}$$

Then

$$\begin{aligned} [f(\sigma(t)) \ominus_H f(t)]^\alpha &= [\min\{1+4t^2-1-t^2, 8+32t^2-8-8t^2\}, \\ &\quad \max\{1+4t^2-1-t^2, 8+32t^2-8-8t^2\}] \end{aligned}$$

$$\begin{aligned} &= [\min\{3t^2, 24t^2\}, \max\{3t^2, 24t^2\}] \\ &= [3t^2, 24t^2], \quad t \in [1,8], \quad \alpha \in [0,1]. \end{aligned}$$

Thus,

$$\begin{aligned} [\delta_H f(t)]^\alpha &= \frac{1}{\mu(t)} \cdot [f(\sigma(t)) \ominus_H f(t)]^\alpha \\ &= \frac{1}{t} \cdot [3t^2, 24t^2] \\ &= [3t, 24t], \quad t \in [1,8], \quad \alpha \in [0,1]. \end{aligned}$$

Example 2.1.34 *Let $\mathbb{T} = 2\mathbb{Z}$ and $f : [0,10] \to F(\mathbb{R})$ be a fuzzy function whose α-levels are given by*

$$[f(t)]^\alpha = [-t, 1+3t], \quad t \in [0,10], \quad \alpha \in [0,1].$$

We will find $\delta_H f(t)$, $t \in [0,10]$. Here

$$\begin{aligned} \sigma(t) &= t+2, \quad t \in \mathbb{T}, \\ [f(\sigma(t))]^\alpha &= [-\sigma(t), 1+3\sigma(t)] \\ &= [-t-2, 1+3(t+2)] \\ &= [-2-t, 7+3t], \quad t \in [0,10], \quad \alpha \in [0,1]. \end{aligned}$$

Hence,

$$\begin{aligned} [f(\sigma(t)) \ominus_H f(t)]^\alpha &= \Big[\min\{-2-t+t, 7+3t-1-3t\}, \\ &\qquad \max\{-2-t+t, 7+3t-1-3t\}\Big] \\ &= [\min\{-2,6\}, \max\{-2,6\}] \\ &= [-2,6], \\ \frac{1}{\mu(t)} \cdot [f(\sigma(t)) \ominus_H f(t)]^\alpha &= \frac{1}{2} \cdot [-2,6] \\ &= [-1,3], \quad t \in [0,10], \quad \alpha \in [0,1]. \end{aligned}$$

Therefore

$$[\delta_H f(t)]^\alpha = [-1,3], \quad t \in [0,10], \quad \alpha \in [0,1].$$

Example 2.1.35 *Let* $\mathbb{T} = 3^{\mathbb{N}_0}$ *and* $f : \mathbb{T} \to F(\mathbb{R})$ *be a fuzzy function whose* α*-levels are given by*

$$[f(t)]^\alpha = g(t) \cdot [1,2], \quad t \in [1,27], \quad \alpha \in [0,1],$$

where

$$g(t) = \begin{cases} \dfrac{t+1}{t+2} & \text{if} \quad t \neq 3 \\ \\ 4 \quad \text{if} \quad t = 3. \end{cases}$$

We will find $[\delta_H f(1)]^\alpha$*. Here*

$$\begin{aligned} \sigma(t) &= 3t, \\ \\ \mu(t) &= 2t, \quad t \in \mathbb{T}, \\ \\ f(1) &= g(1) \cdot [1,2] \\ \\ &= \frac{2}{3} \cdot [1,2] \\ \\ &= \left[\frac{2}{3}, \frac{4}{3}\right], \\ \\ f(\sigma(1)) &= f(3) \\ \\ &= g(3) \cdot [1,2] \\ \\ &= 4 \cdot [1,2] \\ \\ &= [4,8], \end{aligned}$$

$$\begin{aligned} [f(\sigma(1)) \ominus_H f(1)]^\alpha &= \Big[\min\{4 - \frac{2}{3}, 8 - \frac{4}{3}\}, \\ &\quad \max\{4 - \frac{2}{3}, 8 - \frac{4}{3}\}\Big] \\ &= \left[\frac{10}{3}, \frac{20}{3}\right]. \end{aligned}$$

Exercise 2.1.36 *Find* $[\delta_H f(t)]^\alpha$, *where*

1. $\mathbb{T} = \mathbb{Z}$, $[f(t)]^\alpha = g(t) \cdot [-1, 4]$, $g(t) = t + t^2$, $t \in \mathbb{T}$, $\alpha \in [0, 1]$,
2. $\mathbb{T} = 2\mathbb{Z}$, $[f(t)]^\alpha = g(t) \cdot [1, 7]$, $g(t) = 1 + t$, $t \in \mathbb{T}$, $\alpha \in [0, 1]$,
3. $\mathbb{T} = \mathbb{N}_0^2$, $[f(t)]^\alpha = g(t) \cdot [1, 30]$, $g(t) = \dfrac{1 + t + t^2}{1 + 3t}$, $t \in \mathbb{T}$, $\alpha \in [0, 1]$.

Example 2.1.37 *Let* $\mathbb{T} = 3\mathbb{N}_0$, $f, g : \mathbb{T} \to F(\mathbb{R})$ *be fuzzy functions whose* α*-levels are given by*

$$\begin{aligned} [f(t)]^\alpha &= [-3t, 4t^2], \\ [g(t)]^\alpha &= [t + 1, t^2 + t + 1], \quad t \in \mathbb{T}, \quad \alpha \in [0, 1]. \end{aligned}$$

We have

$$\begin{aligned} \sigma(t) &= t + 3, \\ \underline{f}^\alpha(t) &= -3t, \\ \overline{f}^\alpha(t) &= 4t^2, \\ \underline{g}^\alpha(t) &= t + 1, \\ \overline{g}^\alpha(t) &= t^2 + t + 1, \quad t \in \mathbb{T}, \quad \alpha \in [0, 1]. \end{aligned}$$

Then

$$\underline{f}^{\alpha\Delta}(t) = -3,$$

$$\begin{aligned}
\overline{f}^{\alpha\Delta}(t) &= 4(\sigma(t)+t) \\
&= 4(t+3+t) \\
&= 4(2t+3) \\
&= 8t+12, \\
\underline{g}^{\alpha\Delta}(t) &= 1, \\
\overline{g}^{\alpha\Delta}(t) &= \sigma(t)+t+1 \\
&= t+3+t+1 \\
&= 2t+4, \quad t \in \mathbb{T}, \quad \alpha \in [0,1],
\end{aligned}$$

and

$$\begin{aligned}
[\delta_H f(t)]^\alpha &= [-3, 8t+12], \\
[\delta_H g(t)]^\alpha &= [1, 2t+4], \quad t \in \mathbb{T}, \quad \alpha \in [0,1].
\end{aligned}$$

Therefore f and g are δ_H-differentiable in $\mathbb{T}$. Next,

$$\begin{aligned}
[\delta_H f(t)]^\alpha + [\delta_H g(t)]^\alpha &= [-2, `10t+16], \\
[(f+g)(t)]^\alpha &= [-2t+1, 5t^2+t+1], \\
\underline{(f+g)}^\alpha(t) &= -2t+1, \\
\overline{(f+g)}^\alpha(t) &= 5t^2+t+1,
\end{aligned}$$

$$\begin{aligned}
\underline{(f+g)}^{\alpha\Delta}(t) &= -2, \\
\overline{(f+g)}^{\alpha\Delta}(t) &= 5(\sigma(t)+t)+1 \\
&= 5(t+3+t)+1 \\
&= 5(2t+3)+1 \\
&= 10t+15+1 \\
&= 10t+16, \quad t\in\mathbb{T}, \quad \alpha\in[0,1].
\end{aligned}$$

Thus,

$$\begin{aligned}
[\delta_H(f+g)(t)]^\alpha &= [-2, 10t+16] \\
&= [\delta_H f(t)]^\alpha + [\delta_H g(t)]^\alpha, \quad t\in\mathbb{T}, \quad \alpha\in[0,1].
\end{aligned}$$

Exercise 2.1.38 *Check if*

$$[\delta_H(f+g)(t)]^\alpha = [\delta_H f(t)]^\alpha + [\delta_H g(t)]^\alpha, \quad t\in\mathbb{T}, \quad \alpha\in[0,1],$$

where

1. $\mathbb{T}=\mathbb{Z}$ *and*

$$\begin{aligned}
[f(t)]^\alpha &= \left[t^2+t-1, t^4+t^2+t\right], \\
[g(t)]^\alpha &= \left[t^2, t^8+t^6+t^4+t^2\right],
\end{aligned}$$

2. $\mathbb{T}=3\mathbb{Z}$ *and*

$$\begin{aligned}
[f(t)]^\alpha &= \left[t^2-1, t^6+t^4+t^2\right], \\
[g(t)]^\alpha &= \left[t^2-3t-1, t^8+10t^6+t^2-3t-1\right],
\end{aligned}$$

3. $\mathbb{T} = 2^{\mathbb{N}_0}$ *and*

$$\begin{aligned}[f(t)]^\alpha &= \left[\frac{t^2+3t-4}{t^2+t+1}, \frac{2t^4+t^2+3t-4}{t^2+t+1}\right],\\ [g(t)]^\alpha &= \left[t^2-10, t^8+7t^6+6t^4+8t^2-10\right].\end{aligned}$$

Example 2.1.39 *Let* $\mathbb{T} = 3^{\mathbb{N}_0}$, $f : \mathbb{T} \to F(\mathbb{R})$ *be a fuzzy function whose* α*-levels are given by*

$$[f(t)]^\alpha = g(t)\cdot[-1,3], \quad \alpha \in [0,1], \quad t \in \mathbb{T},$$

where

$$g(t) = \frac{t^2+1}{t+3}, \quad t \in \mathbb{T}.$$

We will find $\left[\delta_H f(t)\right]^\alpha$, $\alpha \in [0,1]$, $t \in \mathbb{T}$. *Here*

$$\begin{aligned}\sigma(t) &= 3t,\\ \underline{f}^\alpha(t) &= -\frac{t^2+1}{t+3},\\ \overline{f}^\alpha(t) &= \frac{3t^2+3}{t+3}, \quad t \in \mathbb{T}, \quad \alpha \in [0,1].\end{aligned}$$

We have

$$\begin{aligned}g^\Delta(t) &= \frac{(\sigma(t)+t)(t+3)-(t^2+1)}{(t+3)(\sigma(t)+3)}\\ &= \frac{(3t+t)(t+3)-t^2-1}{(t+3)(3t+3)}\\ &= \frac{4t(t+3)-t^2-1}{3(t+1)(t+3)}\\ &= \frac{4t^2+12t-t^2-1}{3(t+1)(t+3)}\end{aligned}$$

$$= \frac{3t^2+12t-1}{3(t+1)(t+3)}, \quad t \in \mathbb{T}.$$

Therefore

$$\left[\delta_H f(t)\right]^{\alpha} = g^{\Delta}(t)\cdot[-1,3], \quad t \in \mathbb{T}, \quad \alpha \in [0,1].$$

Exercise 2.1.40 *Let* $\mathbb{T} = 4^{\mathbb{N}_0}$, $f : \mathbb{T} \to F(\mathbb{R})$ *be a fuzzy function whose* α*-levels are given by*

$$[f(t)]^{\alpha} = g(t)\cdot[-10,31], \quad \alpha \in [0,1], \quad t \in \mathbb{T},$$

where

$$g(t) = \frac{t^3+2t^2+3t+1}{t^2+4t+3}, \quad t \in \mathbb{T}.$$

Find $\left[\delta_H f(t)\right]^{\alpha}$, $\alpha \in [0,1]$, $t \in \mathbb{T}$.

Before we formulate and prove the next properties of the δ_H derivative, we need the following useful lemmas.

Lemma 2.1.41 *Let* $a,b,c,d,w,r \in F(\mathbb{R})$ *and* $c \ominus_H d$, $w \ominus_H d$ *exist. Then*

$$D(c \ominus_H d, a+b) \;\leq\; D(w \ominus_H d, a) + D(c, w+r) + D(b,r).$$

Proof 2.1.42 *We have*

$$\begin{aligned} D(c \ominus_H d, a+b) &= D(c \ominus_H d + w, a+b+w) \\ &\leq D(w \ominus_H d, a) + D(c, b+w) \\ &= D(w \ominus_H d, a) + D(c+r, b+w+r) \\ &\leq D(w \ominus_H d, a) + D(c, w+r) + D(r,b). \end{aligned}$$

This completes the proof.

Lemma 2.1.43 *Let $a,b,c,d,w,r \in F(\mathbb{R})$ and $c \ominus_H d$, $c \ominus_H w$ exist. Then*

$$D(c \ominus_H d, a+b) \leq D(c \ominus_H w, a) + D(w, d+r) + D(b,r).$$

Proof 2.1.44 *We have*

$$\begin{aligned} D(c \ominus_H d, a+b) &\leq D(c+w, a+b+d+w) \\ &\leq D(c, w+a) + D(w, b+d) \\ &= D(c, w+a) + D(w+r, b+d+r) \\ &\leq D(c \ominus_H w, a) + D(w, d+r) + D(r,b). \end{aligned}$$

This completes the proof.

Theorem 2.1.45 *Let $f : \mathbb{T} \to \mathbb{R}$ is Δ-differentiable at $t \in \mathbb{T}^\kappa$ and $g : \mathbb{T} \to F(\mathbb{R})$ is δ_H-differentiable at t. Let also, for any $h > 0$ sufficiently small there exist $u_1(\sigma(t),h)$, $u_2(\sigma(t),h) \in F(\mathbb{R})$ such that*

$$\begin{aligned} g(t+h) &= g(\sigma(t)) + u_1(\sigma(t),h), \\ g(\sigma(t)) &= g(t-h) + u_2(\sigma(t),h). \end{aligned}$$

If $f(t)f^\Delta(t) > 0$, then

$$\delta_H(f \cdot g)(t) = f^\Delta(t) \cdot g(t) + f(\sigma(t)) \cdot \delta_H g(t).$$

Proof 2.1.46 *Firstly, we note that*

$$\begin{aligned} f(\sigma(t))f^\Delta(t) &= \left(f(t) + \mu(t)f^\Delta(t)\right) f^\Delta(t) \\ &= f(t)f^\Delta(t) + \mu(t)\left(f^\Delta(t)\right)^2 \\ &> 0. \end{aligned}$$

Let $\varepsilon \in (0,1)$. *Define*

$$\varepsilon^* = \varepsilon\left(1 + |f(\sigma(t))| + \|g(t)\| + \left|f^{\Delta}(t)\right|\right)^{-1}.$$

We have $\varepsilon^* \in (0,1)$. *Since* f *is* Δ*-differentiable at* t *and* g *is* δ_H*-differentiable at* t*, there exist neighborhoods* $U_1(\delta,t)$ *and* $U_2(\delta,t)$ *of* t *for some positive* δ *such that for all* $0 < h < \delta$ *with* $t+h \in U_1(t,\delta)$*, we have*

$$\left|f(\sigma(t)) - f(t+h) - f^{\Delta}(t)(\mu(t)-h)\right| \leq \varepsilon^*(h-\mu(t)), \qquad (2.5)$$

and for all $0 < h < \delta$ *with* $t-h, t+h \in U_2(t,\delta)$*, we have*

$$\begin{aligned} D(g(t+h) \ominus_H g(\sigma(t)), (h-\mu(t)) \cdot \delta_H g(t)) &\leq \varepsilon^*(h-\mu(t)), \\ D(g(\sigma(t)) \ominus_H g(t-h), (h+\mu(t)) \cdot \delta_H g(t)) &\leq \varepsilon^*(h+\mu(t)). \end{aligned} \qquad (2.6)$$

Since f *is* Δ*-differentiable at* t*, then* f *is continuous at* t*. Thus, for* $h > 0$ *sufficiently small,* $f(\sigma(t))$ *and* $f(t+h)$ *have the same sign and we have*

$$f(t+h) = f(\sigma(t)) + v_1(\sigma(t),h).$$

Now, using that $f(t)f^{\Delta}(t) > 0$*, we conclude that* $v_1(\sigma(t),h)$ *has the same sign as* $f(\sigma(t))$ *and* $f(t+h)$ *for* $h > 0$ *sufficiently small. From here, we get*

$$\begin{aligned} (f \cdot g)(t+h) &= f(\sigma(t)) \cdot g(\sigma(t)) \\ &\quad + f(\sigma(t)) \cdot u_1(\sigma(t),h) \\ &\quad + v_1(\sigma(t),h) \cdot g(\sigma(t)) \\ &\quad + (v_1 \cdot u_1)(\sigma(t),h), \end{aligned}$$

i.e., the H*-difference*

$$(f(t+h) \cdot g(t+h)) \ominus_H (f(\sigma(t)) \cdot g(\sigma(t))),$$

exists. Next, we have

$$f(\sigma(t)) = f(t-h) + v_2(\sigma(t), h),$$

where $v_2(\sigma(t), h)$ *has the same sign as* $f(\sigma(t))$ *and* $f(t-h)$ *for* $h > 0$ *sufficiently small. Therefore*

$$\begin{aligned}(f \cdot g)(\sigma(t)) &= f(t-h) \cdot g(t-h) \\ &\quad + f(\sigma(t)) \cdot u_2(\sigma(t), h) \\ &\quad + v_2(\sigma(t), h) \cdot g(\sigma(t)) \\ &\quad + (v_2 \cdot u_2)(\sigma(t), h),\end{aligned}$$

for $h > 0$ *sufficiently small, and the H-difference*

$$(f(\sigma(t)) \cdot g(\sigma(t))) \ominus_{gH} (f(t-h) \cdot g(t-h)),$$

exists for $h > 0$ *sufficiently small. Because g is* δ_H*-differentiable at t, we have that g is continuous at t. Then there exists a neighborhood* $U_3(t, \delta)$ *of t such that for all* $0 < h < \delta$ *with* $t-h, t+h \in U_3(t, \delta)$*, we have*

$$D(g(t+h), g(t)) \leq \varepsilon^*,$$

$$\begin{aligned}\|g(t+h)\| &= D\left(g(t+h), \tilde{0}\right) \\ &\leq D\left(g(t), \tilde{0}\right) + 1 \qquad (2.7) \\ &= \|g(t)\| + 1,\end{aligned}$$

$$\begin{aligned}\|g(t-h)\| &= D(g(t-h), \tilde{0}) \\ &\leq D(g(t), \tilde{0}) + 1 \qquad (2.8) \\ &= \|g(t)\| + 1,\end{aligned}$$

$$D(g(t-h), g(t)) \leq \varepsilon^*. \qquad (2.9)$$

Let $U(t,\delta) = U_1(t,\delta)\bigcap U_2(t,\delta)\bigcap U_3(t,\delta)$. *For*

$$\begin{aligned} c &= f(t+h)\cdot g(t+h), \\ d &= f(\sigma(t))\cdot g(\sigma(t)), \\ a &= ((h-\mu(t))f(\sigma(t)))\cdot \delta_H g(t), \\ b &= \left(f^{\Delta}(t)(h-\mu(t))\right)\cdot g(t), \\ w &= f(\sigma(t))\cdot g(t+h), \\ r &= \left(f^{\Delta}(t)(h-\mu(t))\right)\cdot g(t+h), \end{aligned}$$

and for $t-h, t+h \in U(t,\delta)$, *we apply Lemma 2.1.41 and we get*

$$\begin{aligned} & D\Big((f\cdot g)(t+h) \ominus_H (f\cdot g)(\sigma(t)), \\ & \quad (h-\mu(t))\cdot\left(f(\sigma(t))\cdot\delta_H g(t) + f^{\Delta}(t)\cdot g(t)\right)\Big) \\ & \leq D\Big(f(\sigma(t))\cdot(g(t+h)\ominus_H g(\sigma(t))), \\ & \quad ((h-\mu(t))f(\sigma(t)))\cdot\delta_H g(t)\Big) \\ & +D\Big(f(t+h)\cdot g(t+h), \\ & f(\sigma(t))\cdot g(t+h) + \left((h-\mu(t))f^{\Delta}(t)\right)\cdot g(t+h)\Big) \\ & +D\left(\left((h-\mu(t))f^{\Delta}(t)\right)\cdot g(t), \left((h-\mu(t))f^{\Delta}(t)\right)\cdot g(t+h)\right). \end{aligned} \tag{2.10}$$

Since $h-\mu(t)>0$, $f(t)f^{\Delta}(t)>0$, we have that $(h-\mu(t))f^{\Delta}(t)$ has the same sign as $f(\sigma(t))$. Therefore

$$f(\sigma(t))\cdot g(t+h)+\Big((h-\mu(t))f^{\Delta}(t)\Big)\cdot g(t+h)$$

$$= \Big(f(\sigma(t))+f^{\Delta}(t)(h-\mu(t))\Big)\cdot g(t+h).$$

Thus,

$$\begin{aligned} D\Big(& f(t+h)\cdot g(t+h), f(\sigma(t))\cdot g(t+h) \\ &+\Big((h-\mu(t))f^{\Delta}(t)\Big)\cdot g(t+h)\Big) \end{aligned} \tag{2.11}$$

$$=\Big|f(t+h)-f(\sigma(t))-f^{\Delta}(t)(h-\mu(t))\Big|\,\|g(t+h)\|.$$

Hence, from (2.5) *to* (2.11)*, we arrive at*

$$\begin{aligned} D\Big(&(f\cdot g)(t+h)\ominus_H (f\cdot g)(\sigma(t)), \\ &(h-\mu(t))\cdot\Big(f(\sigma(t))\cdot\delta_H g(t)+f^{\Delta}(t)\cdot g(t)\Big)\Big) \end{aligned} \tag{2.12}$$

$$\begin{aligned} \le\ & \varepsilon^*|f(\sigma(t))|(h-\mu(t)) \\ &+\varepsilon^*\|g(t+h)\|(h-\mu(t)) \\ &+(h-\mu(t))\Big|f^{\Delta}(t)\Big|D(g(t+h),g(t)) \\ \le\ & \varepsilon^*(h-\mu(t))\Big(|f(\sigma(t))|+1+\|g(t)\|+\Big|f^{\Delta}(t)\Big|\Big) \\ \le\ & \varepsilon(h-\mu(t)). \end{aligned}$$

Next, there exists a neighborhood $U_4(t,\delta)$ of t such that for all $0<h<\delta$ with $t-h\in U_4(t,\delta)$, we have

$$|f(\sigma(t))-f(t-h)-f^{\Delta}(t)(h+\mu(t))|\le\varepsilon^*(h+\mu(t)). \tag{2.13}$$

For

$$\begin{aligned} d &= f(t-h)\cdot g(t-h), \\ c &= f(\sigma(t))\cdot g(\sigma(t)), \\ a &= ((h+\mu(t))f(\sigma(t)))\cdot \delta_H g(t), \\ b &= \left(f^{\Delta}(t)(h+\mu(t))\right)\cdot g(t), \\ w &= f(\sigma(t))\cdot g(t-h), \\ r &= \left(f^{\Delta}(t)(h+\mu(t))\right)\cdot g(t-h), \end{aligned}$$

we apply Lemma 2.1.43 and we get

$$\begin{aligned} & D\Big((f\cdot g)(\sigma(t))\ominus_H (f\cdot g)(t-h), \\ & (h+\mu(t))\cdot\left(f(\sigma(t))\cdot\delta_H g(t)+f^{\Delta}(t)\cdot g(t)\right)\Big) \\ \leq\ & D\Big((f(\sigma(t))\cdot g(\sigma(t)))\ominus_H (f(\sigma(t))\cdot g(t-h)), \\ & ((h+\mu(t))f(\sigma(t)))\cdot\delta_H g(t)\Big) \\ & +D\Big(f(\sigma(t))\cdot g(t-h), \\ & f(t-h)\cdot g(t-h)+f^{\Delta}(t)(h+\mu(t))\cdot g(t-h)\Big) \\ & +D\left(\left((h+\mu(t))f^{\Delta}(t)\right)\cdot g(t),\left((h+\mu(t))f^{\Delta}(t)\right)\cdot g(t-h)\right). \end{aligned} \tag{2.14}$$

Since $h+\mu(t)>0$ and $f(t)f^{\Delta}(t)>0$, we have that $f^{\Delta}(t)(h+\mu(t))$ has the same sign as $f(\sigma(t))$. Also, we have

$$\begin{aligned}
& f(t-h)\cdot g(t-h)+\Big((h+\mu(t))f^{\Delta}(t)\Big)\cdot g(t-h) \\
&= \Big(f(t-h)+f^{\Delta}(t)(h+\mu(t))\Big)\cdot g(t-h).
\end{aligned}$$

Therefore

$$\begin{aligned}
& D\Big(f(\sigma(t))\cdot g(t-h), f(t-h)\cdot g(t-h) \\
& \quad +\Big((h+\mu(t))f^{\Delta}(t)\Big)\cdot g(t-h)\Big) \qquad (2.15) \\
&= \Big|f(\sigma(t))-f(t-h)-f^{\Delta}(t)(h+\mu(t))\Big|\,\|g(t-h)\|.
\end{aligned}$$

Hence, using (2.6), (2.13), (2.14), (2.15), (2.8), *and* (2.9), *we obtain*

$$\begin{aligned}
& D\Big((f\cdot g)(\sigma(t))\ominus_H (f\cdot g)(t-h), \\
& f(\sigma(t))\cdot \delta_H g(t)+\Big((h+\mu(t))f^{\Delta}(t)\Big)\cdot g(t)\Big) \qquad (2.16) \\
&\leq \varepsilon^*(h+\mu(t))|f(\sigma(t))| \\
& \quad +\varepsilon^*\|g(t-h)\|(h+\mu(t)) \\
& \quad +(h+\mu(t))|f^{\Delta}(t)|D(g(t-h),g(t)) \\
&\leq \varepsilon^*(h+\mu(t))\Big(|f(\sigma(t))|+1+\|g(t)\|+|f^{\Delta}(t)|\Big) \\
&\leq \varepsilon(h+\mu(t)).
\end{aligned}$$

By (2.12) *and* (2.16), *we get the desired result. This completes the proof.*

Example 2.1.47 *Let* $\mathbb{T} = 2^{\mathbb{N}_0}$ *and*

$$\begin{aligned} f(t) &= t^2+t+1, \\ g(t) &= \left(t, t+1, t^2+1\right), \quad t \in \mathbb{T}. \end{aligned}$$

Here

$$\begin{aligned} \sigma(t) &= 2t, \\ \mu(t) &= t, \quad t \in \mathbb{T}. \end{aligned}$$

Then

$$\begin{aligned} f(t)\cdot g(t) &= \left(t(t^2+t+1), (t+1)(t^2+t+1), (t^2+1)(t^2+t+1)\right) \\ &= \left(t^3+t^2+t, t^3+t^2+t+t^2+t+1, t^4+t^3+t^2+t^2+t+1\right) \\ &= \left(t^3+t^2+t, t^3+2t^2+2t+1, t^4+t^3+2t^2+t+1\right), \quad t \in \mathbb{T}. \end{aligned}$$

Let

$$\begin{aligned} h_1(t) &= t^4+t^3+2t^2+t+1, \\ h_2(t) &= t^3+t^2+t, \\ h_3(t) &= t^3+2t^2+2t+1, \quad t \in \mathbb{T}. \end{aligned}$$

Then

$$\begin{aligned} h_1^{\Delta}(t) &= (\sigma(t))^3+t(\sigma(t))^2+t^2\sigma(t)+t^3 \\ &\quad +(\sigma(t))^2+t\sigma(t)+t^2+2(\sigma(t)+t)+1 \end{aligned}$$

$$\begin{aligned}
&= 8t^3+4t^3+2t^3+t^3+4t^2+2t^2+t^2+6t+1\\
&= 15t^3+7t^2+6t+1,\\
h_2^{\Delta}(t) &= (\sigma(t))^2+t\sigma(t)+t^2+\sigma(t)+t+1\\
&= 4t^2+2t^2+t^2+2t+t+1\\
&= 7t^2+3t+1,\\
h_3^{\Delta}(t) &= (\sigma(t))^2+t\sigma(t)+t^2+2(\sigma(t)+t)+2\\
&= 4t^2+2t^2+t^2+6t+2\\
&= 7t^2+6t+2, \quad t\in\mathbb{T}.
\end{aligned}$$

Hence,

$$\begin{aligned}
\delta_H(f\cdot g)(t) &= \left(h_2^{\Delta}(t), h_3^{\Delta}(t), h_1^{\Delta}(t)\right)\\
&= \left(7t^2+3t+1, 7t^2+6t+2, 15t^3+7t^2+6t+1\right),
\end{aligned}$$

$t\in\mathbb{T}$. *Next,*

$$\begin{aligned}
f^{\Delta}(t) &= \sigma(t)+t+1\\
&= 3t+1,\\
f^{\Delta}(t)\cdot g(t) &= \left((3t+1)t, (3t+1)(t+1), (3t+1)(t^2+1)\right)\\
&= \left(3t^2+t, 3t^2+3t+t+1, 3t^3+3t+t^2+1\right)\\
&= \left(3t^2+t, 3t^2+4t+1, 3t^3+t^2+3t+1\right), \quad t\in\mathbb{T}.
\end{aligned}$$

Let

$$
\begin{aligned}
g_1(t) &= t^2+1, \\
g_2(t) &= t, \\
g_3(t) &= t+1, \quad t \in \mathbb{T}.
\end{aligned}
$$

Then

$$
\begin{aligned}
g_1^\Delta(t) &= \sigma(t)+t \\
&= 3t, \\
g_2^\Delta(t) &= 1, \\
g_3^\Delta(t) &= 1, \\
\delta_H g(t) &= \left(g_2^\Delta(t), g_3^\Delta(t), g_1^\Delta(t)\right) \\
&= (1, 1, 3t), \quad t \in \mathbb{T}.
\end{aligned}
$$

Also,

$$
\begin{aligned}
f(\sigma(t)) &= (\sigma(t))^2+\sigma(t)+1 \\
&= 4t^2+2t+1, \quad t \in \mathbb{T}.
\end{aligned}
$$

We get

$$
\begin{aligned}
f(\sigma(t)) \cdot \delta_H g(t) &= \left(4t^2+2t+1, 4t^2+2t+1, 3t(4t^2+2t+1)\right) \\
&= \left(4t^2+2t+1, 4t^2+2t+1, 12t^3+6t^2+3t\right),
\end{aligned}
$$

$t \in \mathbb{T}$, and

$$\begin{aligned}
& f^{\Delta}(t) \cdot g(t) + f(\sigma(t)) \cdot \delta_H g(t) \\
= & \left(3t^2 + t, 3t^2 + 4t + 1, 3t^3 + t^2 + 3t + 1\right) \\
& + \left(4t^2 + 2t + 1, 4t^2 + 2t + 1, 12t^3 + 6t^2 + 3t\right) \\
= & \left(7t^2 + 3t + 1, 7t^2 + 6t + 2, 15t^3 + 7t^2 + 6t + 1\right), \quad t \in \mathbb{T}.
\end{aligned}$$

Therefore

$$\delta_H(f \cdot g)(t) = f^{\Delta}(t) \cdot g(t) + f(\sigma(t)) \cdot \delta_H g(t).$$

Exercise 2.1.48 *Let* $\mathbb{T} = \left\{2 + \frac{1}{4^k}\right\}_{k \in \mathbb{N}} \bigcup \{2\}$,

$$\begin{aligned}
f(t) &= t^3 + 3t^2 + 4t + 7, \\
g(t) &= \left(\frac{t+1}{t+2}, t^2, t^2 + t^3\right), \quad t \in \mathbb{T}.
\end{aligned}$$

Prove that

$$\delta_H(f \cdot g)(2) = f^{\Delta}(2) \cdot g(2) + f(\sigma(2)) \cdot \delta_H g(2).$$

In fact, Theorem 2.1.45 is equivalent to the following theorem.

Theorem 2.1.49 *Let $f : \mathbb{T} \to \mathbb{R}$ be Δ-differentiable at $t \in \mathbb{T}^{\kappa}$ with $f(t) f^{\Delta}(t) > 0$. Let also, $g : \mathbb{T} \to F(\mathbb{R})$ be δ_H-differentiable at t with α-levels*

$$[g(t)]^{\alpha} = \left[\underline{g}^{\alpha}(t), \overline{g}^{\alpha}(t)\right], \quad \alpha \in [0,1].$$

Then $f \cdot g : \mathbb{T} \to F(\mathbb{R})$ is δ_H-differentiable at t and

$$[\delta_H(f \cdot g)(t)]^{\alpha} = \left[f^{\Delta}(t) \cdot g(t)\right]^{\alpha} + [f(\sigma(t)) \cdot \delta_H g(t)]^{\alpha}, \quad \alpha \in [0,1]. \tag{2.17}$$

Proof 2.1.50 *1. Let $f(t) > 0$, $f^{\Delta}(t) > 0$. Then*

$$\begin{aligned}
[f(t)\cdot g(t)]^{\alpha} &= f(t)\cdot [g(t)]^{\alpha} \\
&= f(t)\cdot \left[\underline{g}^{\alpha}(t), \overline{g}^{\alpha}(t)\right] \\
&= \left[f(t)\underline{g}^{\alpha}(t), f(t)\overline{g}^{\alpha}(t)\right], \\
[\delta_H(f\cdot g)(t)]^{\alpha} &= \left[(f\underline{g}^{\alpha})^{\Delta}(t), (f\overline{g}^{\alpha})^{\Delta}(t)\right] \\
&= \Big[f^{\Delta}(t)\underline{g}^{\alpha}(t) + f(\sigma(t))\underline{g}^{\alpha\Delta}(t), \\
&\qquad f^{\Delta}(t)\overline{g}^{\alpha}(t) + f(\sigma(t))\overline{g}^{\alpha\Delta}(t)\Big], \\
\left[f^{\Delta}(t)\cdot g(t)\right]^{\alpha} &= f^{\Delta}(t)\cdot [g(t)]^{\alpha} \\
&= f^{\Delta}(t)\cdot \left[\underline{g}^{\alpha}(t), \overline{g}^{\alpha}(t)\right] \\
&= \left[f^{\Delta}(t)\underline{g}^{\alpha}(t), f^{\Delta}(t)\overline{g}^{\alpha}(t)\right], \\
[f(\sigma(t))\cdot \delta_H g(t)]^{\alpha} &= f(\sigma(t))\cdot [\delta_H g(t)]^{\alpha} \\
&= f(\sigma(t))\cdot \left[\underline{g}^{\alpha\Delta}(t), \overline{g}^{\alpha\Delta}(t)\right] \\
&= \left[f(\sigma(t))\underline{g}^{\alpha\Delta}(t), f(\sigma(t))\overline{g}^{\alpha\Delta}(t)\right].
\end{aligned}$$

Hence,

$$\left[f^{\Delta}(t)\cdot g(t)\right]^{\alpha} + [f(\sigma(t))\cdot \delta_H g(t)]^{\alpha}$$

$$= \left[f^{\Delta}(t)\underline{g}^{\alpha}(t) + f(\sigma(t))\underline{g}^{\alpha\Delta}(t), f^{\Delta}(t)\overline{g}^{\alpha}(t) + f(\sigma(t))\overline{g}^{\alpha\Delta}(t)\right].$$

Therefore, we get (2.17).

2. *Suppose that* $f(t) < 0$, $f^{\Delta}(t) < 0$. *Then* $f(\sigma(t)) < 0$ *and*

$$\begin{aligned}
[f(t)\cdot g(t)]^{\alpha} &= f(t)\cdot [g(t)]^{\alpha} \\
&= f(t)\cdot \left[\underline{g}^{\alpha}(t), \overline{g}^{\alpha}(t)\right] \\
&= \left[f(t)\overline{g}^{\alpha}(t), f(t)\underline{g}^{\alpha}(t)\right], \\
[\delta_H(f\cdot g)(t)]^{\alpha} &= \left[(f\overline{g}^{\alpha})^{\Delta}(t), (f\underline{g}^{\alpha})^{\Delta}(t)\right] \\
&= \Big[f^{\Delta}(t)\overline{g}^{\alpha}(t) + f(\sigma(t))\overline{g}^{\alpha\Delta}(t), \\
&\qquad f^{\Delta}(t)\underline{g}^{\alpha}(t) + f(\sigma(t))\underline{g}^{\alpha\Delta}(t)\Big], \\
\left[f^{\Delta}(t)\cdot g(t)\right]^{\alpha} &= f^{\Delta}(t)\cdot [g(t)]^{\alpha} \\
&= f^{\Delta}(t)\cdot \left[\underline{g}^{\alpha}(t), \overline{g}^{\alpha}(t)\right] \\
&= \left[f^{\Delta}(t)\overline{g}^{\alpha}(t), f^{\Delta}(t)\underline{g}^{\alpha}(t)\right], \\
[f(\sigma(t))\cdot \delta_H g(t)]^{\alpha} &= f(\sigma(t))\cdot [\delta_H g(t)]^{\alpha} \\
&= f(\sigma(t))\cdot \left[\underline{g}^{\alpha\Delta}(t), \overline{g}^{\alpha\Delta}(t)\right] \\
&= \left[f(\sigma(t))\overline{g}^{\alpha\Delta}(t), f(\sigma(t))\underline{g}^{\alpha\Delta}(t)\right].
\end{aligned}$$

From here,

$$\left[f^{\Delta}(t)\cdot g(t)\right]^{\alpha} + [f(\sigma(t))\cdot \delta_H g(t)]^{\alpha}$$

$$= \left[f^{\Delta}(t)\overline{g}^{\alpha}(t) + f(\sigma(t))\overline{g}^{\alpha\Delta}(t), f^{\Delta}(t)\underline{g}^{\alpha}(t) + f(\sigma(t))\underline{g}^{\alpha\Delta}(t) \right].$$

Consequently, we obtain (2.17). *This completes the proof.*

Example 2.1.51 *Let* $\mathbb{T} = 2^{\mathbb{N}_0}$, $f : \mathbb{T} \to \mathbb{R}$ *be given by*

$$f(t) = \frac{t+1}{t+2}, \quad t \in \mathbb{T},$$

and $g : \mathbb{T} \to F(\mathbb{R})$ *be with* α*-levels*

$$[g(t)]^{\alpha} = \left[t+2, t^2+t+2 \right], \quad t \in \mathbb{T}, \quad \alpha \in [0,1].$$

We have

$$\begin{aligned}
f(t) &> 0, \\
\sigma(t) &= 2t, \\
f^{\Delta}(t) &= \frac{t+2-(t+1)}{(t+2)(\sigma(t)+2)} \\
&= \frac{1}{2(t+1)(t+2)} \\
&> 0, \\
f(\sigma(t)) &= \frac{\sigma(t)+1}{\sigma(t)+2} \\
&= \frac{2t+1}{2(t+1)}, \\
\underline{g}^{\alpha}(t) &= t+2, \\
\overline{g}^{\alpha}(t) &= t^2+t+2,
\end{aligned}$$

$$\begin{aligned}
\underline{g}^{\alpha\Delta}(t) &= 1, \\
\overline{g}^{\alpha\Delta}(t) &= \sigma(t)+t+1 \\
&= 3t+1, \\
[\delta_H g(t)]^{\alpha} &= [1, 3t+1], \quad t \in \mathbb{T}, \quad \alpha \in [0,1].
\end{aligned}$$

Therefore

$$\begin{aligned}
\left[f^{\Delta}(t)\cdot g(t)\right]^{\alpha} &= f^{\Delta}(t)\cdot [g(t)]^{\alpha} \\
&= \frac{1}{2(t+1)(t+2)}\cdot\left[t+2, t^2+t+2\right] \\
&= \left[\frac{1}{2(t+1)}, \frac{t^2+t+2}{2(t+1)(t+2)}\right], \\
[f(\sigma(t))\cdot\delta_H g(t)]^{\alpha} &= f(\sigma(t))\cdot[\delta_H g(t)]^{\alpha} \\
&= \frac{2t+1}{2(t+1)}\cdot[1, 3t+1] \\
&= \left[\frac{2t+1}{2(t+1)}, \frac{(2t+1)(3t+1)}{2(t+1)}\right] \\
&= \left[\frac{2t+1}{2(t+1)}, \frac{6t^2+5t+1}{2(t+1)}\right], \quad t \in \mathbb{T}, \quad \alpha \in [0,1].
\end{aligned}$$

Therefore

$$\begin{aligned}
[\delta_H(f\cdot g)(t)]^{\alpha} &= \left[f^{\Delta}(t)\cdot g(t)\right]^{\alpha} + [f(\sigma(t))\cdot\delta_H g(t)]^{\alpha} \\
&= \left[\frac{1}{2(t+1)}, \frac{t^2+t+2}{2(t+1)(t+2)}\right]
\end{aligned}$$

$$
\begin{aligned}
& + \left[\frac{2t+1}{2(t+1)}, \frac{6t^2+5t+1}{2(t+1)} \right] \\
&= \left[\frac{2t+2}{2(t+1)}, \frac{t^2+t+2}{2(t+1)(t+2)} + \frac{6t^2+5t+1}{2(t+1)} \right] \\
&= \left[1, \frac{t^2+t+2+(6t^2+5t+1)(t+2)}{2(t+1)(t+2)} \right] \\
&= \left[1, \frac{t^2+t+2+6t^3+12t^2+5t^2+10t+t+2}{2(t+1)(t+2)} \right] \\
&= \left[1, \frac{6t^3+18t^2+12t+4}{2(t+1)(t+2)} \right] \\
&= \left[1, \frac{3t^3+9t^2+6t+2}{(t+1)(t+2)} \right], \quad t \in \mathbb{T}, \quad \alpha \in [0,1].
\end{aligned}
$$

Exercise 2.1.52 *Let $\mathbb{T} = 3^{\mathbb{N}_0}$, $f : \mathbb{T} \to \mathbb{R}$ be given by*

$$
f(t) = \frac{t^2+t+1}{t^2+3t+2} + t^2+t+1, \quad t \in \mathbb{T},
$$

and $g : \mathbb{T} \to F(\mathbb{R})$ be with α-levels

$$
[g(t)]^\alpha = \left[t+2+\frac{1}{t+1}, t^2+t+2+\frac{2}{t+1} \right], \quad t \in \mathbb{T}, \quad \alpha \in [0,1].
$$

Find

$$
[\delta_H(f \cdot g)(t)]^\alpha, \quad t \in \mathbb{T}, \quad \alpha \in [0,1].
$$

Theorem 2.1.53 *Let $t \in \mathbb{T}^\kappa$, $f : \mathbb{T} \to \mathbb{R}$ be Δ-differentiable at t, $f(\sigma(t))f^\Delta(t) < 0$, $g : \mathbb{T} \to F(\mathbb{R})$ be δ_H-differentiable at t and the H-differences*

$$
(f(t+h) \cdot g(t+h)) \ominus_H (f(\sigma(t)) \cdot g(\sigma(t))),
$$

and

$$
(f(\sigma(t)) \cdot g(\sigma(t))) \ominus_H (f(t-h) \cdot g(t-h)),
$$

exist. Let also, for any $h > 0$ sufficiently small, there exist $u_1(\sigma(t), h)$, $u_2(\sigma(t), h) \in F(\mathbb{R})$ such that

$$\begin{aligned} g(t+h) &= g(\sigma(t)) + u_1(\sigma(t), h), \\ g(\sigma(t)) &= g(t-h) + u_2(\sigma(t), h). \end{aligned}$$

Then

$$\delta_H(f \cdot g)(t) = (f(\sigma(t)) \cdot \delta_H g(t)) \ominus_H \left(\left(-f^{\Delta}(t) \right) \cdot g(t) \right).$$

Proof 2.1.54 *Firstly, note that*

$$\begin{aligned} f(\sigma(t)) f^{\Delta}(t) &= \left(f(t) + \mu(t) f^{\Delta}(t) \right) f^{\Delta}(t) \\ &= f(t) f^{\Delta}(t) + \mu(t) \left(f^{\Delta}(t) \right)^2 \\ &< 0. \end{aligned}$$

Therefore, $f(t)f^{\Delta}(t) < 0$. Without loss of generality, suppose that $f(t) > 0$, $f(\sigma(t)) > 0$ and $f^{\Delta}(t) < 0$. Take $\varepsilon \in (0,1)$ and

$$\varepsilon^* = \varepsilon \left(1 + |f(\sigma(t))| + \|g(t)\| + \left| f^{\Delta}(t) \right| \right)^{-1}.$$

Note that $\varepsilon^* \in (0,1)$. Because f is Δ-differentiable at t and g is δ_H-differentiable at t, there exist neighborhoods $U_1(t,\delta)$ and $U_2(t,\delta)$ of t for some positive δ such that for all $0 < h < \delta$ with $t + h \in U_1(t,\delta)$, we have

$$\left| f(\sigma(t)) - f(t+h) - f^{\Delta}(t)(\mu(t) - h) \right| \leq \varepsilon^*(h - \mu(t)), \tag{2.18}$$

and for all $0 < h < \delta$ with $t-h, t+h \in U_2(t,\delta)$, we have

$$\begin{aligned} D(g(t+h) \ominus_H g(\sigma(t)), (h-\mu(t)) \cdot \delta_H g(t)) &\leq \varepsilon^*(h - \mu(t)), \\ D(g(\sigma(t)) \ominus_H g(t-h), (h+\mu(t)) \cdot \delta_H g(t)) &\leq \varepsilon^*(h + \mu(t)). \end{aligned} \tag{2.19}$$

Because g is δ_H-differentiable at t, we have that g is continuous at t. Then there exists a neighborhood $U_3(t,\delta)$ of t such that for all $0<h<\delta$ with $t-h, t+h \in U_3(t,\delta)$, we have

$$D(g(t+h),g(t)) \leq \varepsilon^*,$$

$$\begin{aligned}\|g(t+h)\| &= D\left(g(t+h),\widetilde{0}\right)\\ &\leq D\left(g(t),\widetilde{0}\right)+1\\ &= \|g(t)\|+1,\end{aligned}$$

and

$$\begin{aligned}D(g(t-h),g(t)) &\leq \varepsilon^*,\\ \|g(t-h)\| &= D\left(g(t-h),\widetilde{0}\right)\\ &\leq D\left(g(t),\widetilde{0}\right)+1\\ &= \|g(t)\|+1.\end{aligned}$$

Let $U(t,\delta)=U_1(t,\delta)\bigcap U_2(t,\delta)\bigcap U_3(t,\delta)$. For

$$\begin{aligned}c &= f(t+h)\cdot g(t+h),\\ d &= f(\sigma(t))\cdot g(\sigma(t)),\\ a &= ((h-\mu(t))f(\sigma(t)))\cdot \delta_H g(t),\\ b &= \left(f^{\Delta}(t)(h-\mu(t))\right)\cdot g(t),\\ w &= f(\sigma(t))\cdot g(t+h),\\ r &= \left(f^{\Delta}(t)(h-\mu(t))\right)\cdot g(t+h).\end{aligned}$$

and for $t-h, t+h \in U(t,\delta)$ we apply Lemma 2.1.41 and we get

$$D\Big((f\cdot g)(t+h)\ominus_H (f\cdot g)(\sigma(t)),$$

$$(h-\mu(t))\cdot\Big(f(\sigma(t))\cdot\delta_H g(t)\ominus_H\Big((-f^{\Delta}(t))\cdot g(t)\Big)\Big)\Big)$$

$$\leq D(f(\sigma(t))\cdot(g(t+h)\ominus_H g(\sigma(t))),((h-\mu(t))f(\sigma(t)))\cdot\delta_H g(t))$$

$$+D\Big(f(t+h)\cdot g(t+h), f(\sigma(t))\cdot g(t+h)+((h-\mu(t))f^{\Delta}(t))\cdot g(t+h)\Big)$$

$$+D\Big(((h-\mu(t))f^{\Delta}(t))\cdot g(t),((h-\mu(t))f^{\Delta}(t))\cdot g(t+h)\Big)$$

$$\begin{aligned}
&\leq \varepsilon^*(h-\mu(t))|f(\sigma(t))|+\varepsilon^*(h-\mu(t))\|g(t+h)\| \\
&\quad +\varepsilon^*(h-\mu(t))|f^{\Delta}(t)| \\
&= \varepsilon^*(h-\mu(t))\Big(1+|f(\sigma(t))|+|f^{\Delta}(t)|+\|g(t)\|\Big) \\
&= \varepsilon(h-\mu(t)).
\end{aligned}$$

Next, there exists a neighborhood $U_4(t,\delta)$ of t such that for all $0 < h < \delta$ with $t-h \in U_4(t,\delta)$, we have

$$|f(\sigma(t))-f(t-h)-f^{\Delta}(t)(h+\mu(t))|\leq\varepsilon^*(h+\mu(t)).$$

For

$$\begin{aligned}
d &= f(t-h)\cdot g(t-h), \\
c &= f(\sigma(t))\cdot g(\sigma(t)),
\end{aligned}$$

$$\begin{aligned}
a &= (h+\mu(t))f(\sigma(t))\cdot\delta_H g(t),\\
b &= (f^\Delta(t)(h+\mu(t)))\cdot g(t),\\
w &= f(\sigma(t))\cdot g(t-h),\\
r &= (f^\Delta(t)(h+\mu(t)))\cdot g(t-h),
\end{aligned}$$

we apply Lemma 2.1.43 and we get

$$\begin{aligned}
& D\Big((f\cdot g)(\sigma(t))\ominus_H(f\cdot g)(t-h),\\
& (h+\mu(t))\cdot\Big(f(\sigma(t))\cdot\delta_H g(t)\ominus_H((-f^\Delta(t))\cdot g(t))\Big)\Big)\\
\leq\ & D\Big((f(\sigma(t))\cdot g(\sigma(t)))\ominus_H(f(\sigma(t))\cdot g(t-h)),\\
& (h+\mu(t))f(\sigma(t))\cdot\delta_H g(t)\Big)\\
& +D\Big(f(\sigma(t))\cdot g(t-h),\\
& f(t-h)\cdot g(t-h)+(f^\Delta(t)(h+\mu(t)))\cdot g(t-h)\Big)\\
& +D\Big(((h+\mu(t))f^\Delta(t))\cdot g(t),((h+\mu(t))f^\Delta(t))\cdot g(t-h)\Big)\\
\leq\ & |f(\sigma(t))|\varepsilon^*(h+\mu(t))+\varepsilon^*(h+\mu(t))\|g(t-h)\|\\
& +\varepsilon^*(h+\mu(t))|f^\Delta(t)|\\
=\ & \varepsilon^*\left(1+|f^\Delta(t)|+|f(\sigma(t))|+\|g(t)\|\right)(h+\mu(t))\\
=\ & \varepsilon(h+\mu(t)).
\end{aligned}$$

This completes the proof.

Example 2.1.55 *Let* $\mathbb{T} = \mathbb{N}$ *and*

$$\begin{aligned} f(t) &= \frac{1}{t}, \\ g(t) &= \left(\frac{t}{t+1}, t, t^2\right), \quad t \in \mathbb{N}. \end{aligned}$$

We have

$$\begin{aligned} \sigma(t) &= t+1, \\ \mu(t) &= 1, \quad t \in \mathbb{T}. \end{aligned}$$

Let

$$\begin{aligned} g_1(t) &= \frac{t}{t+1}, \\ g_2(t) &= t, \\ g_3(t) &= t^2, \quad t \in \mathbb{T}. \end{aligned}$$

Then

$$\begin{aligned} g_1^{\Delta}(t) &= \frac{t+1-t}{(t+1)(\sigma(t)+1)} \\ &= \frac{1}{(t+1)(t+2)}, \\ g_2^{\Delta}(t) &= 1, \\ g_3^{\Delta}(t) &= \sigma(t)+t \\ &= t+1+t \\ &= 2t+1, \end{aligned}$$

$$\begin{aligned} f^{\Delta}(t) &= -\frac{1}{t\sigma(t)} \\ &= -\frac{1}{t(t+1)}, \\ f(\sigma(t)) &= \frac{1}{\sigma(t)} \\ &= \frac{1}{t+1}, \quad t \in \mathbb{T}. \end{aligned}$$

From here,

$$\begin{aligned} \delta_H g(t) &= \left(g_1^{\Delta}(t), g_2^{\Delta}(t), g_3^{\Delta}(t)\right) \\ &= \left(\frac{1}{(t+1)(t+2)}, 1, 2t+1\right), \\ f(\sigma(t)) \cdot \delta_H g(t) &= \left(\frac{1}{(t+1)^2(t+2)}, \frac{1}{t+1}, \frac{2t+1}{t+1}\right), \\ \left(-f^{\Delta}(t)\right) \cdot g(t) &= \left(\frac{1}{(t+1)^2}, \frac{1}{t+1}, \frac{t}{t+1}\right), \quad t \in \mathbb{T}, \end{aligned}$$

and

$$\begin{aligned} &(f(\sigma(t)) \cdot \delta_H g(t)) \ominus_H \left(\left(-f^{\Delta}(t)\right) \cdot g(t)\right) \\ &= \left(\frac{1}{(t+1)^2(t+2)} - \frac{1}{(t+1)^2}, 0, \frac{2t+1}{t+1} - \frac{t}{t+1}\right) \\ &= \left(-\frac{1}{(t+1)(t+2)}, 0, 1\right), \quad t \in \mathbb{T}. \end{aligned}$$

Next,

$$(f \cdot g)(t) = \left(\frac{1}{t+1}, 1, t\right), \quad t \in \mathbb{T}.$$

Let

$$\begin{aligned} h_1(t) &= \frac{1}{t+1}, \\ h_2(t) &= 1, \\ h_3(t) &= 1, \quad t \in \mathbb{T}. \end{aligned}$$

Then

$$\begin{aligned} h_1^{\Delta}(t) &= -\frac{1}{(t+1)(\sigma(t)+1)} \\ &= -\frac{1}{(t+1)(t+2)}, \\ h_2^{\Delta}(t) &= 0, \\ h_3^{\Delta}(t) &= 0, \quad t \in \mathbb{T}. \end{aligned}$$

Therefore

$$\begin{aligned} \delta_H(f \cdot g)(t) &= \left(h_1^{\Delta}(t), h_2^{\Delta}(t), h_3^{\Delta}(t)\right) \\ &= \left(-\frac{1}{(t+1)(t+2)}, 0, 1\right), \quad t \in \mathbb{T}. \end{aligned}$$

Consequently

$$\delta_H(f \cdot g)(t) = (f(\sigma(t)) \cdot \delta_H g(t)) \ominus_H \left(\left(-f^{\Delta}(t)\right) \cdot g(t)\right).$$

Exercise 2.1.56 *Let* $\mathbb{T} = 3^{\mathbb{N}_0}$,

$$\begin{aligned} f(t) &= \frac{1}{t+1}, \\ g(t) &= \left(t^2, t^4, t^4+1\right), \quad t \in \mathbb{T}. \end{aligned}$$

Prove that

$$\delta_H(f\cdot g)(t) = (f(\sigma(t))\cdot\delta_H g(t))\ominus_H\left(\left(-f^{\Delta}(t)\right)\cdot g(t)\right),\quad t\in\mathbb{T}.$$

Exercise 2.1.57 *Let $t\in\mathbb{T}^{\kappa}$, $f:\mathbb{T}\to\mathbb{R}$ be Δ-differentiable at t, $f(\sigma(t))f^{\Delta}(t)<0$, $g:\mathbb{T}\to F(\mathbb{R})$ be δ_H-differentiable at t and the H-differences*

$$(f(t+h)\cdot g(t+h))\ominus_H(f(\sigma(t))\cdot g(\sigma(t))),$$

and

$$(f(\sigma(t))\cdot g(\sigma(t)))\ominus_H(f(t-h)\cdot g(t-h)),$$

exist. Let also, for any $h>0$ sufficiently small, there exist $u_1(\sigma(t),h)$, $u_2(\sigma(t),h)\in F(\mathbb{R})$ such that

$$\begin{aligned} g(t+h) &= g(\sigma(t))+u_1(\sigma(t),h),\\ g(\sigma(t)) &= g(t-h)+u_2(\sigma(t),h). \end{aligned}$$

Prove that

$$\delta_H(f\cdot g)(t) = \left(f^{\Delta}(t)\cdot\delta_H g(t)\right)\ominus_H((-f(\sigma(t)))\cdot g(t)).$$

Theorem 2.1.58 *Let $f:\mathbb{T}\to\mathbb{R}$ be Δ-differentiable at $t\in\mathbb{T}^{\kappa}$ with $f(\sigma(t))f^{\Delta}(t)<0$. Let also, $g:\mathbb{T}\to F(\mathbb{R})$ be δ_H-differentiable at t with α-levels*

$$[g(t)]^{\alpha} = \left[\underline{g}^{\alpha}(t),\overline{g}^{\alpha}(t)\right],\quad \alpha\in[0,1].$$

Then $f\cdot g:\mathbb{T}\to F(\mathbb{R})$ is δ_H-differentiable at t and

$$[\delta_H(f\cdot g)(t)]^{\alpha} = \left[f^{\Delta}(t)\cdot g(t)\right]^{\alpha}\ominus_H[(-f(\sigma(t)))\cdot\delta_H g(t)]^{\alpha},\ \alpha\in[0,1]. \tag{2.20}$$

Proof 2.1.59 *1. Assume $f(\sigma(t))>0$, $f^{\Delta}(t)<0$. Then $f(t)>0$ and*

$$[f(t)\cdot g(t)]^{\alpha} = f(t)\cdot[g(t)]^{\alpha}$$

$$
\begin{aligned}
&= f(t)\cdot\left[\underline{g}^{\alpha}(t),\overline{g}^{\alpha}(t)\right] \\
&= \left[f(t)\underline{g}^{\alpha}(t), f(t)\overline{g}^{\alpha}(t)\right],
\end{aligned}
$$

$$
\begin{aligned}
[\delta_H(f\cdot g)(t)]^{\alpha} &= \left[(f\underline{g}^{\alpha})^{\Delta}(t),(f\overline{g}^{\alpha})^{\Delta}(t)\right] \\
&= \Big[f^{\Delta}(t)\underline{g}^{\alpha}(t)+f(\sigma(t))\underline{g}^{\alpha\Delta}(t), \\
&\qquad f^{\Delta}(t)\overline{g}^{\alpha}(t)+f(\sigma(t))\overline{g}^{\alpha\Delta}(t)\Big],
\end{aligned}
$$

$$
\begin{aligned}
\left[f^{\Delta}(t)\cdot g(t)\right]^{\alpha} &= f^{\Delta}(t)\cdot[g(t)]^{\alpha} \\
&= f^{\Delta}(t)\cdot\left[\underline{g}^{\alpha}(t),\overline{g}^{\alpha}(t)\right] \\
&= \left[f^{\Delta}(t)\overline{g}^{\alpha}(t), f^{\Delta}(t)\underline{g}^{\alpha}(t)\right],
\end{aligned}
$$

$$
\begin{aligned}
[(-f(\sigma(t)))\cdot\delta_H g(t)]^{\alpha} &= (-f(\sigma(t)))\cdot[\delta_H g(t)]^{\alpha} \\
&= (-f(\sigma(t)))\cdot\left[\underline{g}^{\alpha\Delta}(t),\overline{g}^{\alpha\Delta}(t)\right] \\
&= \left[-f(\sigma(t))\overline{g}^{\alpha\Delta}(t), -f(\sigma(t))\underline{g}^{\alpha\Delta}(t)\right].
\end{aligned}
$$

From here,

$$
\left[f^{\Delta}(t)\cdot g(t)\right]^{\alpha}\ominus_H[(-f(\sigma(t)))\cdot\delta_H g(t)]^{\alpha}
$$

$$
= \left[f^{\Delta}(t)\underline{g}^{\alpha}(t)+f(\sigma(t))\underline{g}^{\alpha\Delta}(t), f^{\Delta}(t)\overline{g}^{\alpha}(t)+f(\sigma(t))\overline{g}^{\alpha\Delta}(t)\right].
$$

Therefore, we get (2.20).

2. *Let* $f(\sigma(t)) < 0$, $f^{\Delta}(t) > 0$. *Then* $f(t) < 0$ *and*

$$
\begin{aligned}
[f(t)\cdot g(t)]^{\alpha} &= f(t)\cdot [g(t)]^{\alpha}\\
&= f(t)\cdot \left[\underline{g}^{\alpha}(t), \overline{g}^{\alpha}(t)\right]\\
&= \left[f(t)\overline{g}^{\alpha}(t), f(t)\underline{g}^{\alpha}(t)\right],\\
[\delta_H(f\cdot g)(t)]^{\alpha} &= \left[(f\overline{g}^{\alpha})^{\Delta}(t), (f\underline{g}^{\alpha})^{\Delta}(t)\right]\\
&= \Big[f^{\Delta}(t)\overline{g}^{\alpha}(t) + f(\sigma(t))\overline{g}^{\alpha\Delta}(t),\\
&\qquad f^{\Delta}(t)\underline{g}^{\alpha}(t) + f(\sigma(t))\underline{g}^{\alpha\Delta}(t)\Big],\\
\left[f^{\Delta}(t)\cdot g(t)\right]^{\alpha} &= f^{\Delta}(t)\cdot [g(t)]^{\alpha}\\
&= f^{\Delta}(t)\cdot \left[\underline{g}^{\alpha}(t), \overline{g}^{\alpha}(t)\right]\\
&= \left[f^{\Delta}(t)\underline{g}^{\alpha}(t), f^{\Delta}(t)\overline{g}^{\alpha}(t)\right],\\
[(-f(\sigma(t)))\cdot \delta_H g(t)]^{\alpha} &= (-f(\sigma(t)))\cdot [\delta_H g(t)]^{\alpha}\\
&= (-f(\sigma(t)))\cdot \left[\underline{g}^{\alpha\Delta}(t), \overline{g}^{\alpha\Delta}(t)\right]\\
&= \left[-f(\sigma(t))\underline{g}^{\alpha\Delta}(t), -f(\sigma(t))\overline{g}^{\alpha\Delta}(t)\right].
\end{aligned}
$$

Hence,

$$
\left[f^{\Delta}(t)\cdot g(t)\right]^{\alpha} \ominus_H [(-f(\sigma(t)))\cdot \delta_H g(t)]^{\alpha}
$$

$$
= \left[f^{\Delta}(t)\overline{g}^{\alpha}(t)+f(\sigma(t))\overline{g}^{\alpha\Delta}(t), f^{\Delta}(t)\underline{g}^{\alpha}(t)+f(\sigma(t))\underline{g}^{\alpha\Delta}(t)\right].
$$

Thus, we obtain (2.20). *This completes the proof.*

Example 2.1.60 *Let* $\mathbb{T} = 3^{\mathbb{N}_0}$, $f : \mathbb{T} \to \mathbb{R}$ *be given by*

$$f(t) = \frac{1}{t}, \quad t \in \mathbb{T},$$

and $g : \mathbb{T} \to F(\mathbb{R})$ *be with* α*-levels*

$$[g(t)]^{\alpha} = \left[t, t + t^2\right]^{\alpha}, \quad t \in \mathbb{T}, \quad \alpha \in [0,1].$$

We have

$$\begin{aligned}
\sigma(t) &= 3t, \\
f(t) &> 0, \\
f(\sigma(t)) &= \frac{1}{\sigma(t)} \\
&= \frac{1}{3t} \\
&> 0, \\
f^{\Delta}(t) &= -\frac{1}{t\sigma(t)} \\
&= -\frac{1}{3t^2} \\
&< 0, \\
\underline{g}^{\alpha}(t) &= t, \\
\overline{g}^{\alpha}(t) &= t + t^2, \\
\underline{g}^{\alpha\Delta}(t) &= 1,
\end{aligned}$$

$$\begin{aligned}
\overline{g}^{\alpha\Delta}(t) &= 1+\sigma(t)+t \\
&= 1+4t, \quad t\in\mathbb{T}, \quad \alpha\in[0,1].
\end{aligned}$$

Then

$$\begin{aligned}
\left[f^{\Delta}(t)\cdot g(t)\right]^{\alpha} &= f^{\Delta}(t)\cdot[g(t)]^{\alpha} \\
&= \left(-\frac{1}{3t^2}\right)\cdot\left[t,t+t^2\right] \\
&= \left[-\frac{1+t}{3t},-\frac{1}{3t}\right], \\
[(-f(\sigma(t)))\cdot\delta_H g(t)]^{\alpha} &= (-f(\sigma(t)))\cdot[\delta_H g(t)]^{\alpha} \\
&= \left(-\frac{1}{3t}\right)\cdot[1,1+4t] \\
&= \left[-\frac{1+4t}{3t},-\frac{1}{3t}\right], \quad t\in\mathbb{T}, \quad \alpha\in[0,1].
\end{aligned}$$

Consequently

$$\begin{aligned}
[\delta_H(f\cdot g)(t)]^{\alpha} &= \left[-\frac{1+t}{3t},-\frac{1}{3t}\right] \\
&\quad \ominus_H\left[-\frac{1+4t}{3t},-\frac{1}{3t}\right] \\
&= \left[-\frac{1}{3t}+\frac{1}{3t},-\frac{1+t}{3t}+\frac{1+4t}{3t}\right] \\
&= [0,1], \quad t\in\mathbb{T}, \quad \alpha\in[0,1].
\end{aligned}$$

Exercise 2.1.61 *Let* $\mathbb{T}=2^{\mathbb{N}_0}$, $f:\mathbb{T}\to\mathbb{R}$ *be given by*

$$f(t)=\frac{t}{t^2+t+1}, \quad t\in\mathbb{T},$$

and $g : \mathbb{T} \to F(\mathbb{R})$ *be with* α*-levels*

$$[g(t)]^\alpha = \left[t + t^2, 1 + 2t + 3t^2 + \frac{1}{t+1}\right]^\alpha, \quad t \in \mathbb{T}, \quad \alpha \in [0,1].$$

Find

$$[\delta_H(f \cdot g)(t)]^\alpha, \quad t \in \mathbb{T}, \quad \alpha \in [0,1].$$

2.2 SECOND TYPE FUZZY DELTA DIFFERENTIATION

Definition 2.2.1 *Assume that* $f : \mathbb{T} \to F(\mathbb{R})$ *is a fuzzy function and* $t \in \mathbb{T}^\kappa$. *Then* f *is said to be the second type right fuzzy delta differentiable at* t, *shortly right* δ_{SH}*-differentiable at* t, *if there exists an element* $\delta_{SH}^+ f(t) \in F(\mathbb{R})$ *with the property that for any given* $\varepsilon > 0$ *there exists a neighborhood* $U_\mathbb{T}$ *of* t, *i.e.,* $U_\mathbb{T} = (t - \delta, t + \delta) \bigcap \mathbb{T}$ *for some* $\delta > 0$ *such that for all* $t + h \in U_\mathbb{T}$ *the H-difference* $f(\sigma(t)) \ominus_H f(t+h)$ *exists and*

$$D\big(f(\sigma(t)) \ominus_H f(t+h), \delta_{SH}^+ f(t)(-(h - \mu(t)))\big) \leq \varepsilon |h - \mu(t)|,$$

with $0 \leq h < \delta$. *In this case,* $\delta_{SH}^+ f(t)$ *is said to be a second type right fuzzy delta derivative of* f *at* t, *shortly right* δ_{SH}*-derivative of* f *at* t.

Definition 2.2.2 *Assume that* $f : \mathbb{T} \to F(\mathbb{R})$ *is a fuzzy function and* $t \in \mathbb{T}^\kappa$. *Then* f *is said to be second type left fuzzy delta differentiable at* t, *shortly left* δ_{SH}*-differentiable at* t, *if there exists an element* $\delta_{SH}^- f(t) \in F(\mathbb{R})$ *with the property that for any given* $\varepsilon > 0$ *there exists a neighborhood* $U_\mathbb{T}$ *of* t, *i.e.,* $U_\mathbb{T} = (t - \delta, t + \delta) \bigcap \mathbb{T}$ *for some* $\delta > 0$ *such that for all* $t - h \in U_\mathbb{T}$ *the H-difference* $f(t-h) \ominus_H f(\sigma(t))$ *exists and*

$$D\big(f(t-h) \ominus_H f(\sigma(t)), \delta_{SH}^- f(t)(-(h + \mu(t)))\big) \leq \varepsilon (h + \mu(t)),$$

with $0 \leq h < \delta$.

Definition 2.2.3 *Let* $f : \mathbb{T} \to F(\mathbb{R})$ *be a fuzzy function and* $t \in \mathbb{T}^\kappa$. *Then* f *is said to be second type fuzzy delta differentiable at* t, *shortly*

δ_{SH}-differentiable at t, if f is both second type left and right fuzzy delta differentiable at $t \in \mathbb{T}^\kappa$ and $\delta_{SH}^{-} f(t) = \delta_{SH}^{+} f(t)$, and we will denote it by $\delta_{SH} f(t)$. We call $\delta_{SH} f(t)$ the second type fuzzy delta derivative of f at t, shortly δ_{SH}-derivative of f at t. We say that f is second type fuzzy delta differentiable at t, shortly δ_{SH}-differentiable at t, if its δ_{SH}-derivative exists at t. We say that f is second type fuzzy delta differentiable on $\mathbb{T}^\kappa$, shortly δ_{SH}-differentiable on $\mathbb{T}^\kappa$, if its δ_{SH}-derivative exists at each $t \in \mathbb{T}^\kappa$. The fuzzy function $\delta_{SH} f : \mathbb{T}^\kappa \to F(\mathbb{R})$ is then called the second type fuzzy delta derivative, shortly δ_{SH}-derivative of f in $\mathbb{T}^\kappa$.

The proofs of the following properties of the δ_{SH}-derivative are similar to the proofs of the corresponding properties of the δ_H-derivative. Therefore, we leave them to the reader as exercises.

Theorem 2.2.4 *If the δ_{SH}-derivative of f in $t \in \mathbb{T}^\kappa$ exists, then it is unique. Therefore, the δ_{SH}-derivative is well defined.*

Theorem 2.2.5 *Assume that $f : \mathbb{T} \to F(\mathbb{R})$ is a continuous function at $t_1 \in \mathbb{T}^\kappa$ and t_1 is right-scattered. Then f is δ_{SH}-differentiable at t_1 and*

$$\delta_{SH} f(t_1) = \frac{f(t_1) \ominus_H f(\sigma(t_1))}{-\mu(t_1)}.$$

Example 2.2.6 *Let $\mathbb{T} = 2^{\mathbb{N}_0}$,*

$$f(t) = \left(\frac{1}{t+2}, \frac{1}{t+1}, \frac{1}{t}\right), \quad t \in \mathbb{T}.$$

Here

$$\begin{aligned} \sigma(t) &= 2t, \\ \mu(t) &= t, \\ f(\sigma(t)) &= \left(\frac{1}{\sigma(t)+2}, \frac{1}{\sigma(t)+1}, \frac{1}{\sigma(t)}\right) \end{aligned}$$

$$= \left(\frac{1}{2t+2}, \frac{1}{2t+1}, \frac{1}{2t}\right), \quad t \in \mathbb{T}.$$

Then

$$\begin{aligned} f(t) \ominus_H f(\sigma(t)) &= \left(\frac{1}{t+2} - \frac{1}{2t+2}, \frac{1}{t+1} - \frac{1}{2t+1}, \frac{1}{t} - \frac{1}{2t}\right) \\ &= \left(\frac{2t+2-t-2}{2(t+1)(t+2)}, \frac{2t+1-t-1}{(t+1)(2t+1)}, \frac{1}{2t}\right) \\ &= \left(\frac{t}{2(t+1)(t+2)}, \frac{t}{(t+1)(2t+1)}, \frac{1}{2t}\right), \quad t \in \mathbb{T}, \end{aligned}$$

and

$$\begin{aligned} \delta_{SH} f(t) &= \left(-\frac{1}{t}\right) \cdot \left(\frac{t}{2(t+1)(t+2)}, \frac{t}{(t+1)(2t+1)}, \frac{1}{2t}\right) \\ &= \left(-\frac{1}{2t^2}, -\frac{1}{(t+1)(2t+1)}, -\frac{1}{2(t+1)(t+2)}\right), \quad t \in \mathbb{T}. \end{aligned}$$

Exercise 2.2.7 *Let* $\mathbb{T} = 2\mathbb{Z}$,

$$f(t) = \left(\frac{1}{t^2+3}, \frac{1}{t^2+2}, \frac{1}{t^2+1}\right), \quad t \in \mathbb{T}.$$

Find $\delta_{SH} f(t)$, $t \in \mathbb{T}$.

Theorem 2.2.8 *Assume that* $f : \mathbb{T} \to F(\mathbb{R})$ *is* δ_{SH}*-differentiable at* $t \in \mathbb{T}^\kappa$. *Then* f *is continuous at* t.

Theorem 2.2.9 *Let* $f : \mathbb{T} \to F(\mathbb{R})$ *be a function, and let* $t \in \mathbb{T}^\kappa$ *be right-dense. Then* f *is* δ_{SH}*-differentiable at* t *if and only if the limits*

$$\lim_{h\to 0+} \frac{f(t) \ominus_H f(t+h)}{-h} \quad \text{and} \quad \lim_{h\to 0+} \frac{f(t-h) \ominus_H f(t)}{-h},$$

exist and satisfy the relations

$$\lim_{h\to 0+} \frac{f(t) \ominus_H f(t+h)}{-h} = \lim_{h\to 0+} \frac{f(t-h) \ominus_H f(t)}{-h} = \delta_{SH} f(t).$$

Example 2.2.10 *Let* $\mathbb{T} = \{1\} \bigcup \left\{1 + \frac{1}{12^k}\right\}_{k\in\mathbb{N}}$,

$$f(t) = \left(\frac{1}{t+4}, \frac{1}{t+3}, \frac{1}{t+2}\right), \quad t \in \mathbb{T}.$$

Note that $t = 1$ *is right-dense. Take* $h > 0$ *arbitrarily. Then*

$$\begin{aligned}
f(1) &= \left(\frac{1}{5}, \frac{1}{4}, \frac{1}{3}\right), \\
f(1+h) &= \left(\frac{1}{5+h}, \frac{1}{4+h}, \frac{1}{3+h}\right), \\
f(1-h) &= \left(\frac{1}{5-h}, \frac{1}{4-h}, \frac{1}{3-h}\right), \\
f(1) \ominus_H f(1+h) &= \left(\frac{1}{5} - \frac{1}{5+h}, \frac{1}{4} - \frac{1}{4+h}, \frac{1}{3} - \frac{1}{3+h}\right) \\
&= \left(\frac{h}{5(5+h)}, \frac{h}{4(4+h)}, \frac{h}{3(3+h)}\right), \\
f(1-h) \ominus_H f(1) &= \left(\frac{1}{5-h} - \frac{1}{5}, \frac{1}{4-h} - \frac{1}{4}, \frac{1}{3-h} - \frac{1}{3}\right) \\
&= \left(\frac{h}{5(5-h)}, \frac{h}{4(4-h)}, \frac{h}{3(3-h)}\right), \\
\frac{f(1) \ominus_H f(1+h)}{-h} &= \left(-\frac{1}{3(3+h)}, -\frac{1}{4(4+h)}, -\frac{1}{5(5+h)}\right), \\
\frac{f(1-h) \ominus_H f(1)}{-h} &= \left(-\frac{1}{3(3-h)}, -\frac{1}{4(4-h)}, -\frac{1}{5(5-h)}\right).
\end{aligned}$$

Hence,

$$\lim_{h\to 0} \frac{f(1) \ominus_H f(1+h)}{-h} = \left(-\frac{1}{9}, -\frac{1}{16}, -\frac{1}{25}\right),$$

$$\lim_{h\to 0}\frac{f(1-h)\ominus_H f(1)}{-h} = \left(-\frac{1}{9},-\frac{1}{16},-\frac{1}{25}\right).$$

Therefore

$$\begin{aligned}\delta_{SH}f(1) &= \lim_{h\to 0}\frac{f(1)\ominus_H f(1+h)}{-h}\\ &= \lim_{h\to 0}\frac{f(1-h)\ominus_H f(1)}{-h}\\ &= \left(-\frac{1}{9},-\frac{1}{16},-\frac{1}{25}\right).\end{aligned}$$

Exercise 2.2.11 *Let* $\mathbb{T}=\{0\}\bigcup\left\{\frac{1}{3^k}\right\}_{k\in\mathbb{N}}$,

$$f(t)=\left(\frac{1}{t^2+4},\frac{1}{t^2+2},\frac{1}{t^2+1}\right),\quad t\in\mathbb{T}.$$

Find

$$\delta_{SH}f(0),\quad \delta_{SH}f\left(\frac{1}{3}\right),\quad \delta_{SH}f\left(\frac{1}{9}\right).$$

Theorem 2.2.12 *Let* $f:\mathbb{T}\to F(\mathbb{R})$ *be* δ_{SH}*-differentiable at* $t\in\mathbb{T}^\kappa$. *Then*

$$f(\sigma(t))=f(t)+(-\mu(t))\cdot\delta_{SH}f(t)$$

or

$$f(t)=f(\sigma(t))+(-\mu(t))\cdot\delta_{SH}f(t),\quad t\in\mathbb{T}^\kappa.$$

Theorem 2.2.13 *Let* $f,g:\mathbb{T}\to F(\mathbb{R})$ *be* δ_{SH}*-differentiable at* $t\in\mathbb{T}^\kappa$. *Then* $f+g:\mathbb{T}\to F(\mathbb{R})$ *is* δ_{SH}*-differentiable at* $t\in\mathbb{T}^\kappa$ *and*

$$\delta_{SH}(f+g)(t)=\delta_{SH}f(t)+\delta_{SH}g(t).$$

Example 2.2.14 *Let* $\mathbb{T}=2\mathbb{Z}$,

$$f(t) = \left(\frac{1}{t+3},\frac{1}{t+2},\frac{1}{t+1}\right),$$

$$g(t) = \left(\frac{1}{2t+5}, \frac{1}{2t+3}, \frac{1}{2t+1}\right), \quad t \in \mathbb{T}.$$

We have

$$\begin{aligned} \sigma(t) &= t+2, \\ \mu(t) &= 2, \quad t \in \mathbb{T}, \end{aligned}$$

and

$$\begin{aligned} f(0) &= \left(\frac{1}{3}, \frac{1}{2}, 1\right), \\ f(\sigma(0)) &= f(2) \\ &= \left(\frac{1}{5}, \frac{1}{4}, \frac{1}{3}\right), \\ g(0) &= \left(\frac{1}{5}, \frac{1}{3}, 1\right), \\ g(\sigma(0)) &= g(2) \\ &= \left(\frac{1}{9}, \frac{1}{7}, \frac{1}{5}\right), \\ f(0) \ominus_H f(\sigma(0)) &= \left(\frac{1}{3} - \frac{1}{5}, \frac{1}{2} - \frac{1}{4}, 1 - \frac{1}{3}\right) \\ &= \left(\frac{2}{15}, \frac{1}{4}, \frac{2}{3}\right), \\ g(0) \ominus_H g(\sigma(0)) &= \left(\frac{1}{5} - \frac{1}{9}, \frac{1}{3} - \frac{1}{7}, 1 - \frac{1}{5}\right) \\ &= \left(\frac{4}{45}, \frac{4}{21}, \frac{4}{5}\right), \end{aligned}$$

$$
\begin{aligned}
\delta_{SH} f(0) &= \frac{f(0) \ominus_H f(\sigma(0))}{-2} \\
&= \left(-\frac{1}{3}, -\frac{1}{8}, -\frac{1}{15}\right), \\
\delta_{SH} g(0) &= \frac{g(0) \ominus_H g(\sigma(0))}{-2} \\
&= \left(-\frac{2}{5}, -\frac{2}{21}, -\frac{2}{45}\right), \\
\delta_{SH} f(0) + \delta_{SH} g(0) &= \left(-\frac{1}{3} - \frac{2}{5}, -\frac{1}{8} - \frac{2}{21}, -\frac{1}{15} - \frac{2}{45}\right) \\
&= \left(-\frac{11}{15}, -\frac{37}{168}, -\frac{1}{9}\right), \\
(f+g)(0) &= \left(\frac{1}{3} + \frac{1}{5}, \frac{1}{2} + \frac{1}{3}, 1+1\right) \\
&= \left(\frac{8}{15}, \frac{5}{6}, 2\right), \\
(f+g)(\sigma(0)) &= \left(\frac{1}{5} + \frac{1}{9}, \frac{1}{4} + \frac{1}{7}, \frac{1}{3} + \frac{1}{5}\right) \\
&= \left(\frac{14}{45}, \frac{11}{28}, \frac{8}{15}\right), \\
(f+g)(0) \ominus_H (f+g)(\sigma(0)) &= \left(\frac{8}{15} - \frac{14}{45}, \frac{5}{6} - \frac{11}{28}, 2 - \frac{8}{15}\right) \\
&= \left(\frac{24-14}{45}, \frac{70-33}{84}, \frac{30-8}{15}\right) \\
&= \left(\frac{2}{9}, \frac{37}{84}, \frac{22}{15}\right),
\end{aligned}
$$

$$\begin{aligned}\delta_{SH}(f+g)(0) &= \frac{(f+g)(0)\ominus_H (f+g)(\sigma(0))}{-2}\\ &= \left(-\frac{11}{15},-\frac{37}{168},-\frac{1}{9}\right).\end{aligned}$$

Consequently

$$\delta_{SH}(f+g)(0) = \delta_{SH}f(0)+\delta_{SH}g(0).$$

Exercise 2.2.15 *Let* $\mathbb{T} = \{0\}\bigcup\left\{\frac{1}{2^k}\right\}_{k\in\mathbb{N}}$,

$$\begin{aligned}f(t) &= \left(\frac{1}{t+8},\frac{1}{t+6},\frac{1}{t+4}\right),\\ g(t) &= \left(\frac{1}{t+6},\frac{1}{t+4},\frac{1}{t+2}\right), \quad t\in\mathbb{T}.\end{aligned}$$

Prove that

$$\delta_{SH}(f+g)(0) = \delta_{SH}f(0)+\delta_{SH}g(0).$$

Theorem 2.2.16 *Let* $f:\mathbb{T}\to F(\mathbb{R})$ *be* δ_{SH}*-differentiable at* $t\in\mathbb{T}^\kappa$. *Then for any* $\lambda\in\mathbb{R}$ *the function* $\lambda\cdot f:\mathbb{T}\to F(\mathbb{R})$ *is* δ_{SH}*-differentiable at* $t\in\mathbb{T}^\kappa$ *and*

$$\delta_{SH}(\lambda\cdot f)(t) = \lambda\cdot\delta_{SH}f(t).$$

Theorem 2.2.17 *Let* $t\in\mathbb{T}^\kappa$, $f:\mathbb{T}\to F(\mathbb{R})$ *and* $f_\alpha(t) = [f(t)]^\alpha$, $t\in\mathbb{T}$, $\alpha\in[0,1]$. *If* f *is* δ_{SH}*-differentiable at* t*, then* f_α *is* δ_{SH}*-differentiable at* t *and*

$$[\delta_{SH}f(t)]^\alpha = \delta_{SH}f_\alpha(t), \quad t\in\mathbb{T}, \quad \alpha\in[0,1].$$

Theorem 2.2.18 *Let* $t\in\mathbb{T}^\kappa$, $f:\mathbb{T}\to F(\mathbb{R})$ *be* δ_{SH}*-differentiable at* t. *Let also,*

$$[f(t)]^\alpha = \left[\underline{f}^\alpha(t),\overline{f}^\alpha(t)\right], \quad \alpha\in[0,1].$$

Then $\underline{f}^\alpha$ *and* $\overline{f}^\alpha$ *are* Δ*-differentiable at* t *and*

$$[\delta_{SH}f(t)]^\alpha = \left[\overline{f}^{\alpha\Delta}(t),\underline{f}^{\alpha\Delta}(t)\right], \quad \alpha\in[0,1].$$

Example 2.2.19 *Let* $\mathbb{T} = \left\{\frac{1}{4^k}\right\}_{k\in\mathbb{N}}$, $f : \mathbb{T} \to F(\mathbb{R})$ *and*

$$[f(t)]^\alpha = [e_{-1}(t,1)+t-10, e_{-1}(t,1)-t+10], \quad t \in \mathbb{T}.$$

Set

$$\begin{aligned} g_1(t) &= e_{-1}(t,1)+t-10, \\ g_2(t) &= e_{-1}(t,1)-t+10, \quad t \in \mathbb{T}. \end{aligned}$$

Then

$$\begin{aligned} g_1^\Delta(t) &= -e_{-1}(t,1)+1, \\ g_2^\Delta(t) &= -e_{-1}(t,1)-1, \quad t \in \mathbb{T}, \end{aligned}$$

and

$$[\delta_{SH} f(t)]^\alpha = [-e_{-1}(t,1)-1, -e_{-1}(t,1)+1], \quad t \in \mathbb{T}.$$

Exercise 2.2.20 *Let* $\mathbb{T} = 2^{\mathbb{N}_0}$, $f : \mathbb{T} \to F(\mathbb{R})$ *and*

$$[f(t)]^\alpha = \left[\frac{1}{t+1}, \frac{1}{t}\right], \quad t \in \mathbb{T}.$$

Find $[\delta_{SH} f(t)]^\alpha$.

Example 2.2.21 *Let* $\mathbb{T} = 2^{\mathbb{N}_0}$, $f, g : \mathbb{T} \to F(\mathbb{R})$ *be fuzzy functions whose* α*-levels are given by*

$$\begin{aligned} [f(t)]^\alpha &= [e_{-1}(t,0)-t+2, e_{-1}(t,0)+t+5], \\ [g(t)]^\alpha &= [-e_{-1}(t,0)+t-10, -e_{-1}(t,0)-t+10], \end{aligned}$$

$t \in \mathbb{T}$, $\alpha \in [0,1]$. *Here*

$$\sigma(t) = 2t,$$

$$
\begin{aligned}
\underline{f}^{\alpha}(t) &= e_{-1}(t,0) - t + 2, \\
\overline{f}^{\alpha}(t) &= e_{-1}(t,0) + t + 5, \\
\underline{g}^{\alpha}(t) &= -e_{-1}(t,0) + t - 10, \\
\overline{g}^{\alpha}(t) &= -e_{-1}(t,0) - t + 10, \\
\underline{f}^{\alpha\Delta}(t) &= -e_{-1}(t,0) - 1, \\
\overline{f}^{\alpha\Delta}(t) &= -e_{-1}(t,0) + 1, \\
\underline{g}^{\alpha\Delta}(t) &= e_{-1}(t,0) + 1, \\
\overline{g}^{\alpha\Delta}(t) &= e_{-1}(t,0) - 1, \\
[\delta_H f(t)]^{\alpha} &= \left[\underline{f}^{\alpha\Delta}(t), \overline{f}^{\alpha\Delta}(t)\right] \\
&= [-e_{-1}(t,0) - 1, -e_{-1}(t,0) + 1], \\
[\delta_{SH} g(t)]^{\alpha} &= [e_{-1}(t,0) - 1, e_{-1}(t,0) + 1], \quad t \in \mathbb{T}, \quad \alpha \in [0,1].
\end{aligned}
$$

Thus, f is δ_H-differentiable in $\mathbb{T}$ and g is δ_{SH}-differentiable in $\mathbb{T}$. We have

$$
\begin{aligned}
[\delta_H f(t)]^{\alpha} + [\delta_{SH} g(t)]^{\alpha} &= [-2,2], \\
[f(t) + g(t)]^{\alpha} &= [-8,15], \\
[\delta_H (f+g)(t)]^{\alpha} &= [0,0], \quad t \in \mathbb{T}, \quad \alpha \in [0,1].
\end{aligned}
$$

Note that

$$[\delta_H(f+g)(t)]^\alpha \neq [\delta_H f(t)]^\alpha + [\delta_{SH} g(t)]^\alpha, \quad t \in \mathbb{T}, \quad \alpha \in [0,1].$$

Theorem 2.2.22 *Let $t \in \mathbb{T}^\kappa$, $f : \mathbb{T} \to \mathbb{R}$ be Δ-differentiable at t, $f(\sigma(t))f^\Delta(t) < 0$, $g : \mathbb{T} \to F(\mathbb{R})$ be δ_{SH}-differentiable at t. Let also, there exist $u_1(\sigma(t),h)$, $u_2(\sigma(t),h) \in F(\mathbb{R})$ such that*

$$\begin{aligned} g(\sigma(t)) &= g(t+h) + u_1(\sigma(t),h), \\ g(t-h) &= g(\sigma(t)) + u_2(\sigma(t),h), \end{aligned}$$

for $h > 0$ sufficiently small. Then $f \cdot g$ is δ_{SH}-differentiable at t and

$$\delta_{SH}(f \cdot g)(t) = f^\Delta(t) \cdot g(t) + f(\sigma(t)) \cdot \delta_{SH} g(t).$$

Proof 2.2.23 *Let $\varepsilon \in (0,1)$ and*

$$\varepsilon^* = \varepsilon \left(1 + |f(\sigma(t))| + \|g(t)\| + |f^\Delta(t)|\right)^{-1}.$$

We have $\varepsilon^* \in (0,1)$. Since f is Δ-differentiable at t, there exists a neighborhood $U_T^1(t,\delta)$ of t such that for $0 < h < \delta$ with $t+h, t-h \in U_T^1(t,\delta)$ we have

$$|f(\sigma(t)) - f(t+h) - f^\Delta(t)(\mu(t) - h)| \leq \varepsilon^*(h - \mu(t)), \tag{2.21}$$

$$|f(\sigma(t)) - f(t-h) - f^\Delta(t)(h + \mu(t))| \leq \varepsilon^*(h + \mu(t)). \tag{2.22}$$

Because g is δ_{SH}-differentiable at t, there exists a neighborhood $U_T^2(t,\delta)$ such that for all $0 < h < \delta$ with $t-h, t+h \in U_T^2(t,\delta)$, we have that the H-differences

$$g(\sigma(t)) \ominus_H g(t+h) \quad and \quad g(t-h) \ominus_H g(\sigma(t)),$$

exist and

$$D(g(\sigma(t)) \ominus_H g(t+h), \delta_{SH} g(t)(\mu(t) - h)) \leq \varepsilon^*(h - \mu(t)), \tag{2.23}$$

$$D(g(t-h) \ominus_H g(\sigma(t)), \delta_{SH} g(t)(-(h + \mu(t)))) \leq \varepsilon^*(h + \mu(t)). \tag{2.24}$$

Also,

$$\|g(t-h)\| \leq 1 + \|g(t)\|, \tag{2.25}$$

$$\|g(t+h)\| \leq 1 + \|g(t)\|, \tag{2.26}$$

$$D(g(t), g(t-h)) \leq \varepsilon^*, \tag{2.27}$$

$$D(g(t), g(t+h)) \leq \varepsilon^*, \tag{2.28}$$

for $h > 0$ sufficiently small. Because f is continuous at t and $f(\sigma(t))f^{\Delta}(t) < 0$, we have that $f(t)$, $f(t+h)$, and $f(t-h)$ have the same sign for $h > 0$ sufficiently small and

$$f(\sigma(t)) = f(t+h) + v_1(\sigma(t), h),$$

where $v_1(\sigma(t), h)$ has the same sign as $f(\sigma(t))$ and $f(t+h)$. Consequently

$$\begin{aligned} f(\sigma(t)) \cdot g(\sigma(t)) &= (f(t+h)+v_1(\sigma(t),h)) \cdot (g(t+h)+u_1(\sigma(t),h)) \\ &= f(t+h) \cdot g(t+h) + f(t+h) \cdot u_1(\sigma(t),h) \\ &\quad + v_1(\sigma(t),h) \cdot g(t+h) + v_1(\sigma(t),h) \cdot u_1(\sigma(t),h). \end{aligned}$$

Therefore the H-difference

$$(f(\sigma(t)) \cdot g(\sigma(t))) \ominus_H (f(t+h) \cdot g(t+h)),$$

exists. Next,

$$f(t-h) = f(\sigma(t)) + v_2(\sigma(t), h),$$

where $v_2(\sigma(t), h)$ has the same sign as $f(t-h)$ and $f(\sigma(t))$ for $h > 0$ sufficiently small. From here,

$$\begin{aligned} f(t-h) \cdot g(t-h) &= (f(\sigma(t))+v_2(\sigma(t),h)) \cdot (g(\sigma(t))+u_2(\sigma(t),h)) \\ &= f(\sigma(t)) \cdot g(\sigma(t)) + f(\sigma(t)) \cdot u_2(\sigma(t),h) \\ &\quad + v_2(\sigma(t),h) \cdot g(\sigma(t)) + v_2(\sigma(t),h) \cdot u_2(\sigma(t),h), \end{aligned}$$

and thus, the H-difference

$$(f(t-h)\cdot g(t-h))\ominus_H (f(\sigma(t))\cdot g(\sigma(t))),$$

exists for $h>0$ *sufficiently small. Let*

$$\begin{aligned} c &= f(t-h)\cdot g(t-h),\\ d &= f(\sigma(t))\cdot g(\sigma(t)),\\ a &= f(\sigma(t))\cdot \delta_{SH}g(t)(-(h+\mu(t))),\\ b &= f^{\Delta}(t)\cdot g(t)(-(h+\mu(t))),\\ w &= f(\sigma(t))\cdot g(t-h),\\ r &= f^{\Delta}(t)\cdot g(t-h)(-(h+\mu(t))). \end{aligned}$$

Now we apply Lemma 2.1.41 for a,b,c,d,w,r *and using (2.22), (2.24), (2.25), (2.27), we obtain*

$$\begin{aligned} &D\Big((f\cdot g)(t-h)\ominus_H (f\cdot g)(\sigma(t)),\\ &\Big(f(\sigma(t))\cdot \delta_{SH}g(t)+f^{\Delta}(t)\cdot g(t)\Big)(-(h+\mu(t)))\Big)\\ &\leq D\Big((f(\sigma(t))\cdot g(t-h))\ominus_H (f(\sigma(t))\cdot g(\sigma(t))),\\ &\quad f(\sigma(t))\cdot \delta_{SH}g(t)(-(h+\mu(t)))\Big)\\ &\quad +D\Big(f(t-h)\cdot g(t-h), \end{aligned}$$

$$
\begin{aligned}
& f(\sigma(t)) \cdot g(t-h) + f^{\Delta}(t) \cdot g(t-h)\left(-(h+\mu(t))\right)\Big) \\
& +D\left(f^{\Delta}(t) \cdot g(t)\left(-(h+\mu(t))\right), f^{\Delta}(t) \cdot g(t-h)\left(-(h+\mu(t))\right)\right) \\
\leq\ & |f(\sigma(t))| D\left(g(t-h) \ominus_H g(\sigma(t)), \delta_{SH} g(t)\left(-(h+\mu(t))\right)\right) \\
& + \left| f(t-h) - f(\sigma(t)) - f^{\Delta}(t)\left(-(h+\mu(t))\right)\right| \|g(t-h)\| \\
& + \left| f^{\Delta}(t)\right| D(g(t), g(t-h))(h+\mu(t)) \\
\leq\ & \varepsilon^*(h+\mu(t))|f(\sigma(t))| + \varepsilon^*(h+\mu(t))(\|g(t)\|+1) \\
& +\varepsilon^* \left| f^{\Delta}(t)\right| (h+\mu(t)) \\
=\ & \varepsilon^*\left(1+|f(\sigma(t))|+\|g(t)\|+\left|f^{\Delta}(t)\right|\right)(h+\mu(t)) \\
=\ & \varepsilon(h+\mu(t)),
\end{aligned}
$$

for $h > 0$ *sufficiently small. Let*

$$
\begin{aligned}
c &= f(\sigma(t)) \cdot g(\sigma(t)), \\
d &= f(t+h) \cdot g(t+h), \\
a &= f(\sigma(t)) \cdot \delta_{SH} g(t)\left(\mu(t)-h\right), \\
b &= f^{\Delta}(t) \cdot g(t)\left(\mu(t)-h\right), \\
w &= f(\sigma(t)) \cdot g(t+h), \\
r &= f^{\Delta}(t) \cdot g(t+h)\left(\mu(t)-h\right).
\end{aligned}
$$

We apply Lemma 2.1.43 and using (2.21), (2.23), (2.25), *and* (2.28), *we get*

$$D\Big((f\cdot g)(\sigma(t))\ominus_H(f\cdot g)(t+h),$$

$$\Big(f(\sigma(t))\cdot\delta_{SH}g(t)+f^{\Delta}(t)\cdot g(t)\Big)(\mu(t)-h)\Big)$$

$$\leq D((f\cdot g)(\sigma(t))\ominus_H(f(\sigma(t))\cdot g(t+h)),$$

$$f(\sigma(t))\cdot\delta_{SH}g(t)(\mu(t)-h))$$

$$+D\Big(f(\sigma(t))\cdot g(t+h),$$

$$f(t+h)\cdot g(t+h)+f^{\Delta}(t)\cdot g(t+h)(\mu(t)-h)\Big)$$

$$+D\left(f^{\Delta}(t)\cdot g(t)(\mu(t)-h),f^{\Delta}(t)\cdot g(t+h)(\mu(t)-h)\right)$$

$$=|f(\sigma(t))|D(g(\sigma(t))\ominus_H g(t+h),\delta_{SH}g(t)(\mu(t)-h))$$

$$+\left|f(\sigma(t))-f(t+h)-f^{\Delta}(t)(\mu(t)-h)\right|\|g(t+h)\|$$

$$+\left|f^{\Delta}(t)\right|(h-\mu(t))D(g(t),g(t+h))$$

$$\leq\varepsilon^*|f(\sigma(t))|(h-\mu(t))+\varepsilon^*(1+\|g(t)\|)(h-\mu(t))$$

$$+\varepsilon^*\left|f^{\Delta}(t)\right|(h-\mu(t))$$

$$=\varepsilon^*\left(1+|f(\sigma(t))|+\left|f^{\Delta}(t)\right|+\|g(t)\|\right)(h-\mu(t))$$

$$=\varepsilon(h-\mu(t)),$$

for $h>0$ *sufficiently small. This completes the proof.*

Example 2.2.24 *Let* $\mathbb{T} = 2^{\mathbb{N}_0}$,

$$\begin{aligned} f(t) &= \frac{1}{t}, \\ g(t) &= \left(\frac{1}{3t}, \frac{1}{2t}, \frac{1}{t}\right), \quad t \in \mathbb{T}. \end{aligned}$$

We have

$$\begin{aligned} \sigma(t) &= 2t, \\ f(t) \cdot g(t) &= \left(\frac{1}{3t^2}, \frac{1}{2t^2}, \frac{1}{t^2}\right), \quad t \in \mathbb{T}. \end{aligned}$$

Let

$$h_1(t) = \frac{1}{t^2}, \quad t \in \mathbb{T}.$$

Then

$$\begin{aligned} h_1^\Delta(t) &= -\frac{\sigma(t)+t}{t^2(\sigma(t))^2} \\ &= -\frac{2t+t}{t^2(2t)^2} \\ &= -\frac{3t}{4t^4} \\ &= -\frac{3}{4t^3}, \\ \delta_{SH}(f \cdot g)(t) &= \left(-\frac{3}{4t^3}, -\frac{3}{8t^3}, -\frac{1}{4t^3}\right), \quad t \in \mathbb{T}. \end{aligned}$$

Next,

$$\begin{aligned} f^\Delta(t) &= -\frac{1}{t\sigma(t)} \\ &= -\frac{1}{2t^2}, \end{aligned}$$

$$
\begin{aligned}
\delta_{SH}g(t) &= \left(-\frac{1}{2t^2}, -\frac{1}{4t^2}, -\frac{1}{6t^2}\right), \\
f(\sigma(t)) &= \frac{1}{\sigma(t)} \\
&= \frac{1}{2t}, \\
f(\sigma(t)) \cdot \delta_{SH}g(t) &= \left(-\frac{1}{4t^3}, -\frac{1}{8t^3}, -\frac{1}{12t^3}\right), \\
f^{\Delta}(t) \cdot g(t) &= \left(-\frac{1}{2t^3}, -\frac{1}{4t^3}, -\frac{1}{6t^3}\right), \quad t \in \mathbb{T},
\end{aligned}
$$

and

$$
f^{\Delta}(t) \cdot g(t) + f(\sigma(t)) \cdot \delta_{SH}g(t)
$$

$$
\begin{aligned}
&= \left(-\frac{1}{2t^3}, -\frac{1}{4t^3}, -\frac{1}{6t^3}\right) + \left(-\frac{1}{4t^3}, -\frac{1}{8t^3}, -\frac{1}{12t^3}\right) \\
&= \left(-\frac{1}{2t^3} - \frac{1}{4t^3}, -\frac{1}{4t^3} - \frac{1}{8t^3}, -\frac{1}{6t^3} - \frac{1}{12t^3}\right) \\
&= \left(-\frac{3}{4t^3}, -\frac{3}{8t^3}, -\frac{1}{4t^3}\right), \quad t \in \mathbb{T}.
\end{aligned}
$$

Consequently

$$
\delta_{SH}(f \cdot g)(t) = f^{\Delta}(t) \cdot g(t) + f(\sigma(t)) \cdot \delta_{SH}g(t), \quad t \in \mathbb{T}.
$$

Exercise 2.2.25 *Let* $\mathbb{T} = \left\{\frac{1}{2^k}\right\}_{k \in \mathbb{N}} \bigcup \{0\}$,

$$
\begin{aligned}
f(t) &= \frac{1}{t+1}, \\
g(t) &= \left(\frac{1}{t+7}, \frac{1}{t+6}, \frac{1}{t+5}\right), \quad t \in \mathbb{T}.
\end{aligned}
$$

Prove that

$$\delta_{SH}(f\cdot g)(0)=f^{\Delta}(0)\cdot g(0)+f(\sigma(0))\cdot\delta_{SH}g(0).$$

Lemma 2.2.26 *Let $a,b,c,d,e,w,r\in F(\mathbb{R})$ and H-differences $c\ominus_H d$, $a\ominus_H b$, $r\ominus_H c$ exist. Then*

$$\begin{aligned} D(c\ominus_H d,a\ominus_H b) &\leq D(r,d+w)+D(a,w)\\ &+ D(e,r\ominus_H c)+D(e,b). \end{aligned}$$

Proof 2.2.27 *We have*

$$\begin{aligned} D(c\ominus_H d,a\ominus_H b) &= D(c\ominus_H d+r,a\ominus_H b+r)\\ &= D(b+r+c,a+d+r)\\ &= D(b+r,a+d+r\ominus_H c)\\ &\leq D(r,a+d)+D(b,r\ominus_H c)\\ &= D(r+w,a+d+w)+D(b+e,r\ominus_H c+e)\\ &\leq D(r,d+w)+D(w,a)\\ &\quad +D(e,r\ominus_H c)+D(b,e). \end{aligned}$$

This completes the proof.

Lemma 2.2.28 *Let $a,b,c,d,w,r\in F(\mathbb{R})$ and $c\ominus_H d$, $a\ominus_H b$, $c\ominus_H w$ exist. Then*

$$D(c\ominus_H d,a\ominus_H b) \leq D(c\ominus_H w,a)+D(w+r,d)+D(r,b).$$

Proof 2.2.29 *We have*

$$D(c\ominus_H d,a\ominus_H b) = D(c+b,a+d)$$

$$\begin{aligned} &= D(c+b+w, a+d+w) \\ &\leq D(c, a+w) + D(b+w, d) \\ &= D(c \ominus_H w, a) + D(b+w+r, d+r) \\ &\leq D(c \ominus_H w, a) + D(w+r, d) + D(r, b). \end{aligned}$$

This completes the proof.

Theorem 2.2.30 *Let $t \in \mathbb{T}^{\kappa}$, $f : \mathbb{T} \to \mathbb{R}$ be Δ-differentiable at t, $f(t)f^{\Delta}(t) > 0$, $g : \mathbb{T} \to F(\mathbb{R})$ be δ_{SH}-differentiable at t and the H-differences*

$$((f \cdot g)(t+h)) \ominus_H ((f \cdot g)(\sigma(t)))$$

and

$$((f \cdot g)(\sigma(t))) \ominus_H ((f \cdot g)(t-h))$$

exist for $h > 0$ sufficiently small. Then $f \cdot g$ is δ_H-differentiable at t and

$$\delta_H(f \cdot g)(t) = \left(f^{\Delta}(t) \cdot g(t)\right) \ominus_H ((-f(\sigma(t))) \cdot \delta_{SH} g(t)).$$

Proof 2.2.31 *Let $\varepsilon \in (0,1)$ and*

$$\begin{aligned} \varepsilon^* &= \varepsilon \min\left\{ \left(2 + \|g(t)\| + \left|f^{\Delta}(t)\right| + |f(t)| + \|\delta_{SH} g(t)\|\right)^{-1}, \right. \\ &\quad \left. \left(2 + 2\|g(t)\| + |f^{\Delta}(t)|\right)^{-1} \right\}. \end{aligned}$$

Note that $\varepsilon^* \in (0,1)$. Since f is Δ-differentiable at t, there exists a neighborhood $U_T^1(t,\delta)$ of t such that for $0 < h < \delta$ with $t+h, t-h \in U_T^1(t,\delta)$, we have

$$|f(t+h) - f(\sigma(t)) - f^{\Delta}(t)(h - \mu(t))| \leq \varepsilon^*(h - \mu(t)), \qquad (2.29)$$

and

$$|f(\sigma(t)) - f(t-h) - f^{\Delta}(t)(h+\mu(t))| \leq \varepsilon^*(h+\mu(t)). \quad (2.30)$$

Because g is δ_{SH}*-differentiable at t, there exists a neighborhood* $U_T^2(t,\delta)$ *of t such that for* $0 < h < \delta$ *with* $t-h, t+h \in U_T^2(t,\delta)$, *we have that the H-differences*

$$g(\sigma(t)) \ominus_H g(t+h) \quad \text{and} \quad g(t-h) \ominus_H g(\sigma(t)),$$

exist and

$$D(g(\sigma(t)) \ominus_H g(t+h), \delta_{SH} g(t)(h-\mu(t))) \leq \varepsilon^*(h-\mu(t)), \quad (2.31)$$

$$D(g(t-h) \ominus_H g(\sigma(t)), \delta_{SH} g(t)(-(h+\mu(t)))) \leq \varepsilon^*(h+\mu(t)). \quad (2.32)$$

Since g is δ_{SH}*-differentiable at t, then it is continuous at t. So, there exists a neighborhood* $U_T^3(t,\delta)$ *of t such that*

$$D(g(\sigma(t)), g(t)) \leq \varepsilon^*, \quad (2.33)$$

$$D(g(\sigma(t)), g(t+h)) \leq \varepsilon^* \quad (2.34)$$

and

$$\begin{aligned} \|g(\sigma(t))\| &= D\left(g(\sigma(t)), \widetilde{0}\right) \\ &\leq D(g(\sigma(t)), g(t)) + D(g(t), \widetilde{0}) \\ &\leq \varepsilon^* + \|g(t)\| \\ &< 1 + \|g(t)\|. \end{aligned} \quad (2.35)$$

Because f is Δ*-differentiable at t, then it is continuous at t. Then there exists a neighborhood* $U_T^4(t,\delta)$ *of t so that* $t+h \in U_T^4(t,\delta)$ *and*

$$|f(\sigma(t)) - f(t+h)| \leq \varepsilon^*, \quad 0 < h < \delta. \quad (2.36)$$

Let

$$U_T(t,\delta) = U_T^1(t,\delta) \bigcap U_T^2(t,\delta) \bigcap U_T^3(t,\delta) \bigcap U_T^4(t,\delta).$$

Set

$$\begin{aligned}
c &= f(t+h)\cdot g(t+h),\\
d &= f(\sigma(t))\cdot g(\sigma(t)),\\
a &= f^{\Delta}(t)\cdot g(t)(h-\mu(t)),\\
b &= -f(\sigma(t))\cdot \delta_{SH}g(t)(h-\mu(t)),\\
r &= f(t+h)\cdot g(\sigma(t)),\\
w &= f^{\Delta}(t)\cdot g(\sigma(t))(h-\mu(t)),\\
e &= -f(t+h)\cdot \delta_{SH}g(t)(h-\mu(t)).
\end{aligned}$$

We apply Lemma 2.2.26 for a, b, c, d, e, w, r and using (2.29), (2.31), (2.33), (2.35), *and* (2.36), *we get*

$$\begin{aligned}
&D\Big((f\cdot g)(t+h)\ominus_H (f\cdot g)(\sigma(t)),\\
&\Big(\Big(f^{\Delta}(t)\cdot g(t)\Big)\ominus_H ((-f(\sigma(t))\cdot \delta_{SH}g(t)))\,(h-\mu(t))\Big)\Big)\\
\le\ & D\Big(f(t+h)\cdot g(\sigma(t)), f(\sigma(t))\cdot g(\sigma(t))\\
&+f^{\Delta}(t)\cdot g(\sigma(t))(h-\mu(t))\Big)\\
&+D\Big(f^{\Delta}(t)\cdot g(t)(h-\mu(t)), f^{\Delta}(t)\cdot g(\sigma(t))(h-\mu(t))\Big)\\
&+D\Big(-f(t+h)\cdot \delta_{SH}g(t)(h-\mu(t)),\\
&(f(t+h)\cdot g(\sigma(t)))\ominus_H (f(t+h)\cdot g(t+h))\Big)
\end{aligned}$$

$$
\begin{aligned}
&+D(-f(t+h)\cdot\delta_{SH}g(t)(h-\mu(t)), f(\sigma(t))\cdot\delta_{SH}g(t)(h-\mu(t)))\\
\leq\ & \left|f(t+h)-f(\sigma(t))-f^{\Delta}(t)(h-\mu(t))\right| \|g(\sigma(t))\|\\
&+\left|f^{\Delta}(t)\right|(h-\mu(t))D(g(t),g(\sigma(t)))\\
&+|f(t+h)|D(\delta_{SH}g(t)(h-\mu(t)), g(\sigma(t))\ominus_H g(t+h))\\
&+\|\delta_{SH}g(t)\||f(\sigma(t))-f(t+h)|(h-\mu(t))\\
\leq\ & \varepsilon^*\|g(\sigma(t))\|(h-\mu(t))+\varepsilon^*|f^{\Delta}(t)|(h-\mu(t))\\
&+\varepsilon^*|f(t+h)|(h-\mu(t))+\varepsilon^*\|\delta_{SH}g(t)\|(h-\mu(t))\\
\leq\ & \varepsilon^*\left(1+|f(t)|+|f^{\Delta}(t)|+\|g(\sigma(t))\|+\|\delta_{SH}g(t)\|\right)(h-\mu(t))\\
\leq\ & \varepsilon^*\left(2+|f(t)|+|f^{\Delta}(t)|+\|g(t)\|+\|\delta_{SH}g(t)\|\right)(h-\mu(t))\\
\leq\ & \varepsilon(h-\mu(t)).
\end{aligned}
$$

Let

$$
\begin{aligned}
c &= d\\
&= f(\sigma(t))\cdot g(\sigma(t)),\\
a &= f^{\Delta}(t)\cdot g(\sigma(t))(h+\mu(t)),\\
b &= f^{\Delta}(t)\cdot g(t+h)(h+\mu(t)),
\end{aligned}
$$

$$
\begin{aligned}
r &= f^{\Delta}(t) \cdot g(\sigma(t))(h+\mu(t)), \\
w &= f(t-h) \cdot g(\sigma(t)).
\end{aligned}
$$

Now we apply Lemma 2.2.28 for a, b, c, d, r, w and using (2.30), (2.32), (2.33), (2.34), *and* (2.35), *we obtain*

$$
D\Big((f \cdot g)(\sigma(t)) \ominus_H (f \cdot g)(t-h),
$$

$$
\Big(f^{\Delta}(t) \cdot g(t)\Big) \ominus_H ((-f(\sigma(t))) \cdot \delta_{SH} g(t)(h+\mu(t)))\Big)
$$

$$
\begin{aligned}
&\leq D\Big((f \cdot g)(\sigma(t)) \ominus_H (f(t-h) \cdot g(\sigma(t))), \\
&\quad f^{\Delta}(t) \cdot g(\sigma(t))(h+\mu(t))\Big) \\
&\quad + D\Big(f(t-h) \cdot g(\sigma(t)) + f^{\Delta}(t) \cdot g(\sigma(t))(h+\mu(t)), \\
&\qquad f(\sigma(t)) \cdot g(\sigma(t))\Big) \\
&\quad + D\Big(f^{\Delta}(t) \cdot g(\sigma(t))(h+\mu(t)), f^{\Delta}(t) \cdot g(t+h)(h+\mu(t))\Big) \\
&\leq 2|f(\sigma(t)) - f(t-h) - f^{\Delta}(t)(h+\mu(t))| \|g(\sigma(t))\| \\
&\quad + |f^{\Delta}(t)| D(g(\sigma(t)), g(t+h))(h+\mu(t)) \\
&\leq 2\varepsilon^*(h+\mu(t))\|g(\sigma(t))\| + \varepsilon^*|f^{\Delta}(t)|(h+\mu(t)) \\
&= \varepsilon^* \Big(2\|g(\sigma(t))\| + |f^{\Delta}(t)|\Big)(h+\mu(t)) \\
&\leq \varepsilon^* \Big(2 + 2\|g(t)\| + |f^{\Delta}(t)|\Big)(h+\mu(t))
\end{aligned}
$$

$$\leq \varepsilon(h+\mu(t)).$$

This completes the proof.

Example 2.2.32 *Let* $\mathbb{T} = 2^{\mathbb{N}_0}$,

$$\begin{aligned} f(t) &= t, \\ g(t) &= \left(\frac{1}{4t}, \frac{1}{3t}, \frac{1}{2t}\right), \quad t \in \mathbb{T}. \end{aligned}$$

Here

$$\sigma(t) = 2t, \quad t \in \mathbb{T}.$$

Let

$$h(t) = \frac{1}{t}, \quad t \in \mathbb{T}.$$

Then

$$\begin{aligned} h^{\Delta}(t) &= -\frac{1}{t\sigma(t)} \\ &= -\frac{1}{2t^2}, \quad t \in \mathbb{T}. \end{aligned}$$

Hence,

$$\begin{aligned} \delta_{SH}g(t) &= \left(-\frac{1}{4t^2}, -\frac{1}{6t^2}, -\frac{1}{8t^2}\right), \\ f(\sigma(t)) &= \sigma(t) \\ &= 2t, \\ -f(\sigma(t)) &= -2t, \\ (-f(\sigma(t)) \cdot \delta_{SH}g(t) &= \left(\frac{1}{4t}, \frac{1}{3t}, \frac{1}{2t}\right), \end{aligned}$$

$$\begin{aligned} f^{\Delta}(t) &= 1, \\ f^{\Delta}(t)\cdot g(t) &= \left(\frac{1}{4t},\frac{1}{3t},\frac{1}{2t}\right), \quad t\in\mathbb{T}. \end{aligned}$$

Therefore

$$\left(f^{\Delta}(t)\cdot g(t)\right)\ominus_H(-f(\sigma(t))\cdot\delta_{SH}g(t)) = (0,0,0), \quad t\in\mathbb{T}.$$

Next,

$$\begin{aligned} f(t)\cdot g(t) &= \left(\frac{1}{4},\frac{1}{3},\frac{1}{2}\right), \\ \delta_H(f\cdot g)(t) &= (0,0,0), \quad t\in\mathbb{T}. \end{aligned}$$

Consequently

$$\delta_H(f\cdot g)(t) = \left(f^{\Delta}(t)\cdot g(t)\right)\ominus_H((-f(\sigma(t)))\cdot\delta_{SH}g(t)).$$

Exercise 2.2.33 *Let* $\mathbb{T}=3^{\mathbb{N}_0}$,

$$\begin{aligned} f(t) &= t+1, \\ g(t) &= \left(\frac{1}{t^2+4t},\frac{1}{t^2+3t},\frac{1}{t^2+2t}\right), \quad t\in\mathbb{T}. \end{aligned}$$

Check if

$$\delta_H(f\cdot g)(t) = \left(f^{\Delta}(t)\cdot g(t)\right)\ominus_H((-f(\sigma(t)))\cdot\delta_{SH}g(t)), \quad t\in\mathbb{T}^{\kappa}.$$

Exercise 2.2.34 *Let* $t\in\mathbb{T}^{\kappa}$, $f:\mathbb{T}\to\mathbb{R}$ *be* Δ*-differentiable at* t, $f(t)f^{\Delta}(t)>0$, $g:\mathbb{T}\to F(\mathbb{R})$ *be* δ_{SH}*-differentiable at* t *and the* H*-differences*

$$((f\cdot g)(\sigma(t)))\ominus_H((f\cdot g)(t+h))$$

and

$$((f\cdot g)(t-h))\ominus_H((f\cdot g)(\sigma(t))),$$

exist. Prove that $f \cdot g$ is δ_{SH}-differentiable at t and

$$\delta_{SH}(f \cdot g)(t) = (f(\sigma(t)) \cdot \delta_{SH} g(t)) \ominus_H \left(\left(-f^{\Delta}(t) \right) \cdot g(t) \right).$$

Example 2.2.35 *Let* $\mathbb{T} = 2^{\mathbb{N}_0}$,

$$\begin{aligned} f(t) &= t, \\ g(t) &= \left(\frac{1}{3t^2}, \frac{1}{2t^2}, \frac{1}{t^2} \right), \quad t \in \mathbb{T}. \end{aligned}$$

Here

$$\sigma(t) = 2t, \quad t \in \mathbb{T}.$$

Let

$$\begin{aligned} h_1(t) &= \frac{1}{t^2}, \\ h_2(t) &= \frac{1}{t}, \quad t \in \mathbb{T}. \end{aligned}$$

We have

$$\begin{aligned} h_1^{\Delta}(t) &= -\frac{\sigma(t)+t}{t^2(\sigma(t))^2} \\ &= -\frac{3t}{4t^4} \\ &= -\frac{3}{4t^3}, \\ h_2^{\Delta}(t) &= -\frac{1}{t\sigma(t)} \\ &= -\frac{1}{2t^2}, \quad t \in \mathbb{T}. \end{aligned}$$

Then

$$\delta_{SH} g(t) = \left(-\frac{3}{4t^3}, -\frac{3}{8t^3} - \frac{1}{4t^3} \right),$$

$$\begin{aligned} f(\sigma(t)) &= 2t, \\ f(\sigma(t)) \cdot \delta_{SH} g(t) &= \left(-\frac{3}{2t^2}, -\frac{3}{4t^2}, -\frac{1}{2t^2}\right), \\ f^{\Delta}(t) &= 1, \\ -f^{\Delta}(t) \cdot g(t) &= \left(-\frac{1}{t^2}, -\frac{1}{2t^2}, -\frac{1}{3t^2}\right), \quad t \in \mathbb{T}. \end{aligned}$$

Hence,

$$\begin{aligned} &(f(\sigma(t)) \cdot \delta_{SH} g(t)) \ominus_H \left(\left(-f^{\Delta}(t)\right) \cdot g(t)\right) \\ &= \left(-\frac{3}{2t^2}, -\frac{3}{4t^2}, -\frac{1}{2t^2}\right) \ominus_H \left(-\frac{1}{t^2}, -\frac{1}{2t^2}, -\frac{1}{3t^2}\right) \\ &= \left(-\frac{1}{2t^2}, -\frac{1}{4t^2}, -\frac{1}{6t^2}\right), \quad t \in \mathbb{T}. \end{aligned}$$

Next,

$$\begin{aligned} f(t) \cdot g(t) &= \left(\frac{1}{3t}, \frac{1}{2t}, \frac{1}{t}\right), \\ \delta_{SH}(f \cdot g)(t) &= \left(-\frac{1}{2t^2}, -\frac{1}{4t^2}, -\frac{1}{6t^2}\right), \quad t \in \mathbb{T}. \end{aligned}$$

Consequently

$$\delta_{SH}(f \cdot g)(t) = (f(\sigma(t)) \cdot \delta_{SH} g(t)) \ominus_H \left(\left(-f^{\Delta}(t)\right) \cdot g(t)\right).$$

Exercise 2.2.36 *Let $t \in \mathbb{T}^{\kappa}$, $f : \mathbb{T} \to \mathbb{R}$ be Δ-differentiable at t, $f(\sigma(t)) f^{\Delta}(t) < 0$, $g : \mathbb{T} \to F(\mathbb{R})$ be δ_H-differentiable at t and the H-differences*

$$((f \cdot g)(\sigma(t))) \ominus_H ((f \cdot g)(t+h))$$

and

$$((f \cdot g)(t-h)) \ominus_H ((f \cdot g)(\sigma(t))),$$

exist. Prove that $f \cdot g$ is δ_{SH}-differentiable at t and

$$\delta_{SH}(f \cdot g)(t) = \left(f^{\Delta}(t) \cdot g(t)\right) \ominus_H ((-f(\sigma(t))) \cdot \delta_H g(t)).$$

Example 2.2.37 *Let* $\mathbb{T} = \left(\frac{1}{3}\right)^{\mathbb{N}_0}$,

$$\begin{aligned} f(t) &= \frac{1}{t^2}, \\ g(t) &= \left(t, t^2+1, t^2+t+2\right), \quad t \in \mathbb{T}. \end{aligned}$$

We have

$$\sigma(t) = 3t, \quad t \in \mathbb{T}.$$

Then

$$\begin{aligned} f^{\Delta}(t) &= -\frac{\sigma(t)+t}{t^2(\sigma(t))^2} \\ &= -\frac{3t+t}{t^2(3t)^2} \\ &= -\frac{4t}{9t^4} \\ &= -\frac{4}{9t^3}, \\ \delta_H g(t) &= (1, \sigma(t)+t, \sigma(t)+t+1) \\ &= (1, 3t+t, 3t+t+1) \\ &= (1, 4t, 4t+1), \end{aligned}$$

$$\begin{aligned} f(\sigma(t)) &= \frac{1}{(\sigma(t))^2} \\ &= \frac{1}{9t^2}, \\ -f(\sigma(t)) &= -\frac{1}{9t^2}, \\ (-f(\sigma(t)))\cdot\delta_H g(t) &= \left(-\frac{4t+1}{9t^2}, -\frac{4}{9t}, -\frac{1}{9t^2}\right), \\ f^{\Delta}(t)\cdot g(t) &= \left(-\frac{4\left(t^2+t+2\right)}{9t^3}, \frac{-4(t^2+1)}{9t^3}, -\frac{4}{9t^2}\right), \ t\in\mathbb{T}. \end{aligned}$$

Hence,

$$\left(f^{\Delta}(t)\cdot g(t)\right)\ominus_H((-f(\sigma(t)))\cdot\delta_H g(t))$$

$$=\left(-\frac{4\left(t^2+t+2\right)}{9t^3}, \frac{-4(t^2+1)}{9t^3}, -\frac{4}{9t^2}\right)\ominus_H\left(-\frac{4t+1}{9t^2}, -\frac{4}{9t}, -\frac{1}{9t^2}\right)$$

$$=\left(-\frac{3t+8}{9t^3}, -\frac{4}{9t^3}, -\frac{1}{3t^2}\right), \quad t\in\mathbb{T}.$$

Next,

$$f(t)\cdot g(t) = \left(\frac{1}{t}, 1+\frac{1}{t^2}, 1+\frac{1}{t}+\frac{2}{t^2}\right), \quad t\in\mathbb{T}.$$

Let

$$\begin{aligned} h_1(t) &= \frac{1}{t}, \\ h_2(t) &= 1+\frac{1}{t^2}, \\ h_3(t) &= 1+\frac{1}{t}+\frac{2}{t^2}, \quad t\in\mathbb{T}. \end{aligned}$$

We have

$$\begin{aligned} h_1^{\Delta}(t) &= -\frac{1}{t\sigma(t)} \\ &= -\frac{1}{3t^2}, \\ h_2^{\Delta}(t) &= -\frac{\sigma(t)+t}{t^2(\sigma(t))^2} \\ &= -\frac{4t}{9t^4} \\ &= -\frac{4}{9t^3}, \\ h_3^{\Delta}(t) &= h_1^{\Delta}(t)+2h_2^{\Delta}(t) \\ &= -\frac{1}{3t^2}-\frac{8}{9t^3} \\ &= -\frac{3t+8}{9t^3}, \quad t \in \mathbb{T}, \end{aligned}$$

and

$$\delta_{SH}(f \cdot g)(t) = \left(-\frac{3t+8}{9t^3}, -\frac{4}{9t^3}, -\frac{1}{3t^2}\right), \quad t \in \mathbb{T}.$$

Consequently

$$\delta_{SH}(f \cdot g)(t) = \left(f^{\Delta}(t) \cdot g(t)\right) \ominus_H (-f(\sigma(t)) \cdot \delta_H g(t)).$$

2.3 FIRST TYPE FUZZY DELTA INTEGRATION

In this section, we introduce the concept of the integration of fuzzy deltas of the first type on time scales. We deduce some of its properties. Let $I \subset \mathbb{T}$.

Definition 2.3.1 *A function $f : \mathbb{T} \to \mathbb{R}$ is called a sector of the fuzzy function $F : I \to F(\mathbb{R})$ if $f(t) \in F(t)$ for all $t \in I$. The set of all rd continuous sectors of F in I is denoted by $S_{HF}(I)$.*

Example 2.3.2 *Let* $\mathbb{T} = 2^{\mathbb{N}_0}$, $I = [1,64]$,

$$F(t) = \left\{t, t^2+1, 2, 4, 16, t^3-1, 64, t^4-10\right\},$$

$$f(t) = t, \quad t \in I.$$

Then $f(t) \in F(t)$ *for all* $t \in I$. *Therefore,* f *is a sector of* F.

Example 2.3.3 *Let* $\mathbb{T} = \mathbb{Z}$, $I = [0,16]$,

$$F(t) = \left\{\frac{1}{2}, \frac{1}{3}, \frac{1}{4}, \frac{1}{t^2+1}\right\},$$

$$f(t) = t^2, \quad t \in I.$$

We have $f(t) \notin F(t)$ *for all* $t \in I$. *Therefore,* f *is not a sector of* F.

Definition 2.3.4 *A fuzzy function* $F : I \to F(\mathbb{R})$ *is said to be regulated provided its regulated sectors exist.*

Definition 2.3.5 *A fuzzy function* $F : I \to F(\mathbb{R})$ *is said to be rd-continuous provided its rd-continuous sectors exist. The set of all rd-continuous fuzzy functions* $F : I \to F(\mathbb{R})$ *is denoted by*

$$\mathscr{C}_{frd} = \mathscr{C}_{frd}(I) = \mathscr{C}_{frd}(I, F(\mathbb{R})).$$

The set of fuzzy functions $F : I \to F(\mathbb{R})$ *which are* δ_H*-differentiable whose* δ_H*-derivatives are rd-continuous is denoted by*

$$\mathscr{C}^1_{Hfrd} = \mathscr{C}^1_{Hfrd}(I) = \mathscr{C}^1_{Hfrd}(I, F(\mathbb{R})).$$

Definition 2.3.6 *A continuous fuzzy function* $F : I \to F(\mathbb{R})$ *is said to be* δ_H*-pre-differentiable with (region of differentiation)* D *provided* $D \subset \mathbb{T}^\kappa$, $\mathbb{T}^\kappa \backslash D$ *is countable and contains no right-scattered elements of* $\mathbb{T}$, *and* F *is* δ_H*-differentiable at each* $t \in \mathbb{T}$. *If* F *is rd-continuous, then* F *is regulated and there exists a fuzzy function* f *which is pre-differentiable with region of differentiation* D *such that*

$$\delta_H f(t) = F(t), \quad t \in D.$$

Definition 2.3.7 *A fuzzy function $F : \mathbb{T} \to F(\mathbb{R})$ is said to be δ_H-integrable on I if its sectors are rd-continuous on I. In this case, we define the δ_H-integral of F on I by*

$$\int_I F(s)\delta_H s = \left\{ \int_I f(s)\Delta s : f \in S_{HF}(I) \right\},$$

and defined level wise by

$$\left[\int_I F(s)\delta_H s \right]^\alpha = \left[\int_I \underline{f}^\alpha(s)\Delta s, \int_I \overline{f}^\alpha(s)\Delta s \right],$$

where

$$[F(s)]^\alpha = \left[\underline{f}^\alpha(s), \overline{f}^\alpha(s) \right], \quad s \in I, \quad \alpha \in [0,1].$$

Example 2.3.8 *Let* $\mathbb{T} = \mathbb{N}$,

$$F(t) = \left(2t+2, 2t+4, 3t^2+3t+5\right), \quad t \in \mathbb{T}.$$

We will find

$$\int_1^4 F(s)\delta_H s.$$

We have

$$\sigma(t) = t+1, \quad t \in \mathbb{T}.$$

Also,

$$\int_1^4 F(s)\delta_H s = \left(\int_1^4 (2s+2)\Delta s, \int_1^4 (2s+4)\Delta s, \int_1^4 \left(3s^2+3s+5\right)\Delta s \right).$$

Let

$$h_1(t) = t^2+t,$$

$$h_2(t) = t^2+3t+4,$$

$$h_3(t) = t^3+4t+7, \quad t \in \mathbb{T}.$$

Then

$$\begin{aligned} h_1^\Delta(t) &= \sigma(t)+t+1 \\ &= t+1+t+1 \\ &= 2t+2, \\ h_2^\Delta(t) &= \sigma(t)+t+3 \\ &= t+1+t+3 \\ &= 2t+4, \\ h_3^\Delta(t) &= (\sigma(t))^2+t\sigma(t)+t^2+4 \\ &= (t+1)^2+t(t+1)+t^2+4 \\ &= t^2+2t+1+t^2+t+t^2+4 \\ &= 3t^2+3t+5, \quad t \in \mathbb{T}. \end{aligned}$$

Therefore

$$\int_1^4 F(s)\delta_H s = \left(\int_1^4 h_1^\Delta(s)\Delta s, \int_1^4 h_2^\Delta(s)\Delta s, \int_1^4 h_3^\Delta(s)\Delta s \right)$$

$$= \left(h_1(s)\Big|_{s=1}^{s=4}, h_2(s))\Big|_{s=1}^{s=4}, h_3(s)\Big|_{s=1}^{s=4} \right)$$

$$= \quad (16+4-1-1, 16+12+4-1-3-4, 64+16+7-1-4-7)$$

$$= \quad (18, 24, 75).$$

Exercise 2.3.9 *Let* $\mathbb{T} = 3^{\mathbb{N}_0}$ *and*

$$F(t) = \left(\frac{1}{t^4}, \frac{1}{t^3}, \frac{1}{t^2}\right), \quad t \in \mathbb{T}.$$

Find

$$\int_1^{81} F(s)\delta_H s.$$

Example 2.3.10 *Let* $\mathbb{T} = 2^{\mathbb{N}_0}$, $f : \mathbb{T} \to F(\mathbb{R})$ *be a fuzzy function whose* α*-levels are given by*

$$[f(t)]^\alpha = \left[3t+1, 7t^2\right]^\alpha, \quad t \in \mathbb{T}, \quad \alpha \in [0,1].$$

We will find

$$\left[\int_1^4 f(t)\delta_H t\right]^\alpha, \quad \alpha \in [0,1].$$

Here

$$\begin{aligned} \sigma(t) &= 2t, \\ \underline{f}^\alpha(t) &= 3t+1, \\ \overline{f}^\alpha(t) &= 7t^2, \quad t \in \mathbb{T}, \quad \alpha \in [0,1]. \end{aligned}$$

Let

$$\begin{aligned} f_1(t) &= t^2+t, \\ f_2(t) &= t^3, \quad t \in \mathbb{T}. \end{aligned}$$

Then

$$
\begin{aligned}
f_1^{\Delta}(t) &= \sigma(t)+t+1 \\
&= 2t+t+1 \\
&= 3t+1, \\
f_2^{\Delta}(t) &= (\sigma(t))^2+t\sigma(t)+t^2 \\
&= (2t)^2+2t^2+t^2 \\
&= 4t^2+3t^2 \\
&= 7t^2, \quad t\in\mathbb{T}.
\end{aligned}
$$

Hence,

$$
\begin{aligned}
\int_1^4 \underline{f}^{\alpha}(t)\Delta t &= \int_1^4 (3t+1)\Delta t \\
&= \int_1^4 f_1^{\Delta}(t)\Delta t \\
&= f_1(t)\Big|_{t=1}^{t=4} \\
&= f_1(4)-f_1(1) \\
&= (16+4)-(1+1) \\
&= 20-2 \\
&= 18,
\end{aligned}
$$

$$\begin{aligned}
\int_1^4 \overline{f}^{\alpha}(t)\Delta t &= 7\int_1^4 t^2\Delta t \\
&= \int_1^4 f_2^{\Delta}(t)\Delta t \\
&= f_2(t)\Big|_{t=1}^{t=4} \\
&= f_2(4)-f_2(1) \\
&= 64-1 \\
&= 63.
\end{aligned}$$

Consequently

$$\begin{aligned}
\left[\int_1^4 f(t)\delta_H t\right]^{\alpha} &= \left[\int_1^4 \underline{f}^{\alpha}(t)\Delta t, \int_1^4 \overline{f}^{\alpha}(t)\Delta t\right] \\
&= [18,63].
\end{aligned}$$

Exercise 2.3.11 *Let $\mathbb{T}=3\mathbb{Z}$. Find*

$$\left[\int_{-3}^{9} f(t)\delta_H t\right]^{\alpha}, \quad \alpha\in[0,1],$$

where

$$[f(t)]^{\alpha}=\left[t-3,t^2+t\right]^{\alpha}, \quad t\in\mathbb{T}, \quad \alpha\in[0,1].$$

Theorem 2.3.12 *Let $t_0,T\in\mathbb{T}$, $t_0<T$, $F,G:[t_0,T]\to F(\mathbb{R})$ be δ_H-integrable. Then $F+G:[t_0,T]\to F(\mathbb{R})$ is δ_H-integrable and*

$$\int_{t_0}^{T}(F(s)+G(s))\delta_H s=\int_{t_0}^{T}F(s)\delta_H s+\int_{t_0}^{T}G(s)\delta_H s. \tag{2.37}$$

Proof 2.3.13 *Since F and G are δ_H-integrable on $[t_0,T]$, then their sectors are rd-continuous on $[t_0,T]$. Hence, the sectors of $F+G$ are rd-continuous on $[t_0,T]$. Let u,v,w be rd-continuous sectors of F, G and $F+G$, respectively. Then there exist $f \in S_{HF}([t_0,T])$, $g \in S_{HG}([t_0,T])$ and $h \in S_{H(F+G)}([t_0,T])$ such that*

$$\begin{aligned} u(T) &= \int_{t_0}^{T} f(s)\Delta s, \\ v(T) &= \int_{t_0}^{T} g(s)\Delta s, \\ w(T) &= \int_{t_0}^{T} h(s)\Delta s. \end{aligned}$$

Hence,

$$\begin{aligned} u(T)+v(T) &= \int_{t_0}^{T} f(s)\Delta s + \int_{t_0}^{T} g(s)\Delta s \\ &= \int_{t_0}^{T} (f(s)+g(s))\Delta s \\ &\in S_{H(F+G)}([t_0,T]). \end{aligned}$$

Thus,

$$\begin{aligned} &\left\{ \int_{t_0}^{T} f(s)\Delta s : f \in S_{HF}([t_0,T]) \right\} \\ &\quad + \left\{ \int_{t_0}^{T} g(s)\Delta s : g \in S_{HG}([t_0,T]) \right\} \end{aligned} \tag{2.38}$$

$$\subset \left\{ \int_{t_0}^{T} h(s)\Delta s : h \in S_{H(F+G)}([t_0,T]) \right\}.$$

Next, there exist $h_1 \in S_{HF}([t_0,T])$ *and* $h_2 \in S_{HG}([t_0,T])$ *such that*

$$h(s) = h_1(s) + h_2(s), \quad s \in [t_0,T].$$

Therefore

$$\begin{aligned} w(T) &= \int_{t_0}^{T} h(s)\Delta s \\ &= \int_{t_0}^{T} (h_1(s)+h_2(s))\Delta s \\ &= \int_{t_0}^{T} h_1(s)\Delta s + \int_{t_0}^{T} h_2(s)\Delta s \\ &\in \left\{ \int_{t_0}^{T} f(s)\Delta s : f \in S_{HF}([t_0,T]) \right\} \\ &\quad + \left\{ \int_{t_0}^{T} g(s)\Delta s : g \in S_{HG}([t_0,T]) \right\}. \end{aligned}$$

Consequently

$$\begin{aligned} &\left\{ \int_{t_0}^{T} h(s)\Delta s : h \in S_{H(F+G)}([t_0,T]) \right\} \\ &\subset \left\{ \int_{t_0}^{T} f(s)\Delta s : f \in S_{HF}([t_0,T]) \right\} \\ &\quad + \left\{ \int_{t_0}^{T} g(s)\Delta s : g \in S_{HG}([t_0,T]) \right\}. \end{aligned}$$

From the last relation and from (2.38), we get

$$\begin{aligned}
&\left\{ \int_{t_0}^{T} h(s)\Delta s : h \in S_{H(F+G)}([t_0,T]) \right\} \\
&= \left\{ \int_{t_0}^{T} f(s)\Delta s : f \in S_{HF}([t_0,T]) \right\} \\
&\quad + \left\{ \int_{t_0}^{T} g(s)\Delta s : g \in S_{HG}([t_0,T]) \right\}.
\end{aligned}$$

This completes the proof.

Example 2.3.14 *Let* $\mathbb{T} = 2^{\mathbb{N}_0}$,

$$\begin{aligned}
F(t) &= \{1,t^2\}, \\
G(t) &= \{1,t\}, \quad t \in \mathbb{T}.
\end{aligned}$$

Then

$$F(t)+G(t) = \{2, 1+t, 1+t^2, t+t^2\}, \quad t \in \mathbb{T}.$$

We will compute

$$\int_1^4 F(t)\delta_H t, \quad \int_1^4 G(t)\delta_H t, \quad \int_1^4 (F(t)+G(t))\delta_H t.$$

Let

$$\begin{aligned}
h_1(t) &= \frac{t^2}{3}, \\
h_2(t) &= \frac{t^3}{7}, \\
h_3(t) &= t+\frac{t^2}{3}, \\
h_4(t) &= t+\frac{t^3}{7}, \\
h_5(t) &= \frac{t^2}{3}+\frac{t^3}{7}, \quad t \in \mathbb{T}.
\end{aligned}$$

We have

$$
\begin{aligned}
\sigma(t) &= 2t, \\
h_1^{\Delta}(t) &= \frac{1}{3}(\sigma(t)+t) \\
&= \frac{1}{3}(2t+t) \\
&= t, \\
h_2^{\Delta}(t) &= \frac{1}{7}\left((\sigma(t))^2+t\sigma(t)+t^2\right) \\
&= \frac{1}{7}(4t^2+2t^2+t^2) \\
&= t^2, \\
h_3^{\Delta}(t) &= 1+\frac{1}{3}(\sigma(t)+t) \\
&= 1+\frac{1}{3}(2t+t) \\
&= 1+t, \\
h_4^{\Delta}(t) &= 1+\frac{1}{7}\left((\sigma(t))^2+t\sigma(t)+t^2\right) \\
&= 1+\frac{1}{7}\left(4t^2+2t^2+t^2\right) \\
&= 1+t^2, \\
h_5^{\Delta}(t) &= \frac{1}{3}(\sigma(t)+t)+\frac{1}{7}\left((\sigma(t))^2+t\sigma(t)+t^2\right)
\end{aligned}
$$

$$
\begin{aligned}
&= \frac{1}{3}(2t+t)+\frac{1}{7}(4t^2+2t^2+t^2)\\
&= t+t^2, \quad t\in\mathbb{T}.
\end{aligned}
$$

Hence,

$$
\begin{aligned}
\int_1^4 F(t)\delta_H t &= \left\{\int_1^4 \Delta t, \int_1^4 t^2\Delta t\right\}\\
&= \left\{3, \int_1^4 h_2^{\Delta}(t)\Delta t\right\}\\
&= \left\{3, h_2(t)\Big|_{t=1}^{t=4}\right\}\\
&= \left\{3, \frac{t^3}{7}\Big|_{t=1}^{t=4}\right\}\\
&= \left\{3, \frac{64}{7}-\frac{1}{7}\right\}\\
&= \{3,9\},
\end{aligned}
$$

$$
\begin{aligned}
\int_1^4 G(t)\delta_H t &= \left\{\int_1^4 \Delta t, \int_1^4 t\Delta t\right\}\\
&= \left\{3, \int_1^4 h_1^{\Delta}(t)\Delta t\right\}\\
&= \left\{3, h_1(t)\Big|_{t=1}^{t=4}\right\}\\
&= \left\{3, \frac{t^2}{3}\Big|_{t=1}^{t=4}\right\}
\end{aligned}
$$

$$= \left\{3, \frac{16}{3} - \frac{1}{3}\right\}$$

$$= \{3, 5\},$$

$$\int_1^4 (F(t)+G(t))\delta_H t = \left\{\int_1^4 2\Delta t, \int_1^4 (1+t)\Delta t, \int_1^4 (1+t^2)\Delta t, \int_1^4 (t+t^2)\Delta t\right\}$$

$$= \left\{6, \int_1^4 h_3^\Delta(t)\Delta t, \int_1^4 h_4^\Delta(t)\Delta t, \int_1^4 h_5^\Delta(t)\Delta t\right\}$$

$$= \left\{6, h_3(t)\Big|_{t=1}^{t=4}, h_4(t)\Big|_{t=1}^{t=4}, h_5(t)\Big|_{t=1}^{t=4}\right\}$$

$$= \left\{6, \left(t+\frac{t^2}{3}\right)\Big|_{t=1}^{t=4}, \left(t+\frac{t^3}{7}\right)\Big|_{t=1}^{t=4}, \left(\frac{t^2}{3}+\frac{t^3}{7}\right)\Big|_{t=1}^{t=4}\right\}$$

$$= \left\{6, 4+\frac{1}{3}-1-\frac{1}{3}, 4+\frac{64}{7}-1-\frac{1}{7}, \frac{16}{3}+\frac{64}{7}-\frac{1}{3}-\frac{1}{7}\right\}$$

$$= \{6, 8, 12, 14\}.$$

Therefore

$$\int_1^4 F(t)\delta_H t + \int_1^4 G(t)\delta_H t = \int_1^4 (F(t)+G(t))\delta_H t.$$

Theorem 2.3.15 *Let $t_0, T \in \mathbb{T}$, $t_0 < T$, $F : [t_0, T] \to F(\mathbb{R})$ be δ_H-integrable. Then $\lambda \cdot F : [t_0, T] \to F(\mathbb{R})$ is δ_H-integrable and*

$$\int_{t_0}^T \lambda \cdot F(s)\delta_H s = \lambda \cdot \int_{t_0}^T F(s)\delta_H s,$$

for any $\lambda \in \mathbb{R}$.

Proof 2.3.16 *Let* $f \in S_{HF}([t_0,T])$ *and*

$$u(T) = \int_{t_0}^{T} f(s)\Delta s.$$

Then

$$\begin{aligned}
\lambda u(T) &= \lambda \int_{t_0}^{T} f(s)\Delta s \\
&= \int_{t_0}^{T} \lambda f(s)\Delta s \\
&\in \left\{ \int_{t_0}^{T} \lambda f(s)\Delta s : f \in S_{HF}([t_0,T]) \right\}.
\end{aligned}$$

Therefore

$$\begin{aligned}
&\lambda \cdot \left\{ \int_{t_0}^{T} f(s)\Delta s : f \in S_{HF}([t_0,T]) \right\} \\
&\subset \left\{ \int_{t_0}^{T} g(s)\Delta s : g \in S_{H(\lambda\cdot F)}([t_0,T]) \right\}. \qquad (2.39)
\end{aligned}$$

Next, let $g \in S_{H(\lambda\cdot F)}([t_0,T])$. *Then there exists an* $h \in S_{HF}([t_0,T])$ *such that*

$$g(s) = \lambda h(s), \quad s \in [t_0,T].$$

Hence,

$$\int_{t_0}^{T} \lambda h(s)\Delta s = \lambda \int_{t_0}^{T} h(s)\Delta s,$$

and thus,

$$\left\{ \int_{t_0}^{T} g(s)\Delta s : g \in S_{H(\lambda\cdot F)}([t_0,T]) \right\}$$

$$\subset \lambda \cdot \left\{ \int_{t_0}^{T} f(s)\Delta s : f \in S_{HF}([t_0,T]) \right\}.$$

From the last relation and from (2.39), *it follows that*

$$\lambda \cdot \left\{ \int_{t_0}^{T} f(s)\Delta s : f \in S_{HF}([t_0,T]) \right\}$$

$$= \left\{ \int_{t_0}^{T} g(s)\Delta s : g \in S_{H(\lambda \cdot F)}([t_0,T]) \right\}$$

and

$$\lambda \cdot \int_{t_0}^{T} F(s)\delta_H s = \int_{t_0}^{T} \lambda \cdot F(s)\delta_H s.$$

This completes the proof.

Theorem 2.3.17 *Let $t_0, T \in \mathbb{T}$, $t_0 < T$, and $F : [t_0,T] \to F(\mathbb{R})$ be δ_H-integrable. Then*

$$\int_{t_0}^{T} F(s)\delta_H s = \int_{t_0}^{t} F(s)\delta_H s + \int_{t}^{T} F(s)\delta_H s,$$

for any $t \in [t_0,T]$.

Proof 2.3.18 *Let $f \in S_{HF}([t_0,T])$ and $t \in [t_0,T]$ be fixed. Then*

$$\begin{aligned}\int_{t_0}^{T} f(s)\Delta s &= \int_{t_0}^{t} f(s)\Delta s + \int_{t}^{T} f(s)\Delta s \\ &\subset \left\{ \int_{t_0}^{t} f(s)\Delta s : f \in S_{HF}([t_0,t]) \right\}\end{aligned}$$

$$+\left\{\int_t^T f(s)\Delta s : f \in S_{HF}([t,T])\right\}.$$

Consequently

$$\left\{\int_{t_0}^T f(s)\Delta s : f \in S_{HF}([t_0,T])\right\} \subset \left\{\int_{t_0}^t f(s)\Delta s : f \in S_{HF}([t_0,t])\right\} + \left\{\int_t^T f(s)\Delta s : f \in S_{HF}([t,T])\right\}. \tag{2.40}$$

Let now $g \in S_{HF}([t_0,t])$, $h \in S_{HF([t,T])}$ *and*

$$z = \int_{t_0}^t g(s)\Delta s + \int_t^T h(s)\Delta s.$$

Define

$$h_1(s) = \begin{cases} g(s) & \text{if} \quad t_0 \le s \le t \\ h(s) & \text{if} \quad t \le s \le T. \end{cases}$$

Then

$$\begin{aligned} \int_{t_0}^T h_1(s)\Delta s &= \int_{t_0}^t g(s)\Delta s + \int_t^T h(s)\Delta s \\ &= z. \end{aligned}$$

Thus,

$$\left\{\int_{t_0}^t f(s)\Delta s : f \in S_{HF}([t_0,t])\right\} + \left\{\int_t^T f(s)\Delta s : f \in S_{HF}([t,T])\right\}$$

$$\subset \left\{ \int_{t_0}^{T} f(s)\Delta s : f \in S_{HF}([t_0, T]) \right\}.$$

From the last relation and from (2.40), *we get*

$$\left\{ \int_{t_0}^{t} f(s)\Delta s : f \in S_{HF}([t_0, t]) \right\} + \left\{ \int_{t}^{T} f(s)\Delta s : f \in S_{HF}([t, T]) \right\}$$

$$= \left\{ \int_{t_0}^{T} f(s)\Delta s : f \in S_{HF}([t_0, T]) \right\}$$

and

$$\int_{t_0}^{T} F(s)\delta_H s = \int_{t_0}^{t} F(s)\delta_H s + \int_{t}^{T} F(s)\delta_H s.$$

This completes the proof.

Theorem 2.3.19 *Let* $t_0, T \in \mathbb{T}$, $t_0 < T$, $F : [t_0, T] \to F(\mathbb{R})$ *is rd-continuous. If* $X_0 \in F(\mathbb{R})$ *and*

$$f(t) = X_0 + \int_{t_0}^{t} F(s)\delta_H s, \quad t \in [t_0, T],$$

then f *is* δ_H*-differentiable and*

$$\delta_H f(t) = F(t), \quad t \in [t_0, T].$$

Proof 2.3.20 *Since* $F : [t_0, T] \to F(\mathbb{R})$ *is rd-continuous, there exists a pre-differentiable function* f *with region of differentiation* D*, such that* $\delta_H f(t)$*,* $t \in [t_0, T]$*, exists and*

$$\delta_H f(t) = F(t), \quad t \in D.$$

Let $t \in I^{\kappa} \backslash D$. Then t is a right-dense point of $\mathbb{T}$. Hence, for $0 < h < \delta$ with $t-h, t+h \in [t_0, T]$, we have

$$\begin{aligned} f(t+h) \ominus_H f(\sigma(t)) &= \int_{t_0}^{t+h} F(s)\delta_H s \ominus_H \int_{t_0}^{\sigma(t)} F(s)\delta_H s \\ &= \int_{\sigma(t)}^{t+h} F(s)\delta_H s. \end{aligned}$$

Thus, for $\varepsilon > 0$ sufficiently small, using the rd-continuity of F, we have

$$D\left(\frac{f(t+h)\ominus_H f(\sigma(t))}{h-\mu(t)}, F(t)\right)$$

$$\begin{aligned} &= \frac{1}{h-\mu(t)} D(f(t+h)\ominus_H f(\sigma(t)), (h-\mu(t))\cdot F(t)) \\ &= \frac{1}{h-\mu(t)} D\left(\int_{\sigma(t)}^{t+h} F(s)\delta_H s, \int_{\sigma(t)}^{t+h} F(t)\delta_H s\right) \\ &\leq \frac{1}{h-\mu(t)} \int_{\sigma(t)}^{t+h} D(F(s), F(t))\delta_H s \\ &< \varepsilon, \end{aligned}$$

for all $t-h, t+h \in [t_0, T]$. Hence,

$$\lim_{h\to 0} \frac{f(t+h)\ominus_H f(\sigma(t))}{h-\mu(t)} = F(t)$$

and thus,

$$\delta_H f(t) = F(t).$$

This completes the proof.

Theorem 2.3.21 *If $f : [a,b] \to F(\mathbb{R})$ is δ_H-differentiable on $[a,b]$, then*

$$\int_a^b \delta_H f(t) = f(b) \ominus_H f(a).$$

Proof 2.3.22 *Let the α-levels of the function f are given by*

$$[f(t)]^\alpha = \left[\underline{f}^\alpha(t), \overline{f}^\alpha(t)\right], \quad t \in [a,b], \quad \alpha \in [0,1].$$

Then

$$[\delta_H f(t)]^\alpha = \left[\underline{f}^{\alpha\Delta}(t), \overline{f}^{\alpha\Delta}(t)\right], \quad t \in [a,b], \quad \alpha \in [0,1],$$

and

$$\begin{aligned} \left[\int_a^b \delta_H f(t)\right]^\alpha &= \left[\int_a^b \underline{f}^{\alpha\Delta}(t)\Delta t, \int_a^b \overline{f}^{\alpha\Delta}(t)\Delta t\right] \\ &= \left[\underline{f}^\alpha(b) - \underline{f}^\alpha(a), \overline{f}^\alpha(b) - \overline{f}^\alpha(a)\right] \\ &= \left[\underline{f}^\alpha(b), \overline{f}^\alpha(b)\right] \ominus_H \left[\underline{f}^\alpha(a), \overline{f}^\alpha(a)\right], \alpha \in [0,1]. \end{aligned}$$

This completes the proof.

Example 2.3.23 *Let $\mathbb{T} = 2^{\mathbb{N}_0}$, $f : \mathbb{T} \to F(\mathbb{R})$ be a fuzzy function whose α-levels are given by*

$$[f(t)]^\alpha = \left[e_2(t,1) + t^2, 3e_2(t,1) + 2t^2 + t + 1\right], \quad t \in \mathbb{T}, \quad \alpha \in [0,1].$$

Here

$$\begin{aligned} \sigma(t) &= 2t, \\ \underline{f}^\alpha(t) &= e_2(t,1) + t^2, \\ \overline{f}^\alpha(t) &= 3e_2(t,1) + 2t^2 + t, \quad t \in \mathbb{T}, \quad \alpha \in [0,1]. \end{aligned}$$

We have that f is δ_H-differentiable on $\mathbb{T}$. Then

$$
\begin{aligned}
[\delta_H f(t)]^\alpha &= \left[\underline{f}^{\alpha\Delta}(t), \overline{f}^{\alpha\Delta}(t)\right] \\
&= [2e_2(t,1)+\sigma(t)+t, 6e_2(t,1)+2(\sigma(t)+t)+1] \\
&= [2e_2(t,1)+2t+t, 6e_2(t,1)+6t+1] \\
&= [2e_2(t,1)+3t, 6e_2(t,1)+6t+1], \quad t \in \mathbb{T}, \quad \alpha \in [0,1].
\end{aligned}
$$

Hence,

$$
\begin{aligned}
\left[\int_1^4 \delta_H f(t)\right]^\alpha &= \left[\int_1^4 (2e_2(t,1)+3t)\,\Delta t, \int_1^4 (6e_2(t,1)+6t+1)\,\Delta t\right] \\
&= \left[\left(e_2(t,1)+t^2\right)\Big|_{t=1}^{t=4}, \left(3e_2(t,1)+2t^2+t+1\right)\Big|_{t=1}^{t=4}\right] \\
&= [e_2(4,1)+16-1-1, 3e_2(4,1)+32+4+1-3-2-1-1] \\
&= [e_2(4,1)+14, 3e_2(4,1)+30], \quad \alpha \in [0,1].
\end{aligned}
$$

Theorem 2.3.24 *Let $f : [a,b] \to F(\mathbb{R})$ be δ_H-integrable. Then*

$$
\int_a^b f(t)\delta_H t = (-1) \cdot \int_b^a f(t)\delta_H t.
$$

Proof 2.3.25 *Let $u \in S_{Hf}([a,b])$. Then*

$$
\int_a^b u(t)\Delta t = -\int_b^a u(t)\Delta t
$$

$$\subset (-1)\cdot\left\{\int_b^a u(t)\Delta t : u\in S_{Hf}([a,b])\right\}.$$

Thus,

$$\left\{\int_a^b u(t)\Delta t : u\in S_{Hf}([a,b])\right\}\subset(-1)\cdot\left\{\int_b^a u(t)\Delta t : u\in S_{Hf}([a,b])\right\}. \tag{2.41}$$

For $v\in S_{Hf}([a,b])$, *we have*

$$\begin{aligned}-\int_b^a v(s)\Delta s &= \int_a^b v(s)\Delta s\\ &\subset \left\{\int_a^b u(t)\Delta t : u\in S_{Hf}([a,b])\right\}.\end{aligned}$$

Hence,

$$(-1)\cdot\left\{\int_b^a u(t)\Delta t : u\in S_{Hf}([a,b])\right\}\subset\left\{\int_a^b u(t)\Delta t : u\in S_{Hf}([a,b])\right\}.$$

From the last relation and from (2.41), *we get*

$$\left\{\int_a^b u(t)\Delta t : u\in S_{Hf}([a,b])\right\}=(-1)\cdot\left\{\int_b^a u(t)\Delta t : u\in S_{Hf}([a,b])\right\}.$$

This completes the proof.

Theorem 2.3.26 (Integration by Parts) *Let* $f,g:[a,b]\to F(\mathbb{R})$ *be* δ_H*-differentiable and* $f\circ g$ *is also* δ_H*-differentiable on* $[a,b]$. *If*

$$I^{\alpha,1}_{f,g}(t)\le 0,\quad I^{\alpha,1}_{f^\sigma,\delta_H g}(t)\le 0,\quad I^{\alpha,1}_{\delta_H f,g}(t)\le 0,\quad t\in[a,b]^\kappa, \tag{2.42}$$

then

$$\int_a^b f(\sigma(t)) \circ \delta_H g(t) \delta_H t = (f(b) \circ g(b)) \ominus_H (f(a) \circ g(a)) \ominus_H \int_a^b \delta_H f(t) \circ g(t) \delta_H t. \tag{2.43}$$

Proof 2.3.27 *By Theorem 2.3.21, we get*

$$\int_a^b \delta_H (f \circ g)(t) \delta_H t = (f(b) \circ g(b)) \ominus_H (f(a) \circ g(a)). \tag{2.44}$$

On the other hand, using (2.42), *we have*

$$\delta_H (f \circ g)(t) = f(\sigma(t)) \circ \delta_H g(t) + \delta_H f(t) \circ g(t), \quad t \in [a,b]. \tag{2.45}$$

Hence, (2.44) *and Theorem 2.3.12, we arrive to the following relation*

$$\begin{aligned} & (f(b) \circ g(b)) \ominus_H (f(a) \circ g(a)) \\ = & \int_a^b (f(\sigma(t)) \circ \delta_H g(t) + \delta_H f(t) \circ g(t)) \, \delta_H t \\ = & \int_a^b f(\sigma(t)) \circ \delta_H t + \int_a^b \delta_H f(t) \circ g(t) \delta_H t, \end{aligned}$$

whereupon we get (2.43). *This completes the proof.*

2.4 SECOND TYPE FUZZY DELTA INTEGRATION

The set of fuzzy functions $F : I \to F(\mathbb{R})$ which are δ_{SH}-differentiable whose δ_{SH}-derivatives are rd-continuous is denoted by

$$\mathscr{C}^1_{SH\,frd} = \mathscr{C}^1_{SH\,frd}(I) = \mathscr{C}^1_{SH\,frd}(I, F(\mathbb{R})).$$

Definition 2.4.1 *A continuous fuzzy function $F : I \to F(\mathbb{R})$ is said to be δ_{SH}-pre-differentiable with (region of differentiation) D provided $D \subset \mathbb{T}^{\kappa}$, $\mathbb{T}^{\kappa}\backslash D$ is countable and contains no right-scattered elements of $\mathbb{T}$, and F is δ_{SH}-differentiable at each $t \in \mathbb{T}$. We have*

$$\delta_{SH} f(t) = F(t).$$

Definition 2.4.2 *A fuzzy function $F : I \to F(\mathbb{R})$ is said to be δ_{SH}-integrable on I if its sectors are rd-continuous on I. In this case, we define the δ_{SH}-integral of F on I by*

$$\int_I F(s)\delta_{SH} s = \left\{ \int_I f(s)\Delta s : f \in S_{SHF}(I) \right\}.$$

As in the previous section, one can prove the following theorem.

Theorem 2.4.3 *Let $t_0, T \in \mathbb{T}$, $t_0 < T$, $F, G : [t_0, T] \to F(\mathbb{R})$ be δ_{SH}-integrable and have rd-continuous sectors. Then*

$$\int_{t_0}^{T} (\alpha \cdot F(t) + \beta \cdot G(t))\delta_{SH} t = \alpha \cdot \int_{t_0}^{T} F(t)\delta_{SH} t + \beta \cdot \int_{t_0}^{T} G(t)\delta_{SH} t,$$

for any $\alpha, \beta \in \mathbb{R}$.

Theorem 2.4.4 *Let $t_0, T \in \mathbb{T}$, $t_0 < T$, $F : [t_0, T] \to F(\mathbb{R})$ be rd-continuous, $X_0 \in F(\mathbb{R})$. Then*

$$G(t) = X_0 \ominus_H (-1) \cdot \int_{t_0}^{t} F(s)\delta_{SH} s, \quad t \in [t_0, T],$$

is δ_{SH}-differentiable and

$$\delta_{SH} G(t) = F(t), \quad t \in [t_0, T].$$

Proof 2.4.5 *If t is right-scattered, then $\delta_{SH}G(t)$ exists and*

$$\delta_{SH}G(t) = F(t).$$

Let t is right-dense. We have

$$\begin{aligned}
(-1)\cdot(G(\sigma(t))\ominus_H G(t+h)) &= \int_{t_0}^{t+h} F(s)\delta_{SH}s \ominus_H \int_{t_0}^{\sigma(t)} F(s)\delta_{SH}s \\
&= \int_{\sigma(t)}^{t+h} F(s)\delta_{SH}s, \\
(-1)\cdot(G(t-h)\ominus_H G(\sigma(t))) &= \int_{t_0}^{\sigma(t)} F(s)\delta_{SH}s \ominus_H \int_{t_0}^{t-h} F(s)\delta_{SH}s \\
&= \int_{t-h}^{\sigma(t)} F(s)\delta_{SH}s.
\end{aligned}$$

Take $\varepsilon > 0$ arbitrarily. By the rd-continuity of F, we get

$$\begin{aligned}
&D\left(\frac{-1}{h-\mu(t)}\cdot(G(\sigma(t))\ominus_H G(t+h)), F(t)\right) \\
&= \frac{1}{h-\mu(t)}D((-1)\cdot(G(\sigma(t))\ominus_H G(t+h)), (h-\mu(t))\cdot F(t)) \\
&= \frac{1}{h-\mu(t)}D\left(\int_{\sigma(t)}^{t+h} F(s)\delta_{SH}s, \int_{\sigma(t)}^{t+h} F(t)\delta_{SH}s\right) \\
&\leq \frac{1}{h-\mu(t)}\int_{\sigma(t)}^{t+h} D(F(s),F(t))\Delta s \\
&< \varepsilon,
\end{aligned}$$

and

$$\begin{aligned}
& D\left(-\frac{1}{h+\mu(t)}\cdot(G(t-h)\ominus_H G(\sigma(t))),F(t)\right) \\
&= \frac{1}{h+\mu(t)}D\left((-1)\cdot(G(t-h)\ominus_H G(\sigma(t))),(h+\mu(t))\cdot F(t)\right) \\
&= \frac{1}{h+\mu(t)}D\left(\int_{t-h}^{\sigma(t)} F(s)\delta_{SH}s, \int_{t-h}^{\sigma(t)} F(t)\delta_{SH}s\right) \\
&\leq \frac{1}{h+\mu(t)}\int_{t-h}^{\sigma(t)} D(F(s),F(t))\Delta s \\
&< \varepsilon,
\end{aligned}$$

for $h > 0$ sufficiently small. Therefore

$$\begin{aligned}
& \lim_{h\to 0}\frac{-1}{h-\mu(t)}\cdot(G(\sigma(t))\ominus_H G(t+h)) \\
&= \lim_{h\to 0}\frac{-1}{h+\mu(t)}\cdot(G(t-h)\ominus_H G(\sigma(t))) \\
&= F(t),
\end{aligned}$$

and

$$\delta_{SH}G(t) = F(t).$$

This completes the proof.

Theorem 2.4.6 *Let $c \in (a,b)$. Assume that $f : [a,b] \to F(\mathbb{R})$ is δ_H-differentiable on $[a,c]$ and δ_{SH}-differentiable on $[c,b]$. Then*

$$f(b)\ominus_H f(a) = \int_a^c \delta_H f(t)\ominus_H\left((-1)\cdot\int_c^b \delta_{SH}f(t)\right).$$

Proof 2.4.7 *We have*

$$\begin{aligned}
&\int_a^c \delta_H f(t) \ominus_H \left((-1) \cdot \int_c^b \delta_{SH} f(t) \right) \\
&= (f(c) \ominus_H f(a)) \ominus_H ((-1) \cdot (f(b) \ominus_H f(c))) \\
&= (f(c) \ominus_H f(a)) \ominus_H (f(c) \ominus_H f(b)) \\
&= f(b) \ominus_H f(a).
\end{aligned}$$

This completes the proof.

As in above, one can prove the following theorem.

Theorem 2.4.8 *Let $c \in (a,b)$. Assume that $f : [a,b] \to F(\mathbb{R})$ is δ_{SH}-differentiable on $[a,c]$ and δ_H-differentiable on $[c,b]$. Then*

$$f(b) \ominus_H f(a) = \int_a^c \delta_{SH} f(t) \ominus_H \left((-1) \cdot \int_c^b \delta_H f(t) \right).$$

Example 2.4.9 *Let $\mathbb{T} = 2^{\mathbb{N}_0}$, $f : \mathbb{T} \to F(\mathbb{R})$ be a fuzzy function whose α-levels are given by*

$$[f(t)]^\alpha = \left[t + \frac{8}{t}, t^3 + \frac{224}{t} \right], \quad t \in \mathbb{T}, \quad \alpha \in [0,1].$$

Here

$$\begin{aligned}
\sigma(t) &= 2t, \\
\underline{f}^\alpha(t) &= t + \frac{8}{t}, \\
\overline{f}^\alpha(t) &= t^3 + \frac{224}{t}, \quad t \in \mathbb{T}, \quad \alpha \in [0,1].
\end{aligned}$$

Then

$$\begin{aligned}
\underline{f}^{\alpha\Delta}(t) &= 1-\frac{8}{t\sigma(t)} \\
&= 1-\frac{8}{2t^2} \\
&= 1-\frac{4}{t^2} \\
&= \frac{4-t^2}{t^2} \\
&= \frac{(2-t)(2+t)}{t^2}, \\
\overline{f}^{\alpha\Delta}(t) &= (\sigma(t))^2+t\sigma(t)+t^2-\frac{224}{t\sigma(t)} \\
&= (2t)^2+2t^2+t^2-\frac{112}{t^2} \\
&= 7t^2-\frac{112}{t^2} \\
&= \frac{7(t^4-16)}{t^2} \\
&= \frac{7(t-2)(t+2)(t^2+4)}{t^2}, \quad t\in\mathbb{T}, \quad \alpha\in[0,1].
\end{aligned}$$

Therefore, $t=2$ is a switching point for the function f. We have

$$\begin{aligned}
[f(1)]^\alpha &= [9,225], \\
[f(4)]^\alpha &= [6,120], \\
\left[\int_1^2 \delta_{SH} f(t) \ominus_H \left((-1)\cdot\int_2^4 \delta_H f(t)\right)\right]^\alpha &= [f(4)\ominus_H f(1)]^\alpha \\
&= [3,105].
\end{aligned}$$

2.5 ADVANCED PRACTICAL PROBLEMS

Problem 2.5.1 *Let* $\mathbb{T} = [-1,3]\bigcup\{4^k\}_{k\in\mathbb{N}}$, *where* $[-1,3]$ *is the real-valued interval. Let also,*

$$f(t) = \left(t, t+t^2+\frac{t^2+1}{t^2+4}, t+2t^2+t^4+3\frac{t^2+1}{t^2+4}\right), \quad t \in \mathbb{T}.$$

Find

$$\delta_H f(3), \quad \delta_H f(4), \quad \delta_H f(16).$$

Problem 2.5.2 *Let* $\mathbb{T} = [0,2]\bigcup\left\{2+\frac{3}{2^k}\right\}$, *where* $[0,2]$ *is the real-valued interval. Let also,*

$$f(t) = \left(\frac{t+1}{t^2+t+1}, t+3, t^3+t+3\right), \quad t \in \mathbb{T}.$$

Find $\delta_H f(2)$.

Problem 2.5.3 *Let* $\mathbb{T} = \{0\}\bigcup\left\{\frac{1}{2^k}\right\}_{k\in\mathbb{N}}$,

$$\begin{aligned} f(t) &= \frac{t^2+t+4}{t^3+3t^2+t+1}, \\ g(t) &= \left(\frac{t+1}{3t+1}, \frac{t+1}{3t+1}+t, \frac{t+1}{3t+1}+t+t^2+1\right), \quad t \in \mathbb{T}. \end{aligned}$$

Prove that

$$\delta_H(f\cdot g)(0) = f^{\Delta}(0)\cdot g(0) + f(\sigma(0))\cdot \delta_H g(0).$$

Problem 2.5.4 *Let* $\mathbb{T} = 2^{\mathbb{N}_0}$ *and*

$$\begin{aligned} f(t) &= \frac{1}{t^2}, \\ g(t) &= \left(\frac{t+1}{t+2}, t, t^2+t\right), \quad t \in \mathbb{T}. \end{aligned}$$

Prove that

$$\delta_H(f\cdot g)(t) = (f(\sigma(t))\cdot \delta_H g(t)) \ominus_H \left(\left(-f^{\Delta}(t)\right)\cdot g(t)\right), \quad t \in \mathbb{T}.$$

Problem 2.5.5 *Find* $[\delta_H f(t)]^\alpha$, *where*

1. $\mathbb{T} = 7\mathbb{Z}$, $[f(t)]^\alpha = g(t) \cdot [-10, 41]$, $g(t) = 3 + 4t + t^2$, $t \in \mathbb{T}$, $\alpha \in [0,1]$,
2. $\mathbb{T} = 3\mathbb{Z}$, $[f(t)]^\alpha = g(t) \cdot [-1, 14]$, $g(t) = \dfrac{1+t}{1+2t}$, $t \in \mathbb{T}$, $\alpha \in [0,1]$,
3. $\mathbb{T} = \mathbb{N}_0^4$, $[f(t)]^\alpha = g(t) \cdot [-10, 3]$, $g(t) = \dfrac{1+t^2}{1+3t^4}$, $t \in \mathbb{T}$, $\alpha \in [0,1]$,
4. $\mathbb{T} = \mathbb{Z}$, $[f(t)]^\alpha = g(t) \cdot [-7, 12]$, $g(t) = \dfrac{1+t^8}{1+3t}$, $t \in \mathbb{T}$, $\alpha \in [0,1]$,
5. $\mathbb{T} = 2^{\mathbb{N}_0}$, $[f(t)]^\alpha = g(t) \cdot [-10, 3]$, $g(t) = \dfrac{1+t}{1+7t}$, $t \in \mathbb{T}$, $\alpha \in [0,1]$

Problem 2.5.6 *Let* $\mathbb{T} = [-2,1] \bigcup \{3,4,7\}$, *where* $[-2,1]$ *is the real-valued interval. Consider the function* $f : \mathbb{T} \to F(\mathbb{R})$ *whose* α*-levels are given by*

$$[f(t)]^\alpha = g(t) \cdot [1,4],$$

where

$$g(t) = \begin{cases} \dfrac{t(t+2)}{2t+5} & \text{if} \quad t \in [-2,1] \\ \\ t^2 - 1 & \text{if} \quad t \in \{3,4,7\}. \end{cases}$$

Find

$$[\delta_H f(0)]^\alpha, \quad [\delta_H f(1)]^\alpha, \quad [\delta_H f(4)]^\alpha.$$

Problem 2.5.7 *Let* $\mathbb{T} = 4^{\mathbb{N}_0}$ *and*

$$F(t) = \left(\frac{1}{t^7}, \frac{1}{t^5}, \frac{1}{t^4} \right), \quad t \in \mathbb{T}.$$

Find

$$\int_1^{256} F(s) \delta_H s.$$

Problem 2.5.8 *Check if*

$$[\delta_H(f+g)(t)]^\alpha = [\delta_H f(t)]^\alpha + [\delta_H g(t)]^\alpha, \quad t \in \mathbb{T}, \quad \alpha \in [0,1],$$

where

1. $\mathbb{T} = 7\mathbb{Z}$ *and*

$$\begin{aligned}[f(t)]^\alpha &= \left[-e_{-2}(t,0)+t^2, e_{-2}(t,0)+t^2+7\right],\\ [g(t)]^\alpha &= \left[t^4, t^4+t^2\right],\end{aligned}$$

2. $\mathbb{T} = \mathbb{Z}$ *and*

$$\begin{aligned}[f(t)]^\alpha &= \left[t^2, t^6+t^4+t^2\right],\\ [g(t)]^\alpha &= \left[t^2, t^8+10t^6+t^2\right],\end{aligned}$$

3. $\mathbb{T} = 3^{\mathbb{N}_0}$ *and*

$$\begin{aligned}[f(t)]^\alpha &= \left[-\sin_{-3}(t,0)-2, \sin_{-3}(t,0)+2\right],\\ [g(t)]^\alpha &= \left[t^2, t^{10}+t^6+t^4+t^2\right].\end{aligned}$$

Problem 2.5.9 *Let* $\mathbb{T} = 3^{\mathbb{N}_0}$. *Find*

$$\left[\int_1^{27} f(t)\delta_H t\right]^\alpha, \quad \alpha \in [0,1],$$

where

$$[f(t)]^\alpha = \left[2t+8, t^3+2t^2\right]^\alpha, \quad t \in \mathbb{T}, \quad \alpha \in [0,1].$$

Problem 2.5.10 *Let* $\mathbb{T} = \mathbb{N}$,

$$f(t) = \left(\frac{1}{t+7}, \frac{1}{t+6}, \frac{1}{t+5}\right), \quad t \in \mathbb{T}.$$

Find $\delta_{SH} f(t)$, $t \in \mathbb{T}$.

Problem 2.5.11 *Let* $\mathbb{T} = \{2\} \bigcup \left\{2 + \frac{1}{4^k}\right\}_{k \in \mathbb{N}}$,

$$f(t) = \left(\frac{1}{t+11}, \frac{1}{t+7}, \frac{1}{t+1}\right), \quad t \in \mathbb{T}.$$

Find

$$\delta_{SH} f(2), \quad \delta_{SH} f\left(\frac{9}{4}\right), \quad \delta_{SH} f\left(\frac{33}{16}\right).$$

Problem 2.5.12 *Let* $\mathbb{T} = \{1\} \bigcup \left\{1 + \frac{1}{4^k}\right\}_{k \in \mathbb{N}}$,

$$f(t) = \left(\frac{1}{t^2+4}, \frac{1}{t^2+3}, \frac{1}{t^2+2}\right),$$

$$g(t) = \left(\frac{1}{t+11}, \frac{1}{t+9}, \frac{1}{t+7}\right), \quad t \in \mathbb{T}.$$

Prove that

$$\delta_{SH}(f+g)(1) = \delta_{SH} f(1) + \delta_{SH} g(1).$$

Problem 2.5.13 *Let* $\mathbb{T} = 3^{\mathbb{N}_0}$, $f : \mathbb{T} \to F(\mathbb{R})$ *and*

$$[f(t)]^\alpha = \left[\frac{1}{2t+1}, \frac{1}{t}\right], \quad t \in \mathbb{T}.$$

Find $[\delta_{SH} f(t)]^\alpha$.

Problem 2.5.14 *Let* $\mathbb{T} = 3^{\mathbb{N}_0}$,

$$f(t) = \frac{1}{t^2+t+1},$$

$$g(t) = \left(\frac{1}{4t^3}, \frac{1}{3t^3}, \frac{1}{t^3}\right), \quad t \in \mathbb{T}.$$

Prove that

$$\delta_{SH}(f \cdot g)(t) = f^\Delta(t) \cdot g(t) + f(\sigma(t)) \cdot \delta_{SH} g(t), \quad t \in \mathbb{T}.$$

Problem 2.5.15 *Let* $\mathbb{T} = 2\mathbb{N}$,

$$f(t) = t^2,$$

$$g(t) = \left(\frac{1}{7t^4}, \frac{1}{5t^4}, \frac{1}{3t^4}\right), \quad t \in \mathbb{T}.$$

Prove that

$$\delta_{SH}(f \cdot g)(t) = \left(f^{\Delta}(t) \cdot g(t)\right) \ominus_H ((-f(\sigma(t))) \cdot \delta_{SH} g(t)), \quad t \in \mathbb{T}^{\kappa}.$$

CHAPTER 3

Existence and Stability of First-Order Fuzzy Impulsive Dynamic Equations

Let $\mathbb{T}$ be a time scale with a forward jump operator and a delta differentiation operator σ and Δ, respectively. Suppose that $0, t_i, T \in \mathbb{T}$, $i \in \{1, \ldots, n\}$, $0 < t_i < t_{i+1}$, $i \in \{1, \ldots, n-1\}$, and t_i are dense in $\mathbb{T}$, $i \in \{1, \ldots, n\}$.

3.1 EXISTENCE OF SOLUTIONS FOR FIRST-ORDER NON-LINEAR FUZZY IMPULSIVE DYNAMIC EQUATIONS-I

Consider the following fuzzy impulsive dynamic equation

$$
\begin{aligned}
\delta_H y(t) &= p(t) \cdot y(t) + f(t, y(t)), \quad t \in [0, T], \\
y(0) &= y_0, \\
y(t_i^+) &= y(t_i^-) + I(t_i, y(t_i)), \quad i \in \{1, \ldots, n\},
\end{aligned}
\tag{3.1}
$$

where

$$
y(t_i^{\pm}) = \lim_{t \to t_i^{\pm}} y(t),
$$

DOI: 10.1201/9781003655367-3

$$y(t_i) \;=\; y(t_i^-), \quad i \in \{1, \ldots, n\},$$

(G1) $p : \mathbb{T} \to \mathbb{R}$, $p \in \mathscr{R}$, $y_0 \in F(\mathbb{R})$,

(G2) $f \in \mathscr{C}(\mathbb{T} \times F(\mathbb{R}))$.

(G3) There exist positive functions q and r, defined on $[0,T]$, such that

$$\xi \;=\; \max_{t \in [0,T]} \int_0^t |e_p(t, \sigma(s))| q(s) \Delta s < \infty,$$

$$\eta \;=\; \max_{t \in [0,T]} \int_0^t |e_p(t, \sigma(s))| r(s) \Delta s < \infty,$$

$$D(f(t,x), f(t,y)) \;\leq\; r(t) D(x,y),$$

$$D(f(t,y), \widetilde{0}) \;\leq\; q(t) + r(t) D(y, \widetilde{0}), \quad t \in \mathbb{T}, \quad x, y \in F(\mathbb{R}).$$

(G4) There exists a positive constant A such that

$$D(I(t,x), I(t,y)) \leq AD(x,y), \quad x, y \in F(\mathbb{R}), \quad t \in [0,T].$$

Define

$$t_{n+1} \;=\; T,$$

$$J_0 \;=\; [0, t_1],$$

$$J_k \;=\; (t_k, t_{k+1}], \quad k \in \{1, \ldots, n\},$$

$$PC \;=\; \Big\{ y : [0,T] \to F(\mathbb{R}), \quad y \in \mathscr{C}(J_k), \quad \exists y(t_k^{\pm}),$$

$$y(t_k^-) = y(t_k), \quad k \in \{1,\ldots,n\}\Big\},$$

$$PC^1 = \Big\{y : [0,T] \to F(\mathbb{R}), \quad y \in \mathscr{C}^1(J_k), \quad k \in \{1,\ldots,n\}\Big\}.$$

The set PC is a Banach space endowed with the supremum norm

$$\|u\| = \max_{k\in\{0,\ldots,n\}} \|u\|_k,$$

where

$$\|u\|_k = \sup_{t\in J_k} D(u(t),\widetilde{0}), \quad k \in \{0,\ldots,n\}.$$

For $u, v \in PC$, when we write $\|u-v\|$ we have in mind

$$\|u-v\| = \max_{k\in\{0,\ldots,n\}} \|u-v\|_k,$$

where

$$\|u-v\|_k = \sup_{t\in J_k} D(u(t),v(t)), \quad k \in \{0,\ldots,n\}.$$

Theorem 3.1.1 *The function $y \in PC^1$ is a solution of the equation* (3.1) *if and only if $y \in PC$ satisfies the following integral equation*

$$\begin{aligned} y(t) &= \int_0^t e_p(t,\sigma(\tau)) \cdot f(\tau, y(\tau))\delta_H \tau + e_p(t,0) \cdot y_0 \\ &\quad + \sum_{\{k:t_k<t\}} e_p(t,t_k) \cdot I(t_k, y(t_k)). \end{aligned} \tag{3.2}$$

Proof 3.1.2 *1. Let $y \in PC^1$ be a solution of the equation* (3.1). *For $t \in J_0$, we have*

$$\begin{aligned} \delta_H y(t) &= p(t) \cdot y(t) + f(t, y(t)), \quad t \in J_0, \\ y(0) &= y_0. \end{aligned} \tag{3.3}$$

For its solution, we have

$$y(t) = e_p(t,0) \cdot y_0 + \int_0^t e_p(t,\sigma(\tau)) \cdot f(\tau, y(\tau))\delta_H \tau, \quad t \in J_0,$$

and

$$y(t_1) = e_p(t_1,0) \cdot y_0 + \int_0^{t_1} e_p(t_1,\sigma(\tau)) \cdot f(\tau, y(\tau))\delta_H \tau.$$

For $t \in J_1$, we get the following IVP

$$\begin{aligned} \delta_H y(t) &= p(t) \cdot y(t) + f(t, y(t)), \quad t \in J_1, \\ y(t_1^+) &= e_p(t_1,0) \cdot y_0 + \int_0^{t_1} e_p(t_1,\sigma(\tau)) \cdot f(\tau, y(\tau))\delta_H \tau \\ &\quad + I(t_1, y(t_1)). \end{aligned}$$

For its solution, we have the following representation.

$$\begin{aligned} y(t) &= e_p(t,t_1) \cdot \left(e_p(t_1,0) \cdot y_0 + \int_0^{t_1} e_p(t_1,\sigma(\tau)) \cdot f(\tau, y(\tau))\delta_H \tau \right. \\ &\quad \left. + I(t_1, y(t_1)) \right) + \int_{t_1}^{t} e_p(t,\sigma(\tau)) \cdot f(\tau, y(\tau))\delta_H \tau \\ &= e_p(t,0) \cdot y_0 + \int_0^{t_1} e_p(t,\sigma(\tau)) \cdot f(\tau, y(\tau))\delta_H \tau \\ &\quad + e_p(t,t_1) \cdot I(t_1, y(t_1)) + \int_{t_1}^{t} e_p(t,\sigma(\tau)) \cdot f(\tau, y(\tau))\delta_H \tau \\ &= e_p(t,0) \cdot y_0 + \int_0^{t} e_p(t,\sigma(\tau)) \cdot f(\tau, y(\tau))\delta_H \tau \end{aligned}$$

$$+e_p(t,t_1)\cdot I(t_1,y(t_1)).$$

Continuing this process, we get (3.2).

2. *Let* $y \in PC$ *satisfies* (3.2). *For* $t \in J_0$, *we have*

$$y(t) = e_p(t,0)\cdot y_0 + \int_0^t e_p(t,\sigma(\tau))\cdot f(\tau,y(\tau))\delta_H\tau. \qquad (3.4)$$

We δ_H*-differentiate the equation* (3.4) *and we find*

$$\begin{aligned}
\delta_H y(t) &= (p(t)e_p(t,0))\cdot y_0 + e_p(\sigma(t),\sigma(t))\cdot f(t,y(t)) \\
&\quad + p(t)\cdot \int_0^t e_p(t,\sigma(\tau))\cdot f(\tau,y(\tau))\delta_H\tau \\
&= p(t)\cdot\left(e_p(t,0)\cdot y_0 + \int_0^t e_p(t,\sigma(\tau))\cdot f(\tau,y(\tau))\delta_H\tau\right) \\
&\quad + f(t,y(t)) \\
&= p(t)\cdot y(t) + f(t,y(t)), \quad t\in J_0.
\end{aligned}$$

Moreover, by equation (3.4), *we find*

$$\begin{aligned}
y(t_1^-) &= e_p(t_1^-,0)\cdot y_0 + \int_0^{t_1^-} e_p(t_1^-,\sigma(\tau))\cdot f(\tau,y(\tau))\delta_H\tau \\
&= e_p(t_1^+,0)\cdot y_0 + \int_0^{t_1^+} e_p(t_1^+,\sigma(\tau))\cdot f(\tau,y(\tau))\delta_H\tau.
\end{aligned} \qquad (3.5)$$

For $t \in J_1$, we have

$$y(t) = e_p(t,0)\cdot y_0 + \int_0^t e_p(t,\sigma(\tau))\cdot f(\tau,y(\tau))\delta_H\tau \\ +e_p(t,t_1)\cdot I(t_1,y(t_1)). \tag{3.6}$$

The last equation we δ_H-differentiate, and we arrive at

$$\begin{aligned}\delta_H y(t) &= p(t)\cdot(e_p(t,0)\cdot y_0) + e_p(\sigma(t),\sigma(t))\cdot f(t,y(t)) \\ &\quad +p(t)\cdot\int_0^t e_p(t,\sigma(\tau))\cdot f(\tau,y(\tau))\delta_H\tau \\ &\quad +p(t)\cdot(e_p(t,t_1)\cdot I(t_1,y(t_1))) \\ &= p(t)\cdot y(t) + f(t,y(t)), \quad t\in J_1.\end{aligned}$$

We put $t = t_1^+$ into (3.6) and using (3.5), we obtain

$$\begin{aligned}y(t_1^+) &= e_p(t_1^+,0)\cdot y_0 + \int_0^{t_1^+} e_p(t_1^+,\sigma(\tau))\cdot f(\tau,y(\tau))\delta_H\tau \\ &\quad +I(t_1,y(t_1)) \\ &= y(t_1^-) + I(t_1,y(t_1)).\end{aligned}$$

Continuing this process, we get $y \in PC^1$ and it satisfies (3.1). This completes the proof.

Define the operators $P,Q,R : PC \to PC$ as follows

$$Py(t) = \int_0^t e_p(t,\sigma(\tau))\cdot f(\tau,y(\tau))\delta_H\tau,$$

$$\begin{aligned} Qy(t) &= e_p(t,0)\cdot y_0 + \sum_{\{k:t_k<t\}} e_p(t,t_k)\cdot I(t_k,y(t_k)), \\ Ry(t) &= Py(t)+Qy(t), \quad t\in[0,T]. \end{aligned}$$

Lemma 3.1.3 *Suppose* $(G1)-(G3)$*. Then* $P: PC \to PC$ *is completely continuous.*

Proof 3.1.4 *Note that* $P: PC \to PC$*. Take* $y \in PC$ *arbitrarily. Then, applying* $(G3)$*, we arrive at*

$$\begin{aligned} D(Py(t),\widetilde{0}) &= D\left(\int_0^t e_p(t,\sigma(\tau))\cdot f(\tau,y(\tau))\delta_H\tau,\widetilde{0}\right) \\ &\le \int_0^t |e_p(t,\sigma(\tau))| D(f(\tau,y(\tau)),\widetilde{0})\Delta\tau \\ &\le \int_0^t |e_p(t,\sigma(\tau))| \left(q(\tau)+r(\tau)D(y(\tau),\widetilde{0})\right)\Delta\tau \\ &= \int_0^t |e_p(t,\sigma(\tau))| q(\tau)\Delta\tau \\ &\quad + \int_0^t |e_p(t,\sigma(\tau))| r(\tau)D(y(\tau),\widetilde{0})\Delta\tau \\ &\le \xi + \left(\int_0^t |e_p(t,\sigma(\tau))| r(\tau)\Delta\tau\right)\|y\| \\ &\le \xi+\eta\|y\|, \quad t\in[0,T]. \end{aligned}$$

Hence, we find

$$\|Py\| \le \xi+\eta\|y\|.$$

Let $\{y_n\}_{n\in\mathbb{N}} \subset PC$ *be such that*

$$y_n \to y, \quad as \quad n \to \infty,$$

where $y \in PC$. *Then*

$$\begin{aligned}
& D(Py_n(t), Py(t)) \\
=& D\left(\int_0^t e_p(t,\sigma(\tau))\cdot f(\tau,y_n(\tau))\delta_H\tau, \int_0^t e_p(t,\sigma(\tau))\cdot f(\tau,y(\tau))\delta_H\tau\right) \\
\leq& \int_0^t |e_p(t,\sigma(\tau))| D(f(\tau,y_n(\tau)), f(\tau,y(\tau)))\Delta\tau \\
\leq& \int_0^t |e_p(t,\sigma(\tau))| r(\tau) D(y_n(\tau),y(\tau))\Delta\tau \\
\leq& \left(\int_0^t |e_p(t,\sigma(\tau))| r(\tau)\Delta\tau\right) \|y_n - y\| \\
\leq& \eta \|y_n - y\|, \quad t \in [0,T],
\end{aligned}$$

and

$$\|Py_n - Py\| \leq \eta \|y_n - y\|.$$

Therefore, $P : PC \to PC$ *is continuous. Now, we arbitrarily take* $y \in PC$ *and* $t_1, t_2 \in [0,T]$. *Without loss of generality, suppose that* $t_1 < t_2$. *Then*

$$\begin{aligned}
D(Py(t_2), Py(t_1)) &= D\Big(\int_0^{t_2} e_p(t_2,\sigma(\tau))\cdot f(\tau,y(\tau))\delta_H\tau, \\
&\qquad \int_0^{t_1} e_p(t_1,\sigma(\tau))\cdot f(\tau,y(\tau))\delta_H\tau\Big)
\end{aligned}$$

$$
\begin{aligned}
&= D\Big(\int_0^{t_1} (e_p(t_2,\sigma(\tau)) - e_p(t_1,\sigma(\tau)))\cdot f(\tau,y(\tau))\delta_H\tau, \\
&\quad \int_{t_1}^{t_2} e_p(t_2,\sigma(\tau))\cdot f(\tau,y(\tau))\delta_H\tau\Big) \\
&\leq |e_p(t_2,t_1)-1|\int_0^{t_1} |e_p(t_1,\sigma(\tau))| D(f(\tau,y(\tau)),\widetilde{0})\Delta\tau \\
&\quad + \int_{t_1}^{t_2} |e_p(t_2,\sigma(\tau))| D(f(\tau,y(\tau)),\widetilde{0})\Delta\tau \\
&\to 0, \quad as \quad t_1 \to t_2.
\end{aligned}
$$

Now, using the Arzela-Ascoli theorem, we conclude that $P : PC \to PC$ is completely continuous. This completes the proof.

Lemma 3.1.5 *Suppose that (G1)-(G4) hold. Let also,*

$$
A \max_{t\in[0,T]} \sum_{k=1}^{n} |e_p(t,t_k)| \leq \zeta < 1. \tag{3.7}
$$

Then $Q : PC \to PC$ is a contraction.

Proof 3.1.6 *Note that $e_p(t,t_k)$, $k \in \{1,\ldots,n\}$, is a continuous function. Therefore, $Q : PC \to PC$. Now, let $y, z \in PC$ be arbitrarily chosen. Then*

$$
\begin{aligned}
D(Qy(t),Qz(t)) &= D\Big(e_p(t,0)\cdot y_0 + \sum_{\{k:t_k<t\}} e_p(t,t_k)\cdot I(t_k,y(t_k)), \\
&\quad e_p(t,0)\cdot y_0 + \sum_{\{k:t_k<t\}} e_p(t,t_k)\cdot I(t_k,z(t_k))\Big) \\
&= D\Big(\sum_{\{k:t_k<t\}} e_p(t,t_k)\cdot I(t_k,y(t_k)), \sum_{\{k:t_k<t\}} e_p(t,t_k)\cdot I(t_k,z(t_k))\Big)
\end{aligned}
$$

$$\begin{aligned}
&\leq \sum_{\{k:t_k<t\}} |e_p(t,t_k)| D(I(t_k,y(t_k)),I(t_k,z(t_k)))\\
&\leq A \sum_{\{k:t_k<t\}} |e_p(t,t_k)| D(y(t_k),z(t_k))\\
&\leq A \sum_{\{k:t_k<t\}} |e_p(t,t_k)| \|y-z\|\\
&\leq \zeta \|y-z\|, \quad t\in[0,T],
\end{aligned}$$

whereupon

$$\|Qy-Qz\| \leq \zeta\|y-z\|.$$

Then $Q : PC \to PC$ is a contraction. This completes the proof.

Theorem 3.1.7 *Assume that*

$$\gamma = \max_{t\in[0,T]} \sum_{k=1}^{n} |e_p(t,t_k)| D(I(t_k,0),\widetilde{0}) < \infty. \tag{3.8}$$

Let (*G*1)-(*G*4) *and* (3.7) *hold. Let also, ξ, γ, η, A and S be positive numbers satisfying the inequality*

$$\xi+\gamma+\max_{t\in[0,T]} |e_p(t,0)| D(y_0,\widetilde{0})+S\left(\eta+A+\max_{t\in[0,T]}\sum_{k=1}^{n}|e_p(t,t_k)|\right)\leq S. \tag{3.9}$$

Then the equation (3.1) *has a solution y such that*

$$\|y\| \leq S.$$

Proof 3.1.8 *Define the set*

$$C=\{y\in PC : \|y\|\leq S\}.$$

By Lemma 3.1.3, it follows that $P : PC \to PC$ is completely continuous. By Lemma 3.1.5, it follows that $Q : PC \to PC$ is a contraction. Now, let $y,z \in C$ be arbitrarily chosen. Then

$$\|y\| \leq S,$$

$$\|z\| \leq S,$$

and

$$\begin{aligned}
& D(Py(t)+Qz(t),\widetilde{0}) \\
= \; & D\Bigg(\int_0^t e_p(t,\sigma(s))\cdot f(s,y(s))\delta_H s+\sum_{k=1}^{n} e_p(t,t_k)\cdot I(t_k,z(t_k)) \\
& +e_p(t,0)\cdot y_0,\widetilde{0}\Bigg) \\
\leq \; & D\Bigg(\int_0^t e_p(t,\sigma(s))\cdot f(s,y(s))\delta_H s,\widetilde{0}\Bigg) \\
+ \; & D\Bigg(\sum_{k=1}^{n} e_p(t,t_k)\cdot I(t_k,z(t_k)),\widetilde{0}\Bigg) \\
& +D\Bigg(e_p(t,0)\cdot y_0,\widetilde{0}\Bigg) \\
\leq \; & \int_0^t |e_p(t,\sigma(s))|D(f(s,y(s)),\widetilde{0})\Delta s+\sum_{k=1}^{n} |e_p(t,t_k)|D(I(t_k,z(t_k)),\widetilde{0}) \\
& +|e_p(t,0)|D(y_0,\widetilde{0}) \\
\leq \; & \int_0^t |e_p(t,\sigma(s))|D(f(s,y(s)),\widetilde{0})\Delta s \\
& +\sum_{k=1}^{n} |e_p(t,t_k)|(D(I(t_k,z(t_k)),I(t_k,0))+D(I(t_k,0),\widetilde{0})) \\
& +|e_p(t,0)|D(y_0,\widetilde{0}) \\
\leq \; & \int_0^t |e_p(t,\sigma(s))|\left(q(s)+r(s)D(y(s),\widetilde{0})\right)\Delta s
\end{aligned}$$

$$
\begin{aligned}
&+\sum_{k=1}^{n}|e_p(t,t_k)|D(I(t_k,z(t_k)),I(t_k,0))\\
&+\sum_{k=1}^{n}|e_p(t,t_k)|D(I(t_k,0),\widetilde{0})+|e_p(t,0)|D(y_0,\widetilde{0})\\
\leq\ & \xi+\eta\|y\|+A\sum_{k=1}^{n}|e_p(t,t_k)|D(z(t_k),\widetilde{0})\\
&+\gamma+|e_p(t,0)|D(y_0,\widetilde{0})\\
\leq\ & \xi+\gamma+\eta S+|e_p(t,0)|D(y_0,\widetilde{0})+\left(A\sum_{k=1}^{n}|e_p(t,t_k)|\right)S\\
=\ & \xi+\gamma+|e_p(t,0)|D(y_0,\widetilde{0})+S\left(\eta+A\sum_{k=1}^{n}|e_p(t,t_k)|\right)\\
\leq\ & \xi+\gamma+\max_{t\in[0,T]}|e_p(t,0)|D(y_0,\widetilde{0})\\
&+S\left(\eta+A\max_{t\in[0,T]}\sum_{k=1}^{n}|e_p(t,t_k)|\right)\\
\leq\ & S,\quad t\in[0,T].
\end{aligned}
$$

Hence,

$$
\|Py+Qz\|\leq S,\quad t\in[0,T],
$$

i.e., $Py+Qz\in C$. Hence and the Krasnosel'skii fixed-point theorem,[1] *it follows that the operator $P+Q$ has a fixed point in C. Now, using*

[1] **Theorem 3.1.9 (Krasnosel'skii fixed-point Theorem)** *Let M be a closed convex non-empty subset of a Banach space $(\mathscr{B},\|\cdot\|)$. Suppose that*

1. *$A: M\to\mathscr{B}$ is completely continuous,*
2. *$B: M\to\mathscr{B}$ is a contraction,*
3. *$x,y\in M$ implies $Ax+By\in M$.*

Then the map $A+B$ has a fixed point in M.

Theorem 3.1.1, we conclude that equation (3.1) *has a solution in the set C. This completes the proof.*

Theorem 3.1.10 *Assume* $(G1)$-$(G4)$, (3.7), (3.8), (3.9) *and*

$$\eta + \zeta < 1.$$

Equation (3.1) *has a unique solution.*

Proof 3.1.11 *By Theorem 3.1.7, it follows that equation* (3.1) *has at least one solution in the set C. Suppose that equation* (3.1) *has two solutions y and z. Then*

$$\begin{aligned} D(Ry(t), Rz(t)) &= D\Big(\int_0^t e_p(t, \sigma(\tau)) \cdot f(\tau, y(\tau)) \delta_H \tau + e_p(t, 0) \cdot y_0, \\ &\quad + \sum_{\{k: t_k < t\}} e_p(t, t_k) \cdot I(t_k, y(t_k)), \\ &\quad \int_0^t e_p(t, \sigma(\tau)) \cdot f(\tau, z(\tau)) \delta_H \tau + e_p(t, 0) \cdot y_0 \\ &\quad + \sum_{\{k: t_k < t\}} e_p(t, t_k) \cdot I(t_k, z(t_k)) \Big) \end{aligned}$$

$$\begin{aligned} &= D\Big(\int_0^t e_p(t, \sigma(\tau)) \cdot f(\tau, y(\tau)) \delta_H \tau + \sum_{\{k: t_k < t\}} e_p(t, t_k) \cdot I(t_k, y(t_k)), \\ &\quad \int_0^t e_p(t, \sigma(\tau)) \cdot f(\tau, z(\tau)) \delta_H \tau + \sum_{\{k: t_k < t\}} e_p(t, t_k) \cdot I(t_k, z(t_k)) \Big) \\ &= D\Big(\int_0^t e_p(t, \sigma(\tau)) \cdot (f(\tau, y(\tau)) \ominus_H f(\tau, z(\tau))) \delta_H \tau, \end{aligned}$$

$$
\begin{aligned}
& \sum_{\{k:t_k<t\}} e_p(t,t_k)\cdot(I(t_k,z(t_k))\ominus_H I(t_k,y(t_k)))\Bigg) \\
\le\ & D\Bigg(\int_0^t e_p(t,\sigma(\tau))\cdot(f(\tau,y(\tau))\ominus_H f(\tau,z(\tau)))\delta_H\tau,\widetilde{0}\Bigg) \\
& +D\Bigg(\sum_{\{k:t_k<t\}} e_p(t,t_k)\cdot(I(t_k,y(t_k))\ominus_H I(t_k,z(t_k))),\widetilde{0}\Bigg) \\
=\ & D\Bigg(\int_0^t e_p(t,\sigma(\tau))\cdot f(\tau,y(\tau))\delta_H\tau,\int_0^t e_p(t,\sigma(\tau))\cdot f(\tau,z(\tau))\delta_H\tau\Bigg) \\
& +D\Bigg(\sum_{\{k:t_k<t\}} e_p(t,t_k)\cdot I(t_k,y(t_k)),\sum_{\{k:t_k<t\}} e_p(t,t_k)\cdot I(t_k,z(t_k))\Bigg) \\
\le\ & \int_0^t |e_p(t,\sigma(\tau))| D(f(\tau,y(\tau)),f(\tau,z(\tau)))\Delta\tau \\
& +\sum_{\{k:t_k<t\}} |e_p(t,t_k)| D(I(t_k,y(t_k)),I(t_k,z(t_k))) \\
\le\ & \int_0^t |e_p(t,\sigma(\tau))| r(\tau) D(y(\tau),z(\tau))\Delta\tau \\
& +A\sum_{\{k:t_k<t\}} |e_p(t,t_k)| D(y(t_k),z(t_k)) \\
\le\ & \eta\|y-z\|+\zeta\|y-z\| \\
=\ & (\eta+\zeta)\|y-z\| \\
<\ & \|y-z\|,\quad t\in[0,T].
\end{aligned}
$$

Hence,

$$\begin{aligned} \|y-z\| &= \|Ry-Rz\| \\ &< \|y-z\|, \end{aligned}$$

which is a contradiction. This completes the proof.

Suppose that

(G5) There exist non-decreasing functions $l_k : \mathbb{R}_+ \to \mathbb{R}_+$ such that

$$D(I_k(t_k, y(t_k)), \widetilde{0}) \leq l_k(D(y(t_k), \widetilde{0})), \quad k \in \{0, \ldots, n\},$$

and

$$\lim_{r\to\infty} \frac{l_k(r)}{r} = 0, \quad k \in \{0, \ldots, n\}.$$

(G6) For each $r > 0$ there exists a continuous function $h_r : [0, T] \to (0, \infty)$ such that

$$D(f(t,y), \widetilde{0}) \leq h_k(t) \text{ for each } (t,y) \in [0,T] \times F(\mathbb{R}),\ D(y, \widetilde{0}) \leq r,$$

and

$$\liminf_{r\to\infty} \frac{1}{r} \int_0^T h_r(s) \Delta s = \beta_1 < \infty.$$

Theorem 3.1.12 *Suppose* (G1), (G2), (G4)–(G6) *and*

$$\beta_1 \max_{(t,\tau)\in[0,T]\times[0,T]} |e_p(t, \sigma(\tau))| < 1. \tag{3.10}$$

Then (3.1) *has at least one solution.*

Proof 3.1.13 *We have $R : PC \to PC$.*

1. *We will prove that R is continuous. Let $\{y_m\}_{m\in\mathbb{N}}$ be a sequence of elements of PC such that*

$$y_m \to y, \quad as \quad m \to \infty, \quad y \in PC.$$

Then

$$
\begin{aligned}
& D(Ry_m(t), Ry(t)) \\
= \; & D\Big(\int_0^t e_p(t,\sigma(\tau)) \cdot f(\tau, y_m(\tau))\delta_H \tau + e_p(t,0) \cdot y_0 \\
& + \sum_{\{k:t_k<t\}} e_p(t,t_k) \cdot I(t_k, y_m(t_k)), \\
& \int_0^t e_p(t,\sigma(\tau)) \cdot f(\tau, y(\tau))\delta_H \tau + e_p(t,0) \cdot y_0 \\
& + \sum_{\{k:t_k<t\}} e_p(t,t_k) \cdot I(t_k, y(t_k)) \Big)
\end{aligned}
$$

$$
\begin{aligned}
= \; & D\Big(\int_0^t e_p(t,\sigma(\tau)) \cdot f(\tau, y_m(\tau))\delta_H \tau \\
& + \sum_{\{k:t_k<t\}} e_p(t,t_k) \cdot I(t_k, y_m(t_k)), \\
& \int_0^t e_p(t,\sigma(\tau)) \cdot f(\tau, y(\tau))\delta_H \tau \\
& + \sum_{\{k:t_k<t\}} e_p(t,t_k) \cdot I(t_k, y(t_k)) \Big) \\
= \; & D\Big(\int_0^t e_p(t,\sigma(\tau)) \cdot (f(\tau, y_m(\tau)) \ominus_H f(\tau, y(\tau)))\delta_H \tau, \\
& \sum_{\{k:t_k<t\}} e_p(t,t_k) \cdot (I(t_k, y(t_k)) \ominus_H I(t_k, y_m(t_k))) \Big)
\end{aligned}
$$

$$\begin{aligned}
&\le D\Big(\int_0^t e_p(t,\sigma(\tau))\cdot(f(\tau,y_m(\tau))\ominus_H f(\tau,y(\tau)))\delta_H\tau,\widetilde{0}\Big)\\
&\quad +D\Big(\sum_{\{k:t_k<t\}} e_p(t,t_k)\cdot(I(t_k,y_m(t_k))\ominus_H I(t_k,y(t_k))),\widetilde{0}\Big)\\
&= D\Big(\int_0^t e_p(t,\sigma(\tau))\cdot f(\tau,y_m(\tau))\delta_H\tau,\\
&\quad \int_0^t e_p(t,\sigma(\tau))\cdot f(\tau,y(\tau))\delta_H\tau\Big)\\
&\quad +D\Big(\sum_{\{k:t_k<t\}} e_p(t,t_k)\cdot I(t_k,y_m(t_k)),\\
&\quad \sum_{\{k:t_k<t\}} e_p(t,t_k)\cdot I(t_k,y(t_k))\Big)\\
&\le \int_0^t |e_p(t,\sigma(\tau))|D(f(\tau,y_m(\tau)),f(\tau,y(\tau)))\Delta\tau\\
&\quad +\sum_{\{k:t_k<t\}} |e_p(t,t_k)|D(I(t_k,y_m(t_k)),I(t_k,y(t_k)))\\
&\le \max_{(t,\tau)\in[0,T]\times[0,T]} |e_p(t,\sigma(\tau))|\int_0^t D(f(\tau,y_m(\tau)),f(\tau,y(\tau)))\Delta\tau\\
&\quad +A\sum_{\{k:t_k<t\}}\max_{t\in[0,T]} |e_p(t,t_k)|D(y_m(t_k),y(t_k))\\
&\to 0, \quad as \quad m\to\infty, \quad t\in[0,T].
\end{aligned}$$

Hence, we get

$$\|Ry_m - Ry\| \to 0, \quad as \quad m\to\infty.$$

2. *For each constant $r > 0$, denote*

$$B_r = \{y \in PC : \|y\| \leq r\}.$$

Note that B_r is a bounded, closed and convex set in PC. We will prove that there exists a number $r > 0$ such that $R(B_r) \subseteq B_r$. Suppose the contrary, i.e., for each $r > 0$ there exists a function $y_r \in B_r$ such that

$$D(R(y_r(t)), \widetilde{0}) > r \quad \textit{for} \quad \textit{some} \quad t \in [0, T].$$

Then

$$\begin{aligned}
r &< D(R(y_r(t)), \widetilde{0}) \\
&= D\Big(\int_0^t e_p(t, \sigma(\tau)) \cdot f(\tau, y_r(\tau)) \delta_H \tau + e_p(t, 0) \cdot y_0 \\
&\quad + \sum_{\{k: t_k < t\}} e_p(t, t_k) \cdot I(t_k, y_r(t_k)), \widetilde{0} \Big) \\
&\leq D\Big(\int_0^t e_p(t, \sigma(\tau)) \cdot f(\tau, y_r(\tau)) \delta_H \tau, \widetilde{0} \Big) + D(e_p(t, 0) \cdot y_0, \widetilde{0}) \\
&\quad + D\Big(\sum_{\{k: t_k < t\}} e_p(t, t_k) \cdot I(t_k, y_r(t_k)), \widetilde{0} \Big) \\
&\leq \int_0^t |e_p(t, \sigma(\tau))| D(f(\tau, y_r(\tau)), \widetilde{0}) \Delta \tau + |e_p(t, 0)| D(y_0, \widetilde{0}) \\
&\quad + \sum_{\{k: t_k < t\}} |e_p(t, t_k)| D(I(t_k, y_r(t_k)), \widetilde{0}) \\
&\leq \max_{t \in [0, T]} |e_p(t, 0)| D(y_0, \widetilde{0}) \\
&\quad + \int_0^t |e_p(t, \sigma(\tau))| D(f(\tau, y_r(\tau)), \widetilde{0}) \Delta \tau
\end{aligned}$$

$$
\begin{aligned}
& + \sum_{\{k:t_k<t\}} |e_p(t,t_k)| D(I(t_k, y_r(t_k)), \widetilde{0}) \\
\leq & \max_{t\in[0,T]} |e_p(t,0)| D(y_0, \widetilde{0}) \\
& + \max_{(t,\tau)\in[0,T]\times[0,T]} |e_p(t,\sigma(\tau))| \int_0^t h_r(s)\Delta s \\
& + \sum_{\{k:t_k<t\}} \max_{t\in[0,T]} |e_p(t,t_k)| l_k\left(D(y(t_k),\widetilde{0})\right) \\
\leq & \max_{t\in[0,T]} |e_p(t,0)| D(y_0, \widetilde{0}) \\
& + \max_{(t,\tau)\in[0,T]\times[0,T]} |e_p(t,\sigma(\tau))| \int_0^t h_r(s)\Delta s \\
& + \sum_{\{k:t_k<t\}} \max_{t\in[0,T]} |e_p(t,t_k)| l_k(r),
\end{aligned}
$$

whereupon

$$
\begin{aligned}
1 \; < \; & \max_{t\in[0,T]} |e_p(t,0)| \frac{D(y_0,\widetilde{0})}{r} \\
& + \max_{(t,\tau)\in[0,T]\times[0,T]} |e_p(t,\sigma(\tau))| \frac{1}{r} \int_0^t h_r(s)\Delta s \\
& + \sum_{\{k:t_k<t\}} \max_{t\in[0,T]} |e_p(t,t_k)| \frac{l_k(r)}{r},
\end{aligned}
$$

and taking limits as $r \to \infty$, we find

$$
1 \; \leq \; \max_{t\in[0,T]} |e_p(t,0)| \lim_{r\to\infty} \frac{D(y_0,\widetilde{0})}{r}
$$

$$+ \max_{(t,\tau)\in[0,T]\times[0,T]} |e_p(t,\sigma(\tau))| \lim_{r\to\infty} \frac{1}{r} \int_0^t h_r(s)\Delta s$$

$$+ \sum_{\{k:t_k<t\}} \max_{t\in[0,T]} |e_p(t,t_k)| \lim_{r\to\infty} \frac{l_k(r)}{r}$$

$$= \beta_1 \max_{(t,\tau)\in[0,T]\times[0,T]} |e_p(t,\sigma(\tau))|,$$

which is a contradiction. Consequently, there is a $r_1 > 0$ such that

$$R(B_{r_1}) \subseteq B_{r_1}.$$

3. *We will show that $R(B_{r_1})$ is an equicontinous family of functions. Let $t^1, t^2 \in [0,T]$ be arbitrarily chosen. Without loss of generality, suppose that $t^1 < t^2$. Then*

$$D(R(y)(t^2), R(y)(t^1))$$

$$= D\Big(\int_0^{t^2} e_p(t^2,\sigma(\tau))\cdot f(\tau,y(\tau))\delta_H\tau + e_p(t^2,0)\cdot y_0$$

$$+ \sum_{\{k:t_k<t^2\}} e_p(t^2,t_k)\cdot I(t_k,y(t_k)),$$

$$\int_0^{t^1} e_p(t^1,\sigma(\tau)) f(\tau,y(\tau))\delta_H\tau + e_p(t^1,0)\cdot y_0$$

$$+ \sum_{\{k:t_k<t^1\}} e_p(t^1,t_k)\cdot I(t_k,y(t_k))\Big)$$

$$= D\Big(\int_0^{t^1} \Big(e_p(t^2,\sigma(\tau)) - e_p(t^1,\sigma(\tau))\Big)\cdot f(\tau,y(\tau))\delta_H\tau$$

$$+ \int_{t^1}^{t^2} e_p(t^2,\sigma(\tau))\cdot f(\tau,y(\tau))\delta_H\tau$$

$$
\begin{aligned}
&+\Big(e_p(t^2,0)-e_p(t^1,0)\Big)\cdot y_0 \\
&+\sum_{\{k:t_k<t^1\}}\Big(e_p(t^2,t_k)-e_p(t^1,t_k)\Big)\cdot I(t_k,y(t_k)) \\
&+\sum_{\{k:t^1\le t_k<t^2\}} e_p(t^2,t_k)\cdot I(t_k,y(t_k)),\widetilde{0}\Big) \\
\le\ & D\Big(\int_0^{t^1}\Big(e_p(t^2,\sigma(\tau))-e_p(t^1,\sigma(\tau))\Big)\cdot f(\tau,y(\tau))\delta_H\tau,\widetilde{0}\Big) \\
&+D\Big(\int_{t^1}^{t^2} e_p(t^2,\sigma(\tau))\cdot f(\tau,y(\tau))\delta_H\tau,\widetilde{0}\Big) \\
&+D\Big(\Big(e_p(t^2,0)-e_p(t^1,0)\Big)\cdot y_0,\widetilde{0}\Big) \\
&+D\left(\sum_{\{k:t_k<t^1\}}\Big(e_p(t^2,t_k)-e_p(t^1,t_k)\Big)\cdot I(t_k,y(t_k)),\widetilde{0}\right) \\
&+D\Big(\sum_{\{k:t^1\le t_k<t^2\}} e_p(t^2,t_k)\cdot I(t_k,y(t_k)),\widetilde{0}\Big) \\
\le\ & \int_0^{t^1}\Big|e_p(t^2,\sigma(\tau))-e_p(t^1,\sigma(\tau))\Big|D(f(\tau,y(\tau)),\widetilde{0})\Delta\tau \\
&+\int_{t^1}^{t^2}\Big|e_p(t^2,\sigma(\tau))\Big|D(f(\tau,y(\tau)),\widetilde{0})\Delta\tau \\
&+\Big|e_p(t^2,0)-e_p(t^1,0)\Big|D(y_0,\widetilde{0})
\end{aligned}
$$

$$+ \sum_{\{k:t_k<t^1\}} \left|e_p(t^2,t_k) - e_p(t^1,t_k)\right| D(I(t_k,y(t_k)),\widetilde{0})$$

$$+ \sum_{\{k:t^1\le t_k<t^2\}} \left|e_p(t^2,t_k)\right| D(I(t_k,y(t_k)),\widetilde{0})$$

$$\to 0, \quad as \quad t^1 \to t^2.$$

From this and the Arzela-Ascoli theorem, we conclude that $R : PC \to PC$ is completely continuous, and it is condensing.

In view of Sadovskii's fixed-point theorem,[2] *it follows that R has a fixed point in PC, which is a solution of the equation* (3.1). *This completes the proof.*

Theorem 3.1.15 *Suppose* (*G*1), (*G*2), (*G*4), (*G*6), (3.7) *and*

$$A \sum_{\{k:t_k<t\}} \max_{t\in[0,T]} |e_p(t,t_k)| < 1.$$

Then (3.1) *has at least one solution.*

Proof 3.1.16 *As we have proved in Theorem 3.1.12, there exists a positive number r such that $R(B_r) \subseteq B_r$ and the operator Q is a compact operator in B_r. Now, let $y,z \in PC$. Then*

$$\begin{aligned} D(Py(t),Pz(t)) &= D\Bigg(e_p(t,0)\cdot y_0 + \sum_{\{k:t_k<t\}} e_p(t,t_k)\cdot I(t_k,y(t_k)), \\ &\qquad e_p(t,0)\cdot y_0 + \sum_{\{k:t_k<t\}} e_p(t,t_k)\cdot I(t_k,z(t_k))\Bigg) \end{aligned}$$

$$= D\Bigg(\sum_{\{k:t_k<t\}} e_p(t,t_k)\cdot I(t_k,y(t_k)), \sum_{\{k:t_k<t\}} e_p(t,t_k)\cdot I(t_k,z(t_k))\Bigg)$$

[2] **Theorem 3.1.14 (Sadovskii's fixed-point Theorem)** *Let P be a condensing operator on a Banach space X. If $P(D) \subset D$ for a convex,closed and bounded set D of X, then P has a fixed point in D.*

$$\begin{aligned}
&\leq \sum_{\{k:t_k<t\}} |e_p(t,t_k)| D(I(t_k,y(t_k)), I(t_k,z(t_k)))\\
&\leq A \sum_{\{k:t_k<t\}} \max_{t\in[0,T]} |e_p(t,t_k)| D(y(t_k),z(t_k))\\
&\leq A \sum_{\{k:t_k<t\}} \max_{t\in[0,T]} |e_p(t,t_k)| \|y-z\|, \quad t\in[0,T].
\end{aligned}$$

Hence,

$$\begin{aligned}
\|Py-Pz\| &\leq A \sum_{\{k:t_k<t\}} \max_{t\in[0,T]} |e_p(t,t_k)| \|y-z\|\\
&< \|y-z\|.
\end{aligned}$$

Consequently $R = P+Q$ is a condensing operator on B_r. Hence, by the Krasnoselskii fixed-point theorem, it follows that the equation (3.1) *has a solution in PC. This completes the proof.*

Exercise 3.1.17 *Consider the fuzzy impulsive dynamic equation*

$$\begin{aligned}
\delta_H y(t) &= p(t)\cdot y(t) + f(t), \quad t\in[0,T],\\
y(0) &= y_0,\\
y(t_k^+) - y(t_k^-) &= I_k(y(t_k^-)), \quad k\in\{1,\ldots,n\}.
\end{aligned}$$

Write analogously to Theorem 3.1.1, Theorem 3.1.10, Theorem 3.1.12, Theorem 3.1.15.

Exercise 3.1.18 *Consider the impulsive dynamic equation*

$$\begin{aligned}
\delta_H y(t) &= p(t)\cdot y(t) + f(y(t)) + g(t), \quad t\in[0,T],\\
y(0) &= y_0,\\
y(t_k^+) - y(t_k^-) &= I_k(y(t_k^-)), \quad k\in\{1,\ldots,n\}.
\end{aligned}$$

Write analogously to Theorem 3.1.1, Theorem 3.1.10, Theorem 3.1.12, and Theorem 3.1.15.

Exercise 3.1.19 *Consider the impulsive dynamic equation*

$$\begin{aligned} \delta_H y(t) &= p(t)\cdot y(t) + g(t)\cdot f(y(t)), \quad t\in[0,T], \\ y(0) &= y_0, \\ y(t_k^+) &= I_k(y(t_k^-)), \quad k\in\{1,\ldots,n\}. \end{aligned}$$

Write analogously to Theorem 3.1.1, Theorem 3.1.10, Theorem 3.1.12, and Theorem 3.1.15.

Exercise 3.1.20 *Consider the fuzzy impulsive dynamic equation*

$$\begin{aligned} \delta_{SH} y(t) &= p(t)\cdot y(t) + f(t), \quad t\in[0,T], \\ y(0) &= y_0, \\ y(t_k^+) - y(t_k^-) &= I_k(y(t_k^-)), \quad k\in\{1,\ldots,n\}. \end{aligned}$$

Write analogously to Theorem 3.1.1, Theorem 3.1.10, Theorem 3.1.12, and Theorem 3.1.15.

Exercise 3.1.21 *Consider the impulsive dynamic equation*

$$\begin{aligned} \delta_{SH} y(t) &= p(t)\cdot y(t) + f(y(t)) + g(t), \quad t\in[0,T], \\ y(0) &= y_0, \\ y(t_k^+) - y(t_k^-) &= I_k(y(t_k^-)), \quad k\in\{1,\ldots,n\}. \end{aligned}$$

Write analogously to Theorem 3.1.1, Theorem 3.1.10, Theorem 3.1.12, Theorem 3.1.15.

Exercise 3.1.22 *Consider the impulsive dynamic equation*

$$\begin{aligned} \delta_{SH}y(t) &= p(t)\cdot y(t)+g(t)\cdot f(y(t)), \quad t\in[0,T], \\ y(0) &= y_0, \\ y(t_k^+) &= I_k(y(t_k^-)), \quad k\in\{1,\ldots,n\}. \end{aligned}$$

Write analogously to Theorem 3.1.1, Theorem 3.1.10, Theorem 3.1.12, and Theorem 3.1.15.

3.2 EXISTENCE OF SOLUTIONS FOR FIRST-ORDER NON-LINEAR FUZZY IMPULSIVE DYNAMIC EQUATIONS-II

Consider the following non-linear fuzzy impulsive dynamic equation

$$\begin{aligned} \delta_{H}y(t) &= (-a(t))\cdot y^{\sigma}(t)+f(t,y(t)), \quad t\in[0,T], \\ y(0) &= 0, \\ y(t_i^+) &= y(t_i^-)+I(t_i,y(t_i)), \quad i\in\{1,\ldots,n\}, \end{aligned} \tag{3.11}$$

where

(H1) $a:\mathbb{T}\to\mathbb{R}$, $a\in\mathscr{R}$,

(H2) $f\in\mathscr{C}(\mathbb{T}\times F(\mathbb{R}))$,

(H3) there exist positive functions g and h, defined on $[0,T]$, such that

$$\alpha=\max_{t\in[0,T]}\int_0^t |e_{\ominus a}(t,s)|g(s)\Delta s<\infty,$$

$$\beta=\max_{t\in[0,T]}\int_0^t |e_{\ominus a}(t,s)|h(s)\Delta s<\infty,$$

$$D(f(t,x),f(t,y)) \leq h(t)D(x,y),$$

$$D(f(t,y),\widetilde{0})| \leq g(t)+h(t)D(y,\widetilde{0}), \ \ t \in [0,T], \ \ x,y \in F(\mathbb{R}),$$

(H4) there exists a positive constant E such that

$$D(I(t,x),I(t,y)) \leq ED(x,y), \quad x,y \in F(\mathbb{R}), \quad t \in [0,T].$$

Suppose that PC and PC^1 are the same as in Section 3.1. Now, we will give an integral representation of the solutions of problem (3.11).

Theorem 3.2.1 *The function $y \in PC^1$ is a solution of equation* (3.11) *if and only if $y \in PC$ satisfies the following integral equation.*

$$y(t) = \int_0^t e_{\ominus a}(t,s) \cdot f(s,y(s))\delta_H s + \sum_{\{i:t_i<t\}} e_{\ominus a}(t,t_i) \cdot I(t_i,y(t_i)). \quad (3.12)$$

Proof 3.2.2 *1. Let $y \in PC^1$ be a solution of the equation* (3.11). *For $t \in J_0$, we have that y satisfies the following initial value problem*

$$\begin{aligned} \delta_H y(t) &= (-a(t)) \cdot y^{\sigma}(t) + f(t,y(t)), \quad t \in J_0, \\ y(0) &= \widetilde{0}. \end{aligned}$$

The last problem has a unique solution such that

$$y(t) = \int_0^t e_{\ominus a}(t,s) \cdot f(s,y(s))\delta_H s, \quad t \in J_0.$$

Then

$$y(t_1^-) = y(t_1)$$

$$= \int_0^{t_1} e_{\ominus a}(t_1, s) \cdot f(s, y(s)) \delta_H s,$$

and

$$y(t_1^+) = \int_0^{t_1} e_{\ominus a}(t_1, s) \cdot f(s, y(s)) \delta_H s + I(t_1, y(t_1)).$$

To find a solution of the equation (3.11) *for* $t \in J_1$*, consider the following the initial value problem*

$$\begin{aligned} \delta_H y(t) &= (-a(t)) \cdot y^\sigma(t) + f(t, y(t)), \quad t \in J_1, \\ y(t_1^+) &= \int_0^{t_1} e_{\ominus a}(t_1, s) \cdot f(s, y(s)) \delta_H s + I(t_1, y(t_1)). \end{aligned}$$

For the solution of the last problem, we get

$$\begin{aligned} y(t) &= e_{\ominus a}(t, t_1) \cdot y(t_1^+) + \int_{t_1^+}^{t} e_{\ominus a}(t, s) \cdot f(s, y(s)) \delta_H s \\ &= e_{\ominus a}(t, t_1) \cdot \left(\int_0^{t_1} e_{\ominus a}(t_1, s) \cdot f(s, y(s)) \delta_H s + I(t_1, y(t_1)) \right) \\ &\quad + \int_{t_1^+}^{t} e_{\ominus a}(t, s) \cdot f(s, y(s)) \delta_H s \end{aligned}$$

$$= e_{\ominus a}(t, t_1) \cdot I(t_1, y(t_1)) + \int_0^{t_1} (e_{\ominus a}(t, t_1) e_{\ominus a}(t_1, s)) \cdot f(s, y(s)) \delta_H s$$

$$+ \int_{t_1^+}^{t} e_{\ominus a}(t, s) \cdot f(s, y(s)) \delta_H s$$

$$= e_{\ominus a}(t,t_1)\cdot I(t_1,y(t_1)) + \int_0^t e_{\ominus a}(t,s)\cdot f(s,y(s))\delta_H s, \quad t \in J_1.$$

We proceed inductively to obtain that if $y \in PC^1$ is a solution of (3.11), *then it satisfies* (3.12).

2. *Let $y \in PC$ satisfies* (3.12). *Then $y \in PC^1$. For $t \in J_0$, we have*

$$y(t) = \int_0^t e_{\ominus a}(t,s)\cdot f(s,y(s))\delta_H s.$$

We δ_H-differentiate the last equation, and we find

$$\delta_H y(t)$$

$$= e_{\ominus a}(\sigma(t),t)\cdot f(t,y(t)) + (\ominus a)(t)\cdot \int_0^t e_{\ominus a}(t,s)\cdot f(s,y(s))\delta_H s$$

$$= (e_{\ominus a}(t,t) + \mu(t)(\ominus a)(t)e_{\ominus a}(t,t))\cdot f(t,y(t))$$

$$+ (\ominus a)(t)\cdot \int_0^t e_{\ominus a}(t,s)\cdot f(s,y(s))\delta_H s$$

$$= \left(1 - \frac{\mu(t)a(t)}{1+\mu(t)a(t)}\right)\cdot f(t,y(t))$$

$$+ (\ominus a)(t)\cdot \int_0^t e_{\ominus a}(t,s)\cdot f(s,y(s))\delta_H s$$

$$= \frac{1}{1+\mu(t)a(t)} \cdot f(t,y(t)) + (\ominus a)(t) \cdot \int_0^t e_{\ominus a}(t,s) \cdot f(s,y(s))\delta_H s,$$

$t \in J_0$. *Moreover,*

$$y^{\sigma}(t) = \int_0^{\sigma(t)} e_{\ominus a}(\sigma(t),s) \cdot f(s,y(s))\delta_H s$$

$$= \int_0^t e_{\ominus a}(\sigma(t),s) \cdot f(s,y(s))\delta_H s + \int_t^{\sigma(t)} e_{\ominus a}(\sigma(t),s) \cdot f(s,y(s))\delta_H s$$

$$= \int_0^t (e_{\ominus a}(t,s) + \mu(t)(\ominus a)(t)e_{\ominus a}(t,s)) \cdot f(s,y(s))\delta_H s$$

$$+ (\mu(t)e_{\ominus a}(\sigma(t),t)) \cdot f(t,y(t))$$

$$= \int_0^t \left(\left(1 - \frac{\mu(t)a(t)}{1+\mu(t)a(t)}\right) e_{\ominus a}(t,s)\right) \cdot f(s,y(s))\delta_H s$$

$$+ (\mu(t)\,(e_{\ominus a}(t,t) + \mu(t)(\ominus a)(t)e_{\ominus a}(t,t))) \cdot f(t,y(t))$$

$$= \frac{1}{1+\mu(t)a(t)} \cdot \int_0^t e_{\ominus a}(t,s) \cdot f(s,y(s))\delta_H s$$

$$+ \left(\mu(t)\left(1 - \frac{\mu(t)a(t)}{1+\mu(t)a(t)}\right)\right) \cdot f(t,y(t))$$

$$= \frac{1}{1+\mu(t)a(t)} \cdot \int_0^t e_{\ominus a}(t,s) \cdot f(s,y(s))\delta_H s$$

$$+ \frac{\mu(t)}{1+\mu(t)a(t)} \cdot f(t,y(t)), \quad t \in J_0.$$

Then

$$(-a(t)) \cdot hy^{\sigma}(t) + f(t,y(t))$$

$$= \left(-\frac{a(t)}{1+\mu(t)a(t)}\right) \cdot \int_0^t e_{\ominus a}(t,s) \cdot f(s,y(s))\delta_H s$$

$$+ \left(-\frac{\mu(t)a(t)}{1+\mu(t)a(t)}\right) \cdot f(t,y(t)) + f(t,y(t))$$

$$= (\ominus a)(t) \cdot \int_0^t e_{\ominus a}(t,s) \cdot f(s,y(s))\delta_H s$$

$$+ \left(1 - \frac{\mu(t)a(t)}{1+\mu(t)a(t)}\right) \cdot f(t,y(t))$$

$$= \frac{1}{1+\mu(t)a(t)} \cdot f(t,y(t)) + (\ominus a)(t) \cdot \int_0^t e_{\ominus a}(t,s) \cdot f(s,y(s))\delta_H s$$

$$= \delta_H y(t), \quad t \in J_0,$$

and

$$y(t_1^-) = y(t_1)$$

$$= \int_0^{t_1} e_{\ominus a}(t_1,s)\cdot f(s,y(s))\delta_H s.$$

Hence, we get the following initial value problem

$$\begin{aligned} \delta_H y(t) &= (-a(t))\cdot y^{\sigma}(t) + f(t,y(t)), \quad t\in J_0, \\ y(0) &= 0. \end{aligned}$$

For $t \in J_1$, we have that y satisfies

$$y(t) = \int_0^t e_{\ominus a}(t,s)\cdot f(s,y(s))\delta_H s + e_{\ominus a}(t,t_1)\cdot I(t_1,y(t_1)).$$

We δ_H-differentiate the last equation, and we find

$$\delta_H y(t)$$

$$\begin{aligned} &= e_{\ominus a}(\sigma(t),t)\cdot f(t,y(t)) + (\ominus a)(t)\cdot \int_0^t e_{\ominus a}(t,s)\cdot f(s,y(s))\delta_H s \\ &\quad + ((\ominus a)(t) e_{\ominus a}(t,t_1))\cdot I(t_1,y(t_1)) \\ &= (e_{\ominus a}(t,t) + \mu(t)(\ominus a)(t) e_{\ominus a}(t,t))\cdot f(t,y(t)) \\ &\quad + (\ominus a)(t)\cdot \int_0^t e_{\ominus a}(t,s)\cdot f(s,y(s))\delta_H s \\ &\quad + ((\ominus a)(t) e_{\ominus a}(t,t_1))\cdot I(t_1,y(t_1)) \\ &= \left(1 - \frac{\mu(t)a(t)}{1+\mu(t)a(t)}\right)\cdot f(t,y(t)) \end{aligned}$$

$$+(\ominus a)(t)\cdot\int_0^t e_{\ominus a}(t,s)\cdot f(s,y(s))\delta_H s$$

$$+((\ominus a)(t)e_{\ominus a}(t,t_1))\cdot I(t_1,y(t_1))$$

$$= \frac{1}{1+\mu(t)a(t)}\cdot f(t,y(t))$$

$$+(\ominus a)(t)\cdot\int_0^t e_{\ominus a}(t,s)\cdot f(s,y(s))\delta_H s$$

$$+((\ominus a)(t)e_{\ominus a}(t,t_1))\cdot I(t_1,y(t_1)),$$

$t \in J_1$. *Next,*

$$y^\sigma(t) = \int_0^{\sigma(t)} e_{\ominus a}(\sigma(t),s)\cdot f(s,y(s))\delta_H s$$

$$+e_{\ominus a}(\sigma(t),t_1)\cdot I(t_1,y(t_1))$$

$$= \int_0^t e_{\ominus a}(\sigma(t),s)\cdot f(s,y(s))\delta_H s$$

$$+\int_t^{\sigma(t)} e_{\ominus a}(\sigma(t),s)\cdot f(s,y(s))\delta_H s$$

$$+(e_{\ominus a}(t,t_1)+\mu(t)(\ominus a)(t)e_{\ominus a}(t,t_1))\cdot I(t_1,y(t_1))$$

$$= \int_0^t (e_{\ominus a}(t,s)+\mu(t)(\ominus a)(t)e_{\ominus a}(t,s))\cdot f(s,y(s))\delta_H s$$

$$+(\mu(t)e_{\ominus a}(\sigma(t),t))\cdot f(t,y(t))$$

$$+\left(\left(1-\frac{\mu(t)a(t)}{1+\mu(t)a(t)}\right)e_{\ominus a}(t,t_1)\right)\cdot I(t_1,y(t_1))$$

$$= \int_0^t \left(\left(1-\frac{\mu(t)a(t)}{1+\mu(t)a(t)}\right)e_{\ominus a}(t,s)\right)\cdot f(s,y(s))\delta_H s$$

$$+(\mu(t)\left(e_{\ominus a}(t,t)+\mu(t)(\ominus a)(t)e_{\ominus a}(t,t)\right))\cdot f(t,y(t))$$

$$+\left(\frac{1}{1+\mu(t)a(t)}e_{\ominus a}(t,t_1)\right)\cdot I(t_1,y(t_1))$$

$$= \frac{1}{1+\mu(t)a(t)}\cdot\int_0^t e_{\ominus a}(t,s)\cdot f(s,y(s))\delta_H s$$

$$+\left(\mu(t)\left(1-\frac{\mu(t)a(t)}{1+\mu(t)a(t)}\right)\right)\cdot f(t,y(t))$$

$$+\left(\frac{1}{1+\mu(t)a(t)}e_{\ominus a}(t,t_1)\right)\cdot I(t_1,y(t_1))$$

$$= \frac{1}{1+\mu(t)a(t)}\cdot\int_0^t e_{\ominus a}(t,s)\cdot f(s,y(s))\delta_H s$$

$$+\frac{\mu(t)}{1+\mu(t)a(t)}\cdot f(t,y(t))$$

$$+\left(\frac{1}{1+\mu(t)a(t)}e_{\ominus a}(t,t_1)\right)\cdot I(t_1,y(t_1)),\quad t\in J_1.$$

Then

$$(-a(t))\cdot_h y^\sigma(t)+f(t,y(t))$$

$$= \left(-\frac{a(t)}{1+\mu(t)a(t)}\right)\cdot\int_0^t e_{\ominus a}(t,s)\cdot f(s,y(s))\delta_H s$$

$$
\begin{aligned}
& + \left(- \frac{\mu(t)a(t)}{1+\mu(t)a(t)} \right) \cdot f(t,y(t)) + f(t,y(t)) \\
& + \left(- \frac{a(t)}{1+\mu(t)a(t)} e_{\ominus a}(t,t_1) \right) \cdot I(t_1,y(t_1)) \\
= \;& (\ominus a)(t) \cdot \int_0^t e_{\ominus a}(t,s) \cdot f(s,y(s)) \delta_H s \\
& + \left(1 - \frac{\mu(t)a(t)}{1+\mu(t)a(t)} \right) \cdot f(t,y(t)) \\
& + ((\ominus a)(t) e_{\ominus a}(t,t_1)) \cdot I(t_1,y(t_1)) \\
= \;& \frac{1}{1+\mu(t)a(t)} \cdot f(t,y(t)) \\
& + (\ominus a)(t) \cdot \int_0^t e_{\ominus a}(t,s) \cdot f(s,y(s)) \delta_H s \\
& + ((\ominus a)(t) e_{\ominus a}(t,t_1)) \cdot I(t_1,y(t_1)) \\
= \;& \delta_H y(t), \quad t \in J_1,
\end{aligned}
$$

and

$$
\begin{aligned}
y(t_1^+) &= \int_0^{t_1} e_{\ominus a}(t_1,s) \cdot f(s,y(s)) \delta_H s + e_{\ominus a}(t_1,t_1) \cdot I(t_1,y(t_1)) \\
&= y(t_1^-) + I(t_1,y(t_1)).
\end{aligned}
$$

Continuing inductively, we get that y satisfies (3.11). *This completes the proof.*

For $\phi \in PC$, define the following operators

$$\begin{aligned}(A\phi)(t) &= \int_0^t e_{\ominus a}(t,s)\cdot f(s,\phi(s))\delta_H s,\\ (B\phi)(t) &= \sum_{\{i:t_i<t\}} e_{\ominus a}(t,t_i)\cdot I(t_i,\phi(t_i)),\\ (H\phi)(t) &= (A\phi)(t)+(B\phi)(t), \quad t\in[0,T].\end{aligned}$$

Lemma 3.2.3 *Suppose $(H1)$–$(H3)$. Then $A: PC \to PC$ is completely continuous.*

Proof 3.2.4 *Firstly, note that $A: PC \to PC$. Let $\phi \in PC$ be arbitrarily chosen. Then*

$$\begin{aligned}D(A\phi(t),\widetilde{0}) &= D\left(\int_0^t e_{\ominus a}(t,s)\cdot f(s,\phi(s))\delta_H s,\widetilde{0}\right)\\ &\le \int_0^t |e_{\ominus a}(t,s)|D(f(s,\phi(s)),\widetilde{0})\Delta s\\ &\le \int_0^t |e_{\ominus a}(t,s)|\left(g(s)+h(s)D(\phi(s),\widetilde{0})\right)\Delta s\\ &\le \int_0^t |e_{\ominus a}(t,s)|g(s)\Delta s+\int_0^t |e_{\ominus a}(t,s)|h(s)D(\phi(s),\widetilde{0})\Delta s\\ &\le \alpha+\left(\int_0^t |e_{\ominus a}(t,s)|h(s)\Delta s\right)\|\phi\|\\ &\le \alpha+\beta\|\phi\|, \quad t\in[0,T].\end{aligned}$$

Hence,

$$\|A\phi\| \le \alpha+\beta\|\phi\|.$$

Now, take $\{\phi_n\}_{n\in\mathbb{N}} \subset PC$ *such that*

$$\phi_n \to \phi, \quad as \quad n \to \infty,$$

for some $\phi \in PC$. *Then, using* $(H2)$, *we get*

$$D(A\phi_n(t), A\phi(t))$$

$$= D\Big(\int_0^t e_{\ominus a}(t,s)\cdot f(s,\phi_n(s))\delta_H s, \int_0^t e_{\ominus a}(t,s)\cdot f(s,\phi(s))\delta_H s\Big)$$

$$\leq \int_0^t |e_{\ominus a}(t,s)| D(f(s,\phi_n(s)), f(s,\phi(s)))\Delta s$$

$$\to 0, \quad as \quad n \to \infty.$$

Thus, $A : PC \to PC$ *is continuous. Next, we take* $\phi \in PC$ *arbitrarily, and* $t^1, t^2 \in [0,T]$ *arbitrarily. Without loss of generality, suppose that* $t^1 < t^2$. *Then*

$$D(A\phi(t^2), A\phi(t^1))$$

$$= D\Big(\int_0^{t^2} e_{\ominus a}(t^2,s)\cdot f(s,\phi(s))\delta_H s, \int_0^{t^1} e_{\ominus a}(t^1,s)\cdot f(s,\phi(s))\delta_H s\Big)$$

$$= D\Big(\int_0^{t^1} e_{\ominus a}(t^2,s)\cdot f(s,\phi(s))\delta_H s + \int_{t^1}^{t^2} e_{\ominus a}(t^2,s)\cdot f(s,\phi(s))\delta_H s,$$

$$\int_0^{t^1} e_{\ominus a}(t^1,s)\cdot f(s,\phi(s))\delta_H s\Big)$$

$$\leq \int_0^{t^1} \Big|e_{\ominus a}(t^2,s) - e_{\ominus a}(t^1,s)\Big| D(f(s,\phi(s)), \widetilde{0})\Delta s$$

$$
\begin{aligned}
&+\int_{t^1}^{t^2} |e_{\ominus a}(t^2,s)| D(f(s,\phi(s)),\widetilde{0})\Delta s \\
&\le \left|e_{\ominus a}(t^2,t^1)-1\right| \int_0^{t^1} \left|e_{\ominus a}(t^1,s)\right| D(f(s,\phi(s)),\widetilde{0})\Delta s \\
&+\int_{t^1}^{t^2} |e_{\ominus a}(t^2,s)| D(f(s,\phi(s)),\widetilde{0})\Delta s \\
&\to 0, \quad as \quad t^1 \to t^2.
\end{aligned}
$$

From this and from the Arzela-Ascoli theorem, it follows that $A : PC \to PC$ is completely continuous. This completes the proof.

Lemma 3.2.5 *Let $(H1)$ and $(H4)$ hold. Let also,*

$$
E \max_{t\in[0,T]} \sum_{i=1}^{n} |e_{\ominus a}(t,t_i)| \le \zeta < 1. \tag{3.13}
$$

Then $B : PC \to PC$ is a contraction.

Proof 3.2.6 *Since $e_{\ominus a}(t,t_i)$ is continuous for any $i \in \{1,\ldots,n\}$, we conclude that $B : PC \to PC$. Let $\phi, \psi \in PC$ be arbitrarily chosen. Then*

$$
\begin{aligned}
D(B\phi(t), B\psi(t)) &= D\Bigg(\sum_{\{i:t_i<t\}} e_{\ominus a}(t,t_i)\cdot I(t_i,\phi(t_i)), \\
&\qquad \sum_{\{i:t_i<t\}} e_{\ominus a}(t,t_i)\cdot I(t_i,\psi(t_i))\Bigg) \\
&\le \sum_{\{i:t_i<t\}} |e_{\ominus a}(t,t_i)| D(I(t_i,\phi(t_i)), I(t_i,\psi(t_i))) \\
&\le E \sum_{\{i:t_i<t\}} |e_{\ominus a}(t,t_i)| D(\phi(t_i),\psi(t_i))
\end{aligned}
$$

$$\leq E\left(\max_{t\in[0,T]}\sum_{i=1}^{n}|e_{\ominus a}(t,t_i)|\right)\|\phi-\psi\|$$

$$\leq \zeta\|\phi-\psi\|, \quad t\in[0,T].$$

From here,

$$\|B\phi-B\psi\|\leq\zeta\|\phi-\psi\|.$$

Therefore, $B:PC\to PC$ is a contraction. This completes the proof.

Theorem 3.2.7 *Assume that*

$$\eta=\max_{t\in[0,T]}\sum_{i=1}^{n}|e_{\ominus a}(t,t_i)|D(I(t_i,0),\widetilde{0})<\infty. \tag{3.14}$$

Let $(H1)$-$(H4)$ and (3.13) hold, and J be a positive constant such that

$$\alpha+\eta+\left(\beta+E\max_{t\in[0,T]}\sum_{i=1}^{n}|e_{\ominus a}(t,t_i)|\right)J\leq J. \tag{3.15}$$

Then problem (3.11) has a solution ϕ such that

$$\|\phi\|\leq J.$$

Proof 3.2.8 *Define*

$$M=\{\phi\in PC:\|\phi\|\leq J\}.$$

By Lemma 3.2.3, it follows that $A:PC\to PC$ is completely continuous. From Lemma 3.2.5, it follows that $B:PC\to PC$ is a contraction. Let $\phi,\psi\in M$ be arbitrarily chosen. Then

$$\|\phi\| \leq J,$$

$$\|\psi\| \leq J,$$

and

$$D(A\phi(t)+B\psi(t),\widetilde{0})$$

$$
\begin{aligned}
&= D\left(\int_0^t e_{\ominus a}(t,s)\cdot f(s,\phi(s))\delta_H s + \sum_{i=1}^{n} e_{\ominus a}(t,t_i)\cdot I(t_i,\psi(t_i)),\widetilde{0}\right)\\
&\le D\left(\int_0^t e_{\ominus a}(t,s)\cdot f(s,\phi(s))\delta_H s,\widetilde{0}\right)\\
&\quad +D\left(\sum_{i=1}^{n} e_{\ominus a}(t,t_i)\cdot I(t_i,\psi(t_i)),\widetilde{0}\right)\\
&\le \int_0^t |e_{\ominus a}(t,s)|D(f(s,\phi(s)),\widetilde{0})\Delta s\\
&\quad +\sum_{j=1}^{n} |e_{\ominus a}(t,t_j)|D(I(t_j,\phi(t_j)),\widetilde{0})\\
&= \int_0^t |e_{\ominus a}(t,s)|D(f(s,\phi(s)),\widetilde{0})\Delta s\\
&\quad +\sum_{i=1}^{n} |e_{\ominus a}(t,t_i)|D(I(t_i,0)+I(t_i,\psi(t_i)),I(t_i,0))\\
&\le \int_0^t |e_{\ominus a}(t,s)|\left(g(s)+h(s)D(\phi(s),\widetilde{0})\right)\Delta s\\
&\quad +\sum_{i=1}^{n} |e_{\ominus a}(t,t_i)|D(I(t_i,0),\widetilde{0})\\
&\quad +\sum_{i=1}^{n} |e_{\ominus a}(t,t_i)|D(I(t_i,\psi(t_i)),I(t_i,0))\\
&= \int_0^t |e_{\ominus a}(t,s)|g(s)\Delta s + \left(\int_0^t h(s)\Delta s\right)\|\phi\|
\end{aligned}
$$

$$
\begin{aligned}
&+\sum_{i=1}^{n} |e_{\ominus a}(t,t_i)| D(I(t_i,0),\widetilde{0}) \\
&+\sum_{i=1}^{n} |e_{\ominus a}(t,t_i)| D(I(t_i,\psi(t_i)),I(t_i,0)) \\
\leq\ & \alpha+\eta+\beta\|\phi\| \\
&+E\sum_{i=1}^{n} |e_{\ominus a}(t,t_i)| D(\psi(t_i),\widetilde{0}) \\
\leq\ & \alpha+\eta+\beta J \\
&+E\left(\max_{t\in[0,T]}\sum_{i=1}^{n} |e_{\ominus a}(t,t_i)|\right)\|\psi\| \\
\leq\ & \alpha+\eta+\left(\beta+E\left(\max_{t\in[0,T]}\sum_{i=1}^{n} |e_{\ominus a}(t,t_i)|\right)\right)J \\
\leq\ & J, \quad t\in[0,T].
\end{aligned}
$$

Hence,

$$\|A\phi+B\psi\|\leq J.$$

Therefore, $A\phi+B\psi\in M$. *Applying the Krasnosel'skii fixed-point theorem, it follows that the operator* $A+B$ *has a fixed point in* M. *From here and from Theorem 3.2.1, we find that equation* (3.11) *has a solution in* M. *This completes the proof.*

Theorem 3.2.9 *Assume that* $(H1)$–$(H4)$, (3.13), (3.14), *and* (3.15) *hold, and*

$$\beta+\zeta<1. \tag{3.16}$$

Then problem (3.11) *has a unique solution.*

Proof 3.2.10 *Since* $(H1)$–$(H4)$ *hold,* (3.13), (3.14), *and* (3.15) *hold, by Theorem 3.2.7, it follows that equation* (3.11) *has a solution in* M.

Assume that equation (3.11) *has two solutions* $\phi, \psi \in M$. *Then*

$$D(H\phi(t), H\psi(t))$$

$$\begin{aligned}
&= D\Bigg(\int_0^t e_{\ominus a}(t,s)\cdot f(s,\phi(s))\delta_H s + \sum_{\{i:t_i<t\}} e_{\ominus a}(t,t_i)\cdot I(t_i,\phi(t_i)),\\
&\qquad \int_0^t e_{\ominus a}(t,s)\cdot f(s,\psi(s))\delta_H s + \sum_{\{i:t_i<t\}} e_{\ominus a}(t,t_i)\cdot I(t_i,\psi(t_i))\Bigg)\\
&\leq D\Bigg(\int_0^t e_{\ominus a}(t,s)\cdot f(s,\phi(s))\delta_H s, \int_0^t e_{\ominus a}(t,s)\cdot f(s,\psi(s))\delta_H s\Bigg)\\
&\quad + D\Bigg(\sum_{\{i:t_i<t\}} e_{\ominus a}(t,t_i)\cdot I(t_i,\phi(t_i)), \sum_{\{i:t_i<t\}} e_{\ominus a}(t,t_i)\cdot I(t_i,\psi(t_i))\Bigg)\\
&\leq \int_0^t |e_{\ominus a}(t,s)| D(f(s,\phi(s)), f(s,\psi(s)))\Delta s\\
&\quad + \sum_{\{i:t_i<t\}} |e_{\ominus a}(t,t_i)| D(I(t_i,\phi(t_i)), I(t_i,\psi(t_i)))\\
&\leq \int_0^t |e_{\ominus a}(t,s)| h(s) D(\phi(s),\psi(s))\Delta s\\
&\quad + E\sum_{\{i:t_i<t\}} |e_{\ominus a}(t,t_i)| D(\phi(t_i),\psi(t_i))\\
&\leq \left(\int_0^t |e_{\ominus a}(t,s)| h(s)\Delta s\right)\|\phi-\psi\|\\
&\quad + \left(E\sum_{\{i:t_i<t\}} |e_{\ominus a}(t,t_i)|\right)\|\phi-\psi\|
\end{aligned}$$

$$\leq (\beta+\zeta)\|\phi-\psi\|, \quad t\in[0,T].$$

Hence,

$$\|H\phi - H\psi\| \leq (\beta+\zeta)\|\phi-\psi\|.$$

Then

$$\begin{aligned}\|\phi-\psi\| &= \|H\phi - H\psi\| \\ &\leq (\beta+\zeta)\|\phi-\psi\| \\ &< \|\phi-\psi\|,\end{aligned}$$

whereupon $\phi\equiv\psi$. *This completes the proof.*

Exercise 3.2.11 *Find conditions for the function* $f\in\mathscr{C}(\mathbb{T}\times\mathbb{R})$ *so that the equation*

$$\begin{aligned}\delta_H y(t) &= f(t,y(t)), \quad t\in(0,T], \\ y(0) &= 0, \\ y(t_i^+) &= y(t_i^-)+I(t_i,y(t_i)), \quad i\in\{1,\ldots,n\},\end{aligned}$$

has a unique solution in M.

Hint. Take a function $a\in\mathscr{R}$ such that

$$1-\mu(t)a(t)\neq 0, \quad t\in\mathbb{T}.$$

Then represent the equation in the following form

$$\begin{aligned}\delta_H y(t) &= f(t,y(t)) \\ &= (-a(t))\cdot y^{\sigma}(t)+a(t)\cdot y^{\sigma}(t)+f(t,y(t)) \\ &= (-a(t))\cdot y^{\sigma}(t)+a(t)y(t)+(\mu(t)a(t))\cdot\delta_H y(t)+f(t,y(t)),\end{aligned}$$

$t \in (0,T]$, whereupon

$$(1-\mu(t)a(t))\cdot\delta_H y(t) = (-a(t))\cdot y^\sigma(t) + a(t)y(t) + f(t,y(t)),$$

$t \in (0,T]$, or

$$\begin{aligned}\delta_H y(t) &= \left(-\frac{a(t)}{1-\mu(t)a(t)}\right)\cdot y^\sigma(t) + \left(\frac{a(t)}{1-\mu(t)a(t)}\right)\cdot y(t)\\ &\quad + \left(\frac{1}{1-\mu(t)a(t)}\right)\cdot f(t,y(t)),\end{aligned}$$

$t \in [0,T]$. Then find conditions for the functions

$$\begin{aligned}\tilde{a}(t) &= -\frac{a(t)}{1-\mu(t)a(t)},\\ \tilde{f}(t,y(t)) &= \left(\frac{a(t)}{1-\mu(t)a(t)}\right)\cdot y(t) + \left(\frac{1}{1-\mu(t)a(t)}\right)\cdot f(t,y(t)),\end{aligned}$$

$t \in \mathbb{T}$, so that $(H1)$–$(H4)$, (3.13), (3.14), (3.15), and (3.16) hold. Assume

(H5) There exist non-decreasing functions $l_k : \mathbb{R}_+ \to \mathbb{R}_+$ such that

$$D(I_k(t_k, y(t_k)), \widetilde{0}) \le l_k(D(y(t_k), \widetilde{0})), \quad k \in \{0,\dots,n\},$$

and

$$\lim_{r\to\infty} \frac{l_k(r)}{r} = 0, \quad k \in \{0,\dots,n\},$$

(H6) For each $r > 0$ there exists a continuous function $h_r : [0,T] \to (0,\infty)$ such that

$$D(f(t,y), \widetilde{0}) \le h_k(t) \quad \text{for each} \quad (t,y) \in [0,T]\times F(\mathbb{R}),$$
$$D(y, \widetilde{0}) \le r,$$

and

$$\liminf_{r\to\infty} \frac{1}{r}\int_0^T h_r(s)\Delta s = \beta_1 < \infty.$$

Theorem 3.2.12 *Suppose* $(H1)$, $(H2)$, $(H4)$–$(H6)$ *and*

$$\beta_1 \max_{(t,\tau)\in[0,T]\times[0,T]} |e_{\ominus a}(t,\tau)| < 1. \tag{3.17}$$

Then (3.11) *has at least one solution.*

Proof 3.2.13 *We have* $H : PC \to PC$.

1. *We will prove that H is continuous. Let* $\{y_m\}_{m\in\mathbb{N}}$ *be a sequence of elements of PC such that*

$$y_m \to y, \quad as \quad m \to \infty, \quad y \in PC.$$

Then

$$\begin{aligned}
& D(Hy_m(t), Hy(t)) \\
= \; & D\Bigg(\int_0^t e_{\ominus a}(t,\tau)\cdot f(\tau, y_m(\tau))\,\delta_H \tau \\
& + \sum_{\{k:t_k<t\}} e_{\ominus a}(t,t_k)\cdot I(t_k, y_m(t_k)), \\
& \int_0^t e_{\ominus a}(t,\tau)\cdot f(\tau, y(\tau))\,\delta_H \tau \\
& + \sum_{\{k:t_k<t\}} e_{\ominus a}(t,t_k)\cdot I(t_k, y(t_k))\Bigg) \\
\le \; & D\Bigg(\int_0^t e_{\ominus a}(t,\tau)\cdot f(\tau, y_m(\tau))\,\delta_H \tau, \\
& \int_0^t e_{\ominus a}(t,\tau)\cdot f(\tau, y(\tau))\,\delta_H \tau\Bigg) \\
+ \; & D\Bigg(\sum_{\{k:t_k<t\}} e_{\ominus a}(t,t_k)\cdot I(t_k, y_m(t_k)),
\end{aligned}$$

$$\sum_{\{k:t_k<t\}} e_{\ominus a}(t,t_k)\cdot I(t_k,y(t_k))\Big)$$

$$\leq \int_0^t |e_{\ominus a}(t,\tau)| D(f(\tau,y_m(\tau)),f(\tau,y(\tau)))\Delta\tau$$

$$+ \sum_{\{k:t_k<t\}} |e_{\ominus a}(t,t_k)| D(I(t_k,y_m(t_k)),I(t_k,y(t_k)))$$

$$\leq \max_{(t,\tau)\in[0,T]\times[0,T]} |e_{\ominus a}(t,\tau)| \int_0^t D(f(\tau,y_m(\tau)),f(\tau,y(\tau)))\Delta\tau$$

$$+E \sum_{\{k:t_k<t\}} \max_{t\in[0,T]} |e_{\ominus a}(t,t_k)| D(y_m(t_k),y(t_k))$$

$$\to 0, \quad as \quad m\to\infty, \quad t\in[0,T].$$

Hence,

$$\|Hy_m - Hy\| \to 0, \quad as \quad m\to\infty.$$

2. *We will prove that there exists a number $r>0$ such that $H(B_r)\subset B_r$. Suppose the contrary, i.e., for each $r>0$ there exists a function $y_r\in B_r$ such that*

$$D(H(y_r(t)),\widetilde{0}) > r \quad for \quad some \quad t\in[0,T].$$

Then

$$r < D(H(y_r(t)),\widetilde{0})$$

$$= D\Big(\int_0^t e_{\ominus a}(t,\tau)\cdot f(\tau,y_r(\tau))\delta_H\tau$$

$$+ \sum_{\{k:t_k<t\}} e_{\ominus a}(t,t_k)\cdot I(t_k,y_r(t_k)),\widetilde{0}\Big)$$

$$
\begin{aligned}
&\leq D\left(\int_0^t e_{\ominus a}(t,\tau)\cdot f(\tau,y_r(\tau))\delta_H\tau,\widetilde{0}\right)\\
&\quad +D\left(\sum_{\{k:t_k<t\}} e_{\ominus a}(t,t_k)\cdot I(t_k,y_r(t_k)),\widetilde{0}\right)\\
&\leq \int_0^t |e_{\ominus a}(t,\tau)|D(f(\tau,y_r(\tau)),\widetilde{0})\Delta\tau\\
&\quad +\sum_{\{k:t_k<t\}} |e_{\ominus a}(t,t_k)|D(I(t_k,y_r(t_k)),\widetilde{0})\\
&\leq \max_{(t,\tau)\in[0,T]\times[0,T]} |e_{\ominus a}(t,\tau)|\int_0^t h_r(s)\Delta s\\
&\quad +\sum_{\{k:t_k<t\}} \max_{t\in[0,T]} |e_{\ominus a}(t,t_k)|l_k(r),
\end{aligned}
$$

whereupon

$$
\begin{aligned}
1 &< \max_{(t,\tau)\in[0,T]\times[0,T]} |e_{\ominus a}(t,\tau)|\frac{1}{r}\int_0^t h_r(s)\Delta s\\
&\quad +\sum_{\{k:t_k<t\}} \max_{t\in[0,T]} |e_{\ominus a}(t,t_k)|\frac{l_k(r)}{r}
\end{aligned}
$$

and taking limits as $r\to\infty$, *we find*

$$
\begin{aligned}
1 &\leq \max_{(t,\tau)\in[0,T]\times[0,T]} |e_{\ominus a}(t,\tau)|\lim_{r\to\infty}\frac{1}{r}\int_0^t h_r(s)\Delta s\\
&\quad +\sum_{\{k:t_k<t\}} \max_{t\in[0,T]} |e_{\ominus a}(t,t_k)|\lim_{r\to\infty}\frac{l_k(r)}{r}\\
&= \beta_1 \max_{(t,\tau)\in[0,T]\times[0,T]} |e_{\ominus a}(t,\tau)|,
\end{aligned}
$$

which is a contradiction. Consequently, there is a $r_1 > 0$ such that

$$H(B_{r_1}) \subset B_{r_1}.$$

3. *We will show that $H(B_{r_1})$ is an equicontinous family of functions. Let $t^1, t^2 \in [0,T]$. Without loss of generality, suppose that $t^1 < t^2$. Then*

$$\begin{aligned}
& D(H(y)(t^2), H(y)(t^1)) \\
= \; & D\Bigg(\int_0^{t^2} e_{\ominus a}(t^2,\tau) \cdot f(\tau, y(\tau))\delta_H \tau \\
& + \sum_{\{k: t_k < t^2\}} e_{\ominus a}(t^2, t_k) \cdot I(t_k, y(t_k)), \\
& \int_0^{t^1} e_{\ominus a}(t^1,\tau) \cdot f(\tau, y(\tau))\delta_H \tau \\
& + \sum_{\{k: t_k < t^1\}} e_{\ominus a}(t^1, t_k) \cdot I(t_k, y(t_k)) \Bigg) \\
\leq \; & D\Bigg(\int_0^{t^2} e_{\ominus a}(t^2,\tau) \cdot f(\tau, y(\tau))\delta_H \tau, \\
& \int_0^{t^1} e_{\ominus a}(t^1,\tau) \cdot f(\tau, y(\tau))\delta_H \tau \Bigg) \\
+ \; & D\Bigg(\sum_{\{k: t_k < t^2\}} e_{\ominus a}(t^2, t_k) \cdot I(t_k, y(t_k)), \\
& \sum_{\{k: t_k < t^1\}} e_{\ominus a}(t^1, t_k) \cdot I(t_k, y(t_k)) \Bigg)
\end{aligned}$$

$$
\begin{aligned}
&= D\Big(\int_0^{t^1} e_{\ominus a}(t^2,\tau)\cdot f(\tau,y(\tau))\delta_H\tau \\
&\quad + \int_{t^1}^{t^2} e_{\ominus a}(t^2,\tau)\cdot f(\tau,y(\tau))\delta_H\tau, \\
&\quad \int_0^{t^1} e_{\ominus a}(t^1,\tau)\cdot f(\tau,y(\tau))\delta_H\tau\Big) \\
&\quad + D\Big(\sum_{\{k:t_k<t^1\}} e_{\ominus a}(t^2,t_k)\cdot I(t_k,y(t_k)) \\
&\quad + \sum_{\{k:t^1\le t_k<t^2\}} e_{\ominus a}(t^2,t_k)\cdot I(t_k,y(t_k)), \\
&\quad \sum_{\{k:t_k<t^1\}} e_{\ominus a}(t^1,t_k)\cdot I(t_k,y(t_k))\Big) \\
&= D\Big(\int_0^{t^1}\Big(e_{\ominus a}(t^2,\tau)-e_{\ominus a}(t^1,\tau)\Big)\cdot f(\tau,y(\tau))\delta_H\tau \\
&\quad + \int_{t^1}^{t^2} e_{\ominus a}(t^2,\tau)\cdot f(\tau,y(\tau))\delta_H\tau,\widetilde{0}\Big) \\
&\quad + D\Big(\sum_{\{k:t_k<t^1\}}\Big(e_{\ominus a}(t^2,t_k)-e_{\ominus a}(t^1,t_k)\Big)\cdot I(t_k,y(t_k)) \\
&\quad + \sum_{\{k:t^1\le t_k<t^2\}} e_{\ominus a}(t^2,t_k)\cdot I(t_k,y(t_k)),\widetilde{0}\Big) \\
&\le \int_0^{t^1}\Big|e_{\ominus a}(t^2,\tau)-e_{\ominus a}(t^1,\tau)\Big| D(f(\tau,y(\tau)),\widetilde{0})\Delta\tau
\end{aligned}
$$

$$+\int_{t^1}^{t^2}\left|e_{\ominus a}(t^2,\tau)\right|D(f(\tau,y(\tau)),\widetilde{0})\Delta\tau$$

$$+\sum_{\{k:t_k<t^1\}}\left|e_{\ominus a}(t^2,t_k)-e_{\ominus a}(t^1,t_k)\right|D(I(t_k,y(t_k)),\widetilde{0})$$

$$+\sum_{\{k:t^1\leq t_k<t^2\}}\left|e_{\ominus a}(t^2,t_k)\right|D(I(t_k,y(t_k)),\widetilde{0})$$

$$\to 0, \quad as \quad t^1\to t^2.$$

Hence, and by the Arzela-Ascoli theorem, we conclude that H : $PC \to PC$ is completely continuous and it is condensing.

In view of Sadovskii's fixed-point theorem, it follows that H has a fixed point in PC, which is a solution of the problem (3.11). *This completes the proof.*

Theorem 3.2.14 *Suppose* $(H1)$, $(H2)$, $(H4)$, $(H6)$, (3.17) *and*

$$E\sum_{\{k:t_k<t\}}\max_{t\in[0,T]}|e_{\ominus a}(t,t_k)|<1.$$

Then problem (3.11) *has at least one solution.*

Proof 3.2.15 *As we have proved in Theorem 3.2.12, there exists a positive number r such that $H(B_r)\subseteq B_r$ and the operator A is a compact operator in B_r. Now, let $y,z\in PC$. Then*

$$\begin{aligned}D(By(t),Bz(t)) &= D\Bigg(\sum_{\{k:t_k<t\}}e_{\ominus a}(t,t_k)I(t_k,y(t_k)),\\ &\qquad \sum_{\{k:t_k<t\}}e_{\ominus a}(t,t_k)I(t_k,z(t_k))\Bigg)\\ &\leq \sum_{\{k:t_k<t\}}|e_{\ominus a}(t,t_k)|D(I(t_k,y(t_k)),I(t_k,z(t_k)))\end{aligned}$$

$$\leq E \sum_{\{k:t_k<t\}} \max_{t\in[0,T]} |e_{\ominus a}(t,t_k)| D(y(t_k),z(t_k))$$

$$\leq E \sum_{\{k:t_k<t\}} \max_{t\in[0,T]} |e_{\ominus a}(t,t_k)| \|y-z\|, \quad t\in[0,T].$$

Hence,

$$\begin{aligned} \|By-Bz\| &\leq E \sum_{\{k:t_k<t\}} \max_{t\in[0,T]} |e_{\ominus a}(t,t_k)| \|y-z\| \\ &< \|y-z\|. \end{aligned}$$

Consequently $H = A+B$ is a condensing operator on B_r. Hence and the Krasnoselskii fixed-point theorem, it follows that the problem (3.11) *has a solution in PC. This completes the proof.*

Exercise 3.2.16 *Consider the impulsive dynamic equation*

$$\begin{aligned} \delta_H y(t) &= (-a(t))\cdot y^{\sigma}(t) + f_1(t) + (f_2(t))^2, \quad t\in[0,T], \\ y(0) &= y_0, \\ y(t_k^+) - y(t_k^-) &= 5\cdot I_k(y(t_k^-)), \quad k\in\{1,\ldots,n\}. \end{aligned}$$

Write analogously to Theorem 3.2.7, Theorem 3.2.9, Theorem 3.2.12, and Theorem 3.2.14.

Exercise 3.2.17 *Consider the impulsive dynamic equation*

$$\delta_H y(t) = (-a(t))\cdot y^{\sigma}(t)y(t) + f(y(t)) + (-7)\cdot g(t), \quad t\in[0,T],$$

$$y(0) = y_0,$$

$$y(t_k^+) - y(t_k^-) = I_k(y(t_k^-)), \quad k\in\{1,\ldots,n\}.$$

Write analogously to Theorem 3.2.7, Theorem 3.2.9, Theorem 3.2.12, and Theorem 3.2.14.

Exercise 3.2.18 *Consider the impulsive dynamic equation*

$$\delta_H y(t) = (-a(t)) \cdot y^{\sigma}(t) + g(t) \cdot f(y(t)), \quad t \in [0,T],$$

$$y(0) = y_0,$$

$$y(t_k^+) = I_k(y(t_k^-)), \quad k \in \{1,\ldots,n\}.$$

Write analogously to Theorem 3.2.7, Theorem 3.2.9, Theorem 3.2.12, and Theorem 3.2.14.

3.3 STABILITY OF THE SOLUTIONS OF FIRST-ORDER FUZZY IMPULSIVE DYNAMIC EQUATIONS-I

Suppose that $\mathbb{T}$ is above unbounded. In this section, we investigate the equation (3.1), where $y_0 \in \mathbb{R}$, t_i, $i \in \{1,\ldots,n\}$, are as in Section 3.1, p satisfies $(G1)$ and

(G7) $e_p(t,0) \to 0$, as $t \to \infty$,

f satisfies $(G2)$ and

(G8) $f(t,\widetilde{0}) = \widetilde{0}$, there exists positive continuous function h such that

$$\beta = \max_{t \in \mathbb{T}} \int_0^t |e_p(t,\sigma(s))| \cdot h(s)\Delta s < \infty,$$

$$\int_0^t |e_p(t,\sigma(s))| \cdot h(s)\Delta s \to 0, \quad \text{as} \quad t \to \infty,$$

$$D(f(t,y_1), f(t,y_2)) \leq h(t)D(y_1,y_2), \quad t \in \mathbb{T}, \quad y_1, y_2 \in F(\mathbb{R}),$$

$$D(y_1,\widetilde{0}) \leq J, \quad D(y_2,\widetilde{0}) \leq J,$$

for some positive constant J,

I satisfies $(G4)$ and

(G9) $I(t,0)=\widetilde{0}$, $t\in\mathbb{T}$.

By $(G7)$, it follows that

$$e_p(t,t_i)\to 0, \quad \text{as} \quad t\to\infty, \quad i\in\{1,\dots,n\}.$$

Define the set S as follows.

$$S = \Big\{\phi\in PC:\phi(0)=y_0, \quad \phi(t)\to\widetilde{0}, \quad \text{as} \quad t\to\infty,$$
$$\phi \quad \text{is} \quad \text{bounded}\Big\}.$$

Note that S is a completely normed space endowed with the norm

$$\|y\| = \sup_{t\in\mathbb{T}} D(y(t),\widetilde{0}).$$

For $\phi\in S$, consider the operator R, defined in Section 3.1.

Definition 3.3.1 *We say that the zero solution of the equation* (3.1) *is stable if for any* $\varepsilon>0$ *there exist a* $\delta=\delta(\varepsilon)>0$ *and a* $t^*>0$ *such that if* $D(y_0,\widetilde{0})<\delta$*, then*

$$D(y(t),\widetilde{0})<\varepsilon \quad \textit{for} \quad \textit{any} \quad t>t^*.$$

Theorem 3.3.2 *Assume* $(G1)$, $(G2)$, $(G4)$, $(G7)$–$(G9)$ *and*

$$\beta + A\max_{t\in\mathbb{T}} \sum_{\{i:t_i<t\}} |e_p(t,t_i)| < 1.$$

Then any solution y *of the equation* (3.1) *with a small initial value* y_0 *is bounded and goes to* 0*, as* $t\to\infty$*. Moreover, the zero solution is stable.*

Proof 3.3.3 *Let* $\phi\in S$ *and*

$$\|\phi\|\le K,$$

for some positive constant K. Then

$$R\phi(0) = y_0,$$

and

$$\begin{aligned}
D(R\phi(t),\widetilde{0}) &= D\Bigg(e_p(t,0)\cdot y_0 + \int_0^t e_p(t,\sigma(s))\cdot f(s,\phi(s))\delta_H s \\
&\quad + \sum_{\{i:t_i<t\}} e_p(t,t_i)\cdot I(t_i,\phi(t_i)),\widetilde{0}\Bigg) \\
&\leq |e_p(t,0)|D(y_0,\widetilde{0}) + D\left(\int_0^t e_p(t,\sigma(s))\cdot f(s,\phi(s))\delta_H s,\widetilde{0}\right) \\
&\quad + D\left(\sum_{\{i:t_i<t\}} e_p(t,t_i)\cdot I(t_i,\phi(t_i)),\widetilde{0}\right) \\
&\leq |e_p(t,0)|D(y_0,\widetilde{0}) + \int_0^t |e_p(t,\sigma(s))|D(f(s,\phi(s)),\widetilde{0})\Delta s \\
&\quad + \sum_{\{i:t_i<t\}} |e_p(t,t_i)|D(I(t_i,\phi(t_i)),\widetilde{0}) \\
&\leq |e_p(t,0)|D(y_0,\widetilde{0}) + \int_0^t |e_p(t,\sigma(s))|h(s)D(\phi(s),\widetilde{0})\Delta s \\
&\quad + A\sum_{\{i:t_i<t\}} |e_p(t,t_i)|D(\phi(t_i),\widetilde{0}) \\
&\leq |e_p(t,0)|D(y_0,\widetilde{0}) + K\int_0^t |e_p(t,\sigma(s))|h(s)\Delta s
\end{aligned}$$

$$+AK \sum_{\{i:t_i<t\}} |e_p(t,t_i)|$$

$$\to 0, \quad as \quad t \to \infty.$$

Therefore $R\phi$ is bounded and $R\phi \in S$. Hence, $R : S \to S$. Let now, $\phi, \psi \in S$. Then

$$\begin{aligned}
D(R\phi(t), R\psi(t)) &= D\Bigg(e_p(t,0)y_0 + \int_0^t e_p(t,\sigma(s)) \cdot f(s,\phi(s))\delta_H s \\
&\quad + \sum_{\{i:t_i<t\}} e_p(t,t_i) \cdot I(t_i,\phi(t_i)), \\
&\quad e_p(t,0) \cdot y_0 + \int_0^t e_p(t,\sigma(s)) \cdot f(s,\phi(s))\delta_H s \\
&\quad + \sum_{\{i:t_i<t\}} e_p(t,t_i) \cdot I(t_i,\psi(t_i)) \Bigg) \\
&= D\Bigg(\int_0^t e_p(t,\sigma(s)) \cdot f(s,\phi(s))\delta_H s \\
&\quad + \sum_{\{i:t_i<t\}} e_p(t,t_i) \cdot I(t_i,\phi(t_i)), \\
&\quad + \int_0^t e_p(t,\sigma(s)) \cdot f(s,\phi(s))\delta_H s \\
&\quad + \sum_{\{i:t_i<t\}} e_p(t,t_i) \cdot I(t_i,\psi(t_i)) \Bigg) \\
&\le D\Bigg(\int_0^t e_p(t,\sigma(s)) \cdot f(s,\phi(s))\delta_H s, \int_0^t e_p(t,\sigma(s)) \cdot f(s,\phi(s))\delta_H s \Bigg)
\end{aligned}$$

$$
\begin{aligned}
& +D\left(\sum_{\{i:t_i<t\}} e_p(t,t_i)\cdot I(t_i,\phi(t_i)), \sum_{\{i:t_i<t\}} e_p(t,t_i)\cdot I(t_i,\psi(t_i))\right) \\
\le & \int_0^t |e_p(t,\sigma(s))| D(f(s,\phi(s)), f(s,\psi(s)))\Delta s \\
& + \sum_{\{i:t_i<t\}} |e_p(t,t_i)| D(I(t_i,\phi(t_i)), I(t_i,\psi(t_i))) \\
\le & \int_0^t |e_p(t,\sigma(s))| h(s) D(\phi(s),\psi(s))\Delta s \\
& +A \sum_{\{i:t_i<t\}} |e_p(t,t_i)| D(\phi(t_i),\psi(t_i)) \\
\le & \ \beta\|\phi-\psi\| \\
& +A \max_{t\in\mathbb{T}} \sum_{\{i:t_i<t\}} |e_p(t,t_i)| \|\phi-\psi\| \\
= & \left(\beta + A\max_{t\in\mathbb{T}} \sum_{\{i:t_i<t\}} |e_p(t,t_i)|\right)\|\phi-\psi\|, \quad t>0.
\end{aligned}
$$

From here,

$$
\begin{aligned}
\|R\phi - R\psi\| \ \le & \left(\beta + A\max_{t\in\mathbb{T}} \sum_{\{i:t_i<t\}} |e_p(t,t_i)|\right)\|\phi-\psi\| \\
< & \ \|\phi-\psi\|.
\end{aligned}
$$

Consequently $R: S \to S$ is a contraction. Then R has a fixed point in S and from here, the equation (3.1) *has a solution in S. Since*

$$
R\phi(t) \to 0, \quad as \quad t\to\infty,
$$

we conclude that the zero solution is stable. This completes the proof.

Exercise 3.3.4 *Investigate for stability the zero solution of the equation*

$$\begin{aligned}\delta_H y(t) &= \frac{t+1}{t^2+1}\cdot y(t)+(y(t))^4, \quad t>0,\\ y(0) &= 0,\\ y(t_i^+) &= y(t_i^-)+(y(t_i))^6, \quad i\in\{1,\ldots,n\}.\end{aligned}$$

3.4 STABILITY OF THE SOLUTIONS OF FIRST-ORDER FUZZY IMPULSIVE DYNAMIC EQUATIONS-II

Suppose that $\mathbb{T}$ is above unbounded. In this section, we investigate the following first-order impulsive dynamic equation

$$\begin{aligned}\delta_H y(t) &= (-a(t))\cdot y^\sigma(t)+f(t,y(t)), \quad t>0,\\ y(0) &= y_0, \qquad\qquad (3.18)\\ y(t_i^+) &= y(t_i^-)+I(t_i,y(t_i)),\end{aligned}$$

where $y_0\in F(\mathbb{R})$, t_i, $i\in\{1,\ldots,n\}$, are as in the previous section, a satisfies $(H1)$ and

(H7) $e_{\ominus a}(t,0)\to 0$, as $t\to\infty$,

f satisfies $(H2)$ and

(H8) $f(t,\widetilde{0})=\widetilde{0}$, there exists a non-negative continuous function h such that

$$\begin{aligned}\beta &= \max_{t\in\mathbb{T}}\int_0^t |e_{\ominus a}(t,s)|\cdot h(s)\Delta s\\ &< \infty,\end{aligned}$$

$$\int_0^t |e_{\ominus a}(t,s)|h(s)\Delta s \to 0, \quad \text{as} \quad t\to\infty,$$

$$D(f(t,y_1),f(t,y_2)) \leq h(t)D(y_1,y_2), \quad t \in \mathbb{T}, \quad y_1,y_2 \in F(\mathbb{R}),$$

$$D(y_1,\widetilde{0}), \quad D(y_2,\widetilde{0}) \leq J,$$

for some positive constant J,

I satisfies $(H4)$ and

(H9) $I(t,0) = \widetilde{0}$, $t \in \mathbb{T}$.

By $(H7)$, it follows that

$$e_{\ominus a}(t,t_i) \to 0, \quad \text{as} \quad t \to \infty, \quad i \in \{1,\ldots,n\}.$$

Define the set S as in Section 3.3. Let the operator H be the same as in Section 3.2.

Definition 3.4.1 *We say that the zero solution of equation* (3.18) *is stable if for any* $\varepsilon > 0$, *there exists a* $\delta = \delta(\varepsilon) > 0$ *and a* $t^* > 0$ *such that if* $D(y_0,\widetilde{0}) < \delta$, *then*

$$D(y(t),\widetilde{0}) < \varepsilon \quad \textit{for} \quad \textit{any} \quad t > t^*.$$

Theorem 3.4.2 *Assume* $(H1)$, $(H2)$, $(H4)$, $(H7)$–$(H9)$ *and*

$$\beta + E \max_{t \in \mathbb{T}} \sum_{\{i:t_i<t\}} |e_{\ominus a}(t,t_i)| < 1.$$

Then any solution y *of* (3.18) *with a small initial value* y_0 *is bounded and goes to* 0, *as* $t \to \infty$. *Moreover, the zero solution is stable.*

Proof 3.4.3 *Let* $\phi \in S$ *and*

$$\|\phi\| \leq K,$$

for some positive constant K. *Then*

$$H\phi(0) = y_0,$$

and

$$
\begin{aligned}
D(H\phi(t),\widetilde{0}) &= D\Big(e_{\ominus a}(t,0)\cdot y_0+\int_0^t e_{\ominus a}(t,s)\cdot f(s,\phi(s))\delta_H s \\
&\quad + \sum_{\{i:t_i<t\}} e_{\ominus a}(t,t_i)\cdot I(t_i,\phi(t_i)),\widetilde{0}\Big) \\
&\le |e_{\ominus a}(t,0)|D(y_0,\widetilde{0})+\int_0^t |e_{\ominus a}(t,s)|D(f(s,\phi(s)),\widetilde{0})\Delta s \\
&\quad + \sum_{\{i:t_i<t\}} |e_{\ominus a}(t,t_i)|D(I(t_i,\phi(t_i)),\widetilde{0}) \\
&\le |e_{\ominus a}(t,0)|D(y_0,\widetilde{0})+\int_0^t |e_{\ominus a}(t,s)|h(s)D(\phi(s),\widetilde{0})\Delta s \\
&\quad +E\sum_{\{i:t_i<t\}} |e_{\ominus a}(t,t_i)|D(\phi(t_i),\widetilde{0}) \\
&\le |e_{\ominus a}(t,0)|D(y_0,\widetilde{0})+K\int_0^t |e_{\ominus a}(t,s)|h(s)\delta_H s \\
&\quad +EK\sum_{\{i:t_i<t\}} |e_{\ominus a}(t,t_i)| \\
&\to 0, \quad as \quad t\to\infty.
\end{aligned}
$$

Therefore, $H\phi$ *is bounded and* $H\phi\in S$. *Hence,* $H:S\to S$. *Let now* $\phi,\psi\in S$. *Then*

$$
\begin{aligned}
D(H\phi(t),H\psi(t)) &= D\Big(e_{\ominus a}(t,0)y_0+\int_0^t e_{\ominus a}(t,s)\cdot f(s,\phi(s))\delta_H s \\
&\quad + \sum_{\{i:t_i<t\}} e_{\ominus a}(t,t_i)\cdot I(t_i,\phi(t_i)),
\end{aligned}
$$

$$
\begin{aligned}
& e_{\ominus a}(t,0)\cdot y_0+\int_0^t e_{\ominus a}(t,s)\cdot f(s,\psi(s))\delta_H s \\
& + \sum_{\{i:t_i<t\}} e_{\ominus a}(t,t_i)\cdot I(t_i,\psi(t_i))\Big) \\
= {} & D\Big(\int_0^t e_{\ominus a}(t,s)\cdot f(s,\phi(s))\delta_H s \\
& + \sum_{\{i:t_i<t\}} e_{\ominus a}(t,t_i)\cdot I(t_i,\phi(t_i)), \\
& \int_0^t e_{\ominus a}(t,s)\cdot f(s,\psi(s))\delta_H s \\
& + \sum_{\{i:t_i<t\}} e_{\ominus a}(t,t_i)\cdot I(t_i,\psi(t_i))\Big) \\
\leq {} & D\Big(\int_0^t e_{\ominus a}(t,s)\cdot f(s,\phi(s))\delta_H s, \int_0^t e_{\ominus a}(t,s)\cdot f(s,\psi(s))\delta_H s\Big) \\
& +D\Big(\sum_{\{i:t_i<t\}} e_{\ominus a}(t,t_i)\cdot I(t_i,\phi(t_i)), \sum_{\{i:t_i<t\}} e_{\ominus a}(t,t_i)\cdot I(t_i,\psi(t_i))\Big) \\
\leq {} & \int_0^t |e_{\ominus a}(t,s)| D(f(s,\phi(s)), f(s,\psi(s)))\Delta_H s \\
& + \sum_{\{i:t_i<t\}} |e_{\ominus a}(t,t_i)| D(I(t_i,\phi(t_i)), I(t_i,\psi(t_i))) \\
\leq {} & \int_0^t |e_{\ominus a}(t,s)| h(s) D(\phi(s),\psi(s))\Delta s
\end{aligned}
$$

$$
\begin{aligned}
& +E \sum_{\{i:t_i<t\}} |e_{\ominus a}(t,t_i)| D(\phi(t_i), \psi(t_i)) \\
\leq\ & \beta \|\phi - \psi\|\|| \\
& +E \max_{t\in\mathbb{T}} \sum_{\{i:t_i<t\}} |e_{\ominus a}(t,t_i)| \|\phi - \psi\| \\
=\ & \left(\beta + E \max_{t\in\mathbb{T}} \sum_{\{i:t_i<t\}} |e_{\ominus a}(t,t_i)|\right) \|\phi - \psi\|, \quad t > 0.
\end{aligned}
$$

From here,

$$
\begin{aligned}
\|H\phi - H\psi\| &\leq \left(\beta + E \max_{t\in\mathbb{T}} \sum_{\{i:t_i<t\}} |e_{\ominus a}(t,t_i)|\right) \|\phi - \psi\| \\
&< \|\phi - \psi\|.
\end{aligned}
$$

Consequently, $H : S \to S$ is a contraction. Then H has a fixed point in S and from here, the equation (3.18) *has a solution in S. Since*

$$H\phi(t) \to 0, \quad as \quad t \to \infty,$$

we conclude that the zero solution is stable. This completes the proof.

Exercise 3.4.4 *Investigate for stability of the zero solution of the equation*

$$
\begin{aligned}
\delta_H y(t) &= \left(-\frac{1}{3t^2+7}\right) \cdot y^\sigma(t) + (y(t))^2, \quad t > 0, \\
y(0) &= \widetilde{0}, \\
y(t_i^+) &= y(t_i^-) + (y(t_i) + 1)^2, \quad i \in \{1, \ldots, n\}.
\end{aligned}
$$

3.5 ADVANCED PRACTICAL PROBLEMS

Problem 3.5.1 *Consider the impulsive dynamic equation*

$$\delta_H y(t) = p(t) \cdot y(t) + (f(y(t)) + (-2) \cdot g(t))^3, \quad t \in [0, T],$$

$$
\begin{aligned}
y(0) &= y_0, \\
y(t_k^+) - y(t_k^-) &= I_k(y(t_k^-)), \quad k \in \{1,\ldots,n\}.
\end{aligned}
$$

Write analogously to Theorem 3.1.1, Theorem 3.1.10, Theorem 3.1.12, Theorem 3.1.15.

Problem 3.5.2 *Consider the impulsive dynamic equation*

$$
\begin{aligned}
\delta_H y(t) &= (-a(t)) \cdot y^{\sigma}(t) + (g(t) + f(y(t)))^2, \quad t \in [0,T], \\
y(0) &= y_0, \\
y(t_k^+) &= I_k(y(t_k^-)), \quad k \in \{1,\ldots,n\}.
\end{aligned}
$$

Write analogously to Theorem 3.2.7, Theorem 3.2.9, Theorem 3.2.12, Theorem 3.2.14.

Problem 3.5.3 *Investigate for stability of the zero solution of the equation*

$$
\begin{aligned}
\delta_H y(t) &= \frac{t^2+1}{t^4+t^2+1} \cdot y(t) + (y(t))^3, \quad t > 0, \\
y(0) &= \widetilde{0}, \\
y(t_i^+) &= y(t_i^-) + (y(t_i))^6, \quad i \in \{1,\ldots,n\}.
\end{aligned}
$$

Problem 3.5.4 *Investigate for stability the zero solution of the equation*

$$
\begin{aligned}
\delta_H y(t) &= t \cdot y(t) + y(t), \quad t > 0, \\
y(0) &= \widetilde{0}, \\
y(t_i^+) &= y(t_i^-) + (y(t_i))^2, \quad i \in \{1,\ldots,n\}.
\end{aligned}
$$

Problem 3.5.5 *Investigate for stability the zero solution of the equation*

$$\begin{aligned} \delta_H y(t) &= (t+1)\cdot y^{\sigma}(t) + (y(t))^3, \quad t > 0, \\ y(0) &= \widetilde{0}, \\ y(t_i^+) &= y(t_i^-) + (y(t_i))^2, \quad i \in \{1,\dots,n\}. \end{aligned}$$

CHAPTER 4

Boundary Value Problems for First-Order Fuzzy Impulsive Dynamic Equations

4.1 LOWER AND UPPER SOLUTIONS FOR FIRST-ORDER FUZZY IMPULSIVE DYNAMIC EQUATIONS

Suppose that $\mathbb{T}$ is a time scale with the forward jump operator, backward jump operator, and delta differentiation operator σ, ρ and Δ, respectively. Let also $0, t_1, T \in \mathbb{T}$, $0 < t_1 < T$, t_1 be right-dense. Set $J = [0,T]$ and consider the fuzzy impulsive dynamic equation with a single impulse.

$$\begin{aligned} \delta_H y(t) &= g(t, y(t)), \quad t \in J \backslash \{t_1\}, \\ y(t_1^+) &= I(y(t_1)), \qquad (4.1) \\ B(y(0), y(T)) &= \widetilde{0}. \end{aligned}$$

Let

$$J_1 = [0, t_1], \quad J_2 = (t_1, T].$$

With y_i, $i \in \{1,2\}$, will be denoted the restrictions of $y : J \to F(\mathbb{R})$ on J_i, $i \in \{1,2\}$. Define

$$\mathscr{C}(J_1) = \{y : J_1 \to F(\mathbb{R}), \quad y \in \mathscr{C}(J_1)\},$$

DOI: 10.1201/9781003655367-4

$$\begin{aligned}\mathscr{C}(J_2) &= \{y : J_2 \to F(\mathbb{R}), \quad y \in \mathscr{C}(J_2), \quad \exists y(t_1^+)\},\\ A &= \{y : J \to F(\mathbb{R}), \quad y_1 \in \mathscr{C}(J_1), \quad y_2 \in \mathscr{C}(J_2)\}.\end{aligned}$$

For $y \in A$, define

$$\|y\| = \sup_{t \in J} D(y(t), \widetilde{0}).$$

Then $(A, \|\cdot\|)$ is a Banach space. For $u, v \in A$, define

$$[u, v] = \{w \in A : u(t) \leq w(t) \leq v(t), \quad t \in J\}.$$

Definition 4.1.1 *We say that* $y : \mathbb{T} \to F(\mathbb{R})$ *is a solution of* (4.1) *if*

1. $y \in A$,
2. $$\begin{aligned}y(t) &= y(0) + \int_0^t g(s, y(s))\delta_H s, \quad t \in J_1,\\ y(t) &= I(y(t_1)) + \int_{t_1}^t g(s, y(s))\delta_H s, \quad t \in J_2,\end{aligned}$$
3. $B(y(0), y(T)) = 0$.

Definition 4.1.2 *We say that* $\alpha : J \to F(\mathbb{R})$ *is a lower solution of the equation* (4.1) *if*

1. $\alpha \in A$,
2. $$\alpha(b) \ominus_H \alpha(a) \leq \int_a^b g(s, \alpha(s))\delta_H s$$

   for $a \leq b$, $a, b \in J_1$ *or* $a, b \in J_2$,

3. $\alpha(t_1^+) \leq I(\alpha(t_1))$,

4. $B(\alpha(0), \alpha(T)) \leq \widetilde{0}$.

Definition 4.1.3 *We say that $\beta : J \to F(\mathbb{R})$ is an upper solution of the equation* (4.1) *if*

1. $\beta \in A$,

2. $$\beta(b) \ominus_H \beta(a) \leq \int_a^b g(s, \beta(s))\delta_H s$$

   for $a \leq b$, $a, b \in J_1$ or $a, b \in J_2$,

3. $\beta(t_1^+) \geq I(\beta(t_1))$,

4. $B(\beta(0), \beta(T)) \geq \widetilde{0}$.

Let

$$p(t,x) = \max\{\alpha(t), \quad \min\{x, \beta(t)\}\}.$$

Suppose that

(I1) $g : J \times F(\mathbb{R}) \to F(\mathbb{R})$ is continuous,

(I2) there is an $h : J \to [0, \infty)$ such that

$$\int_0^T h(s)\Delta s \ < \ \infty,$$

$$D(g(t, p(t,x)), \widetilde{0}) \ \leq \ h(t) \quad (t,x) \in J \times F(\mathbb{R}),$$

(I3) $I : F(\mathbb{R}) \to F(\mathbb{R})$ is continuous and non-decreasing,

(I4) $B : F(\mathbb{R}) \times F(\mathbb{R}) \to F(\mathbb{R})$ is continuous and for each $x \in [\alpha(0), \beta(0)]$, $B(x, \cdot)$ is non-increasing.

Define an operator $G : A \to A$ as follows.

$$Gy(t) = p(0, \bar{y}(0)) + \int_0^t g(s, p(s, y(s)))\delta_H s, \quad t \in J_1, \tag{4.2}$$

$$Gy(t) = I(p(t_1, y(t_1))) + \int_{t_1}^t g(s, p(s, y(s)))\delta_H s, \quad t \in J_2, \tag{4.3}$$

where

$$\bar{y}(0) = y(0) \ominus_H B(y(0), y(T)).$$

Lemma 4.1.4 *Assume* $(I1)$–$(I4)$ *and* $y \in A$ *satisfy*

$$y(t) = p(0, \bar{y}(0)) + \int_0^t g(s, p(s, y(s)))\delta_H s, \quad t \in J_1, \tag{4.4}$$

$$y(t) = I(p(t_1, y(t_1))) + \int_{t_1}^t g(s, p(s, y(s)))\delta_H s, \quad t \in J_2. \tag{4.5}$$

Then

$$y(t) \in [\alpha(t), \beta(t)], \quad t \in J.$$

Proof 4.1.5 *Take* $t = 0$. *From* (4.4), *we get*

$$\begin{aligned} y(0) &= p(0, \bar{y}(0)) \\ &= \max\{\alpha(0), \min\{\bar{y}(0), \beta(0)\}\} \\ &= \max\{\alpha(0), \quad \min\{y(0) \ominus_H B(y(0), y(T)), \beta(0)\}\}. \end{aligned}$$

If $y(0) = \alpha(0)$, *we get* $y(0) \in [\alpha(0), \beta(0)]$. *If*

$$y(0) = \min\{y(0) \ominus_H B(y(0), y(T)), \quad \beta(0)\},$$

then

$$\alpha(0) \le y(0) \le \beta(0).$$

Suppose that there is a $t_2 \in (0, t_1]$ *such that*

$$\alpha(t_2) > y(t_2).$$

Since $\alpha(0) \le y(0)$, *there is a* $t_3 \in [0, t_2)$ *so that*

$$\alpha(t) \le y(t), \quad t \in [0, t_3],$$

and

$$\alpha(t) > y(t), \quad t \in (t_3, t_2].$$

Then

$$\begin{aligned} p(t, y(t)) &= \max\{\alpha(t), \quad \min\{y(t), \beta(t)\}\} \\ &= \alpha(t), \quad t \in (t_3, t_2], \end{aligned}$$

and

$$g(t, p(t, y(t))) = g(t, \alpha(t)), \quad t \in (t_3, t_2].$$

Hence, for $t \in (t_3, t_2]$, *we obtain*

$$\begin{aligned} \int_0^t g(s, p(s, y(s)))\delta_H s &= y(t) \ominus_H p(0, \bar{y}(0)) \\ &= y(t_3) \ominus_H p(0, \bar{y}(0)) + y(t) \ominus_H y(t_3) \\ &= \int_0^{t_3} g(s, p(s, y(s)))\delta_H s + y(t) \ominus_H y(t_3), \end{aligned}$$

and

$$\begin{aligned} y(t) \ominus_H y(t_3) &= \int_0^t g(s, p(s, y(s)))\delta_H s \ominus_H \int_0^{t_3} g(s, p(s, y(s)))\delta_H s \\ &= \int_{t_3}^t g(s, p(s, y(s)))\delta_H s \end{aligned}$$

$$
\begin{aligned}
&= \int_{t_3}^{t} g(s,\alpha(s))\delta_H s \\
&\geq \alpha(t) \ominus_H \alpha(t_3).
\end{aligned}
$$

Since $y(t_3) \geq \alpha(t_3)$, we get $y(t) > \alpha(t)$, $t \in (t_3, t_2]$. This is a contradiction. Therefore, $y(t) \geq \alpha(t)$, $t \in [0, t_1]$. Similarly, $y(t) \leq \beta(t)$, $t \in [0, t_1]$. In particular, we have

$$\alpha(t_1) \leq y(t_1) \leq \beta(t_1).$$

Hence,

$$
\begin{aligned}
\alpha(t_1^+) &\leq I(\alpha(t_1)) \\
&\leq I(y(t_1)) \\
&\leq I(\beta(t_1)) \\
&\leq \beta(t_1^+).
\end{aligned}
$$

Since

$$
\begin{aligned}
I(y(t_1)) &= I(p(t_1, y(t_1))) \\
&= y(t_1^+),
\end{aligned}
$$

we obtain

$$\alpha(t_1^+) \leq y(t_1^+) \leq \beta(t_1^+).$$

Proceeding as above, we arrive at

$$\alpha(t) \leq y(t) \leq \beta(t), \quad t \in J_2.$$

This completes the proof.

Remark 4.1.6 *Suppose that all the conditions of Lemma 4.1.4 are met. Then*

$$\begin{aligned} y(t) &= y(0) + \int_0^t g(s, y(s))\delta_H s, \quad t \in J_1, \\ y(t) &= I(y(t_1)) + \int_{t_1}^t g(s, y(s))\delta_H s, \quad t \in J_2. \end{aligned} \tag{4.6}$$

Lemma 4.1.7 *Assume* (*I*1)-(*I*4) *and* $y \in A$ *satisfy* (4.4), (4.5). *Then*

$$\bar{y}(0) \in [\alpha(0), \beta(0)].$$

Proof 4.1.8 *Suppose that*

$$\alpha(0) > \bar{y}(0).$$

Thus,

$$\begin{aligned} y(0) &= p(0, \bar{y}(0)) \\ &= \max\{\alpha(0), \min\{\bar{y}(0), \beta(0)\}\} \\ &= \max\{\alpha(0), \bar{y}(0)\} \\ &= \alpha(0). \end{aligned}$$

Hence,

$$B(y(0), y(T)) > \tilde{0}$$

and

$$\begin{aligned} B(\alpha(0), \alpha(T)) &= B(y(0), \alpha(T)) \\ &\geq B(y(0), y(T)) \\ &> \tilde{0}, \end{aligned}$$

which is a contradiction. Therefore, $\alpha(0) \leq \bar{y}(0)$. *Now, assume that* $\bar{y}(0) > \beta(0)$. *Then*

$$\begin{aligned} y(0) \ominus_H B(y(0), y(T)) &= \bar{y}(0) \\ &> \beta(0), \end{aligned}$$

and

$$\begin{aligned} y(0) &= p(0, \bar{y}(0)) \\ &= \max\{\alpha(0), \min\{\bar{y}(0), \beta(0)\}\} \\ &= \max\{\alpha(0), \beta(0)\} \\ &= \beta(0), \end{aligned}$$

and

$$B(y(0), y(T)) < \tilde{0}.$$

Hence,

$$\begin{aligned} \tilde{0} &> B(y(0), y(T)) \\ &= B(\beta(0), y(T)) \\ &\geq B(\beta(0), \beta(T)), \end{aligned}$$

which is a contradiction. Consequently $\bar{y}(0) \leq \beta(0)$. *This completes the proof.*

Lemma 4.1.9 *Assume* $(I1)$-$(I4)$ *and* $y \in A$ *satisfies* (4.4), (4.5). *If* y *is a fixed point of operator* G, *then* y *is a solution of* (4.1).

Proof 4.1.10 *By Lemma 4.1.4, we get*

$$y(t) \in [\alpha(t), \beta(t)], \quad t \in J.$$

By Remark 4.1.6, it follows that y satisfies (4.6). *By Lemma 4.1.7, we obtain*

$$\bar{y}(0) \in [\alpha(0), \beta(0)],$$

and

$$\begin{aligned} y(0) &= p(0, \bar{y}(0)) \\ &= \max\{\alpha(0), \min\{\bar{y}(0), \beta(0)\}\} \\ &= \max\{\alpha(0), \bar{y}(0)\} \\ &= \bar{y}(0) \\ &= y(0) \ominus_H B(y(0), y(T)), \end{aligned}$$

whereupon

$$B(y(0), y(T)) = \widetilde{0}.$$

Thus, y is a solution of (4.1). *This completes the proof.*

Define

$$K = \|\alpha\| + \|\beta\|,$$

and $w : J \to F(\mathbb{R})$ as follows

$$w(t) = K + \int_0^t h(s)\Delta s.$$

Let

$$\begin{aligned} S &= \Big\{ y \in A : D(y(0), \widetilde{0}) \leq K, \quad D(y(t_1^+), \widetilde{0}) \leq K, \\ &\quad D(y(b), y(a)) \leq w(b) - w(a) \\ &\quad \text{on} \quad 0 \leq a \leq b \leq t_1 \quad \text{or} \quad t_1 < a \leq b \leq T, \quad a, b \in J \Big\}. \end{aligned}$$

Exercise 4.1.11 *Prove that* S *is a convex and compact subset of* $(A, \|\cdot\|)$.

Lemma 4.1.12 *Let* $(I1)$–$(I4)$ *hold. Then* $G(S) \subset S$.

Proof 4.1.13 *Take* $y \in S$ *arbitrarily. Then*

$$\begin{aligned} D(Gy(0), \widetilde{0}) &= D(p(0, \overline{y}(0)), \widetilde{0}) \\ &= D\left(\max\{\alpha(0), \min\{\overline{y}(0), \beta(0)\}\}, \widetilde{0}\right) \\ &\leq \max\{D(\alpha(0), \widetilde{0}), D(\beta(0), \widetilde{0})\} \\ &\leq K. \end{aligned}$$

Note that

$$\begin{aligned} \beta(t_1) &\geq p(t_1, y(t_1)) \\ &= \max\{\alpha(t_1), \min\{y(t_1), \beta(t_1)\}\} \\ &\geq \alpha(t_1). \end{aligned}$$

Hence,

$$\begin{aligned} \alpha(t_1^+) &\leq I(\alpha(t_1)) \\ &\leq I(p(t_1, y(t_1))) \\ &\leq I(\beta(t_1)) \\ &\leq \beta(t_1^+), \end{aligned}$$

and

$$\begin{aligned} D(I(p(t_1, y(t_1))), \widetilde{0}) &\leq \max\{D(\alpha(t_1^+), \widetilde{0}), D(\beta(t_1^+), \widetilde{0})\} \\ &\leq K. \end{aligned}$$

Let $t \to t_1$, $t > t_1$, *in* (4.3) *and we get*

$$\begin{aligned} D(Gy(t_1^+),\widetilde{0}) &= D(I(p(t_1,y(t_1))),\widetilde{0}) \\ &\leq K. \end{aligned}$$

Let $a,b \in J$, $0 \leq a \leq b \leq t_1$. *Then*

$$\begin{aligned} D(Gy(b),Gy(a)) &= D\Bigg(p(0,\overline{y}(0)) + \int_0^b g(s,p(s,y(s)))\delta_H s, \\ &\qquad p(0,\overline{y}(0)) + \int_0^a g(s,p(s,y(s)))\delta_H s\Bigg) \\ &= D\left(\int_a^b g(s,p(s,y(s)))\delta_H s, \widetilde{0}\right) \\ &\leq \int_a^b D(g(s,p(s,y(s))),\widetilde{0})\Delta s \\ &\leq \int_a^b h(s)\Delta s \\ &= w(b) - w(a). \end{aligned}$$

Now, take $t_1 < a \leq b \leq T$. *Then*

$$\begin{aligned} D(Gy(b),Gy(a)) &= D\Bigg(I(p(t_1,y(t_1))) + \int_{t_1}^b g(s,p(s,y(s)))\delta_H s, \\ &\qquad I(p(t_1,y(t_1))) + \int_{t_1}^a g(s,p(s,y(s)))\delta_H s\Bigg) \end{aligned}$$

$$
\begin{aligned}
&= D\left(\int_a^b g(s,p(s,y(s)))\delta_H s, \widetilde{0}\right) \\
&\le \int_a^b D(g(s,p(s,y(s))),\widetilde{0})\Delta s \\
&\le \int_a^b h(s)\Delta s \\
&= w(b)-w(a).
\end{aligned}
$$

Thus, $G: S \to S$. This completes the proof.

Lemma 4.1.14 *Let $(I1)$-$(I4)$ hold. Then $G: S \to S$ is continuous.*

Proof 4.1.15 *Let $\{y_n\}_{n\in\mathbb{N}} \subset S$ and $y_n \to y$, as $n \to \infty$, $y \in S$, in the space $(A, \|\cdot\|)$. Note that $y_n \to y$, as $n \to \infty$, uniformly in compact subsets of J. Then, for $t \in J_1$, we get*

$$
\begin{aligned}
&D(Gy(t), Gy_n(t)) \\
&\quad = D\Big(Gy(t) \ominus_H p(0,\overline{y}(0)) \ominus_H (Gy_n(t) \ominus_H p(0,\overline{y}_n(0))) \\
&\quad\quad + p(0,\overline{y}(0)) \ominus_H p(0,\overline{y}_n(0)), \widetilde{0}\Big) \\
&\quad = D\Big(\int_0^t g(s,p(s,y(s)))\delta_H s \ominus_H \int_0^t g(s,p(s,y_n(s)))\delta_H s \\
&\quad\quad + p(0,\overline{y}(0)) \ominus_H p(0,\overline{y}_n(0)), \widetilde{0}\Big) \\
&\quad \le \int_0^t D(g(s,p(s,y(s))), g(s,p(s,y_n(s))))\Delta s
\end{aligned}
$$

$$+D(p(0,\overline{y}(0)),p(0,\overline{y}_n(0))$$

$$\to 0, \quad as \quad n \to \infty.$$

For $t \in J_2$, we obtain

$$D(Gy(t),Gy_n(t))$$

$$= D\Big(Gy(t) \ominus_H I(p(t_1,y(t_1))) \ominus_H (Gy_n(t) \ominus_H I(p(t_1,y_n(t_1))))$$

$$+I(p(t_1,y(t_1))) \ominus_H I(p(t_1,y_n(t_1))),\widetilde{0}\Big)$$

$$= D\Big(\int_{t_1}^{t} g(s,p(s,y(s)))\delta_H s \ominus_H \int_{t_1}^{t} g(s,p(s,y_n(s)))\delta_H s$$

$$+I(p(t_1,y(t_1))) \ominus_H I(p(t_1,y_n(t_1))),\widetilde{0}\Big)$$

$$= D\Big(\int_{t_1}^{t} (g(s,p(s,y(s))) \ominus_H g(s,p(s,y_n(s))))\delta_H s$$

$$+I(p(t_1,y(t_1))) \ominus_H I(p(t_1,y_n(t_1))),\widetilde{0}\Big)$$

$$\leq \int_{t_1}^{t} D(g(s,p(s,y(s))),g(s,p(s,y_n(s))))\Delta s$$

$$+D\Big(I(p(t_1,y(t_1))),I(p(t_1,y_n(t_1)))\Big)$$

$$\to 0, \quad as \quad n \to \infty.$$

This completes the proof.

Lemma 4.1.16 *Suppose* $(I1)$–$(I4)$*. Then* G *has a fixed point in* S*.*

Proof 4.1.17 *By Lemma 4.1.12, it follows that $G : S \to S$. From Lemma 4.1.14, we have $G : S \to S$ is continuous. Hence, by Schauder's fixed-point theorem, we conclude that G has a fixed point in S. This completes the proof.*

Theorem 4.1.18 *Suppose (I1)–(I4). Then equation (4.1) has at least one solution $y \in S$ such that $y(t) \in [\alpha(t), \beta(t)]$, $t \in J$.*

Proof 4.1.19 *By Lemma 4.1.16, we see that the operator G has a fixed point $y \in S$. By Lemma 4.1.9, we see that y solves (4.1) and $y(t) \in [\alpha(t), \beta(t)]$, $t \in J$. This completes the proof.*

Exercise 4.1.20 *Let $\mathbb{T} = [0,2] \bigcup [3,4]$, where $[0,2]$ and $[3,4]$ are the real value intervals. Investigate the existence of solutions to the following fuzzy impulsive dynamic equation*

$$\begin{aligned}
\delta_H y(t) &= (t^4, t, t+1) + 2 \cdot y(t), \quad t \in [0,2] \bigcup (3,4], \\
y(3^+) &= y(3) + (-1, -2, 3), \\
B(y(0), y(4)) &= \widetilde{0}.
\end{aligned}$$

4.2 PERIODIC BOUNDARY VALUE PROBLEMS FOR FIRST-ORDER LINEAR FUZZY DYNAMIC EQUATIONS WITH IMPULSES

Let $\mathbb{T}$ be a time scale with forward jump operator and delta differentiation operator σ and Δ, respectively. Let also $0, T \in \mathbb{T}$, $T > 0$, $J = [0, T]$. Assume that

$$0 = t_0 < t_1 < \ldots < t_m < t_{m+1} = T,$$

t_k, $k \in \{1, \ldots, m\}$, are right-dense. Set $J_0 = J \backslash \{t_k\}_{k=1}^m$. Consider the following a linear fuzzy impulsive dynamic equation

$$\begin{aligned}
\delta_H x(t) + p(t) \cdot x^\sigma(t) &= q(t), \quad t \in J_0, \\
x(t_k^+) &= x(t_k^-) + I_k(x(t_k)), \quad k \in \{1, \ldots, m\},
\end{aligned} \tag{4.7}$$

with periodic boundary condition

$$x(0) = x(\sigma(T)), \tag{4.8}$$

(H1) $p, q \in \mathscr{C}([0, \sigma(T)])$, p is regressive on $[0, \sigma(T)]$,

(H2) $I_k \in \mathscr{C}(F(\mathbb{R}))$, $k \in \{1, \ldots, m\}$.

Define

$$\begin{aligned} PC([0,\sigma(T)]) \quad = \quad & \Big\{ x : [0,\sigma(T)] \to \mathbb{R}, \quad x \quad \text{is} \quad \text{continuous} \\ & \text{on} \quad J_0, \quad \text{ld-continuous} \quad \text{at} \quad t_k, \\ & \exists \lim_{t \to t_k + 0} x(t), \quad k \in \{1, \ldots, m\} \Big\}. \end{aligned}$$

$PC([0, \sigma(T)])$ is a Banach space with the norm

$$\|x\| = \max_{j \in \{0, \ldots, m\}} \sup_{t \in [t_j, t_{j+1}]} D(x(t), 0).$$

Now, we define another Banach space

$$\begin{aligned} PC^{\Delta}([0,\sigma(T)]) \quad = \quad & \Big\{ x : x \in PC([0,\sigma(T)]), \quad \exists \delta_H x \quad \text{and} \\ & \text{it} \quad \text{is} \quad \text{continuous} \quad \text{on} \quad (t_k, t_{k+1}), \\ & \exists \lim_{t \to t_k + 0} \delta_H x(t) \quad \text{and} \quad \lim_{t \to t_k -} \delta_H x(t) \quad \text{if} \\ & t_k \quad \text{is} \quad \text{left-scattered}, \quad k \in \{1, \ldots, m\} \Big\} \end{aligned}$$

with the norm

$$\|x\|_{\Delta} = \max\{\|x\|, \sup_{t \in J_0} D\left(\delta_H x(t), \widetilde{0}\right)\}.$$

Definition 4.2.1 *A solution of the problem* (4.7), (4.8), *we mean a function* $x \in PC^{\Delta}([0,\sigma(T)])$ *such that* (4.7) *and* (4.8) *are satisfied.*

Lemma 4.2.2 *Suppose* $e_p(\sigma(T),0) \neq 1$. *If* $x \in PC^{\Delta}([0,\sigma(T)])$ *is a solution of the problems* (4.7), (4.8), *then*

$$x(t) = \int_0^{\sigma(T)} g(t,s) \cdot q(s)\delta_H s + \sum_{j=1}^{m} g(t,t_j) \cdot I_j(x(t_j)), \quad t \in [0,\sigma(T)], \tag{4.9}$$

where

$$\begin{aligned} g(t,s) &= \frac{1}{1 - e_{\ominus p}(\sigma(T),0)} \\ &\quad \times \begin{cases} e_{\ominus p}(t,s) & 0 \le s < t \le \sigma(T) \\ e_{\ominus p}(\sigma(T),0)e_{\ominus p}(t,s) & 0 \le t \le s \le \sigma(T), \end{cases} \end{aligned} \tag{4.10}$$

$t,s \in [0,\sigma(T)]$.

Proof 4.2.3 *Let*

$$y(t) = e_p(t,0) \cdot x(t), \quad t \in [0,\sigma(T)].$$

Then

$$\begin{aligned} \delta_H y(t) &= (p(t)e_p(t,0)) \cdot x^{\sigma}(t) + e_p(t,0) \cdot \delta_H x(t) \\ &= e_p(t,0) \cdot (p(t) \cdot x^{\sigma}(t) + \delta_H x(t)) \\ &= e_p(t,0) \cdot q(t), \quad t \in J_0, \\ y(\sigma(T)) &= e_p(\sigma(T),0) \cdot x(\sigma(T)) \\ &= e_p(\sigma(T),0) \cdot x(0) \end{aligned}$$

$$
\begin{aligned}
&= e_p(\sigma(T),0)\cdot y(0),\\
y(t_j^+) &= e_p(t_j^+,0)\cdot x(t_j^+)\\
&= e_p(t_j,0)\cdot (x(t_j)+I_j(x(t_j)))\\
&= e_p(t_j,0)\cdot x(t_j)+e_p(t_j,0)\cdot I_j(x(t_j))\\
&= (e_p(t_j,0)e_p(0,t_j))\cdot y(t_j)+e_p(t_j,0)\cdot I_j(x(t_j))\\
&= y(t_j)+e_p(t_j,0)\cdot I_j(x(t_j)), \quad j\in\{1,\ldots,m\},
\end{aligned}
$$

i.e., y satisfies the impulsive periodic boundary value problem

$$
\begin{aligned}
\delta_H y(t) &= e_p(t,0)\cdot q(t), \quad t\in J_0,\\
y(t_j^+) &= y(t_j)+e_p(t_j,0)\cdot I_j(x(t_j)), \quad j\in\{1,\ldots,m\},
\end{aligned} \tag{4.11}
$$

$$
y(\sigma(T)) = e_p(\sigma(T),0)\cdot y(0). \tag{4.12}
$$

By the first equation of (4.11), *we derive*

$$
y(t) = y(0) + \int_0^t e_p(s,0)\cdot q(s)\delta_H s, \quad t\in[0,t_1],
$$

and

$$
y(t) = y(t_j^+) + \int_{t_j}^t e_p(s,0)\cdot q(s)\delta_H s, \quad t\in(t_j,t_{j+1}], \quad j\in\{1,\ldots,m\}.
$$

Thus,

$$
y(t_{j+1}) = y(t_j^+) + \int_{t_j}^{t_{j+1}} e_p(s,0)\cdot q(s)\delta_H s
$$

$$
\begin{aligned}
&= \int_{t_j}^{t_{j+1}} e_p(s,0)\cdot q(s)\delta_H s + y(t_j) + e_p(t_j,0)\cdot I_j(x(t_j))\\
&= \int_{t_j}^{t_{j+1}} e_p(s,0)\cdot q(s)\delta_H s + \int_{t_{j-1}}^{t_j} e_p(s,0)\cdot q(s)\delta_H s\\
&\quad + e_p(t_{j-1},0)\cdot I_{j-1}(x(t_{j-1})) + e_p(t_j,0)\cdot I_j(x(t_j))\\
&= \int_{t_{j-1}}^{t_{j+1}} e_p(s,0)\cdot q(s)\delta_H s + e_p(t_{j-1},0)\cdot I_{j-1}(x(t_{j-1}))\\
&\quad + e_p(t_j,0)\cdot I_j(x(t_j))\\
&= \cdots\\
&= \int_{0}^{t_{j+1}} e_p(s,0)\cdot q(s)\delta_H s + \sum_{k=1}^{j} e_p(t_k,0)\cdot I_k(x(t_k)) + y(0).
\end{aligned}
$$

Hence,

$$
\begin{aligned}
y(t) &= y(t_j^+) + \int_{t_j}^{t} e_p(s,0)\cdot q(s)\delta_H s\\
&= y(t_j) + \int_{t_j}^{t} e_p(s,0)\cdot q(s)\delta_H s + e_p(t_j,0)\cdot I_j(x(t_j))\\
&= \int_{0}^{t_j} e_p(s,0)\cdot q(s)\delta_H s + \sum_{k=1}^{j-1} e_p(t_k,0)\cdot I_k(x(t_k)) + y(0)
\end{aligned}
$$

$$
\begin{aligned}
&+\int_{t_j}^{t} e_p(s,0)\cdot q(s)\delta_H s + e_p(t_j,0)\cdot I_j(x(t_j))\\
&= y(0)+\int_{0}^{t} e_p(s,0)\cdot q(s)\delta_H s+\sum_{k=1}^{j} e_p(t_k,0)\cdot I_k(x(t_k)),
\end{aligned}
$$

$t\in(t_j,t_{j+1}]$, $\quad j\in\{1,\dots,m\}$. *Therefore*

$$
y(t)=y(0)+\int_{0}^{t} e_p(s,0)\cdot q(s)\delta_H s+\sum_{j:t_j<t} e_p(t_j,0)\cdot I_j(x(t_j)), \quad (4.13)
$$

$t\in[0,\sigma(T)]$. *Hence, using the periodic boundary value condition* (4.12), *we get*

$$
\begin{aligned}
e_p(\sigma(T),0)\cdot y(0) &= y(\sigma(T))\\
&= y(0)+\int_{0}^{\sigma(T)} e_p(s,0)\cdot q(s)\delta_H s\\
&\quad+\sum_{j:t_j<\sigma(T)} e_p(t_j,0)\cdot I_j(x(t_j))\\
&= y(0)+\int_{0}^{\sigma(T)} e_p(s,0)\cdot q(s)\delta_H s\\
&\quad+\sum_{j=1}^{m} e_p(t_j,0)\cdot I_j(x(t_j)),
\end{aligned}
$$

whereupon

$$
(e_p(\sigma(T),0)-1)\cdot y(0)=\int_{0}^{\sigma(T)} e_p(s,0)\cdot q(s)\delta_H s+\sum_{j=1}^{m} e_p(t_j,0)\cdot I_j(x(t_j)),
$$

or

$$y(0) = \frac{1}{e_p(\sigma(T),0)-1} \cdot \left(\int_0^{\sigma(T)} e_p(s,0) \cdot q(s)\delta_H s + \sum_{j=1}^{m} e_p(t_j,0) \cdot I_j(x(t_j)) \right).$$

From the last expression and from (4.13), *we obtain*

$$e_p(t,0) \cdot x(t)$$

$$= \frac{1}{e_p(\sigma(T),0)-1} \cdot \left(\int_0^{\sigma(T)} e_p(s,0) \cdot q(s)\delta_H s + \sum_{j=1}^{m} e_p(t_j,0) \cdot I_j(x(t_j)) \right)$$

$$+ \int_0^{t} e_p(s,0) \cdot q(s)\delta_H s + \sum_{j:t_j<t} e_p(t_j,0) \cdot I_j(x(t_j)),$$

or

$$x(t)$$

$$= \frac{e_{\ominus p}(t,0)}{e_p(\sigma(T),0)-1} \cdot \left(\int_0^{\sigma(T)} e_p(s,0) \cdot q(s)\delta_H s + \sum_{j=1}^{m} e_p(t_j,0) \cdot I_j(x(t_j)) \right)$$

$$+ e_{\ominus p}(t,0) \cdot \left(\int_0^{t} e_p(s,0) \cdot q(s)\delta_H s + \sum_{j:t_j<t} e_p(t_j,0) \cdot I_j(x(t_j)) \right)$$

$$= \frac{e_{\ominus p}(t,0)}{e_p(\sigma(T),0)-1} \cdot \Bigg(\int_0^{\sigma(T)} e_p(s,0) \cdot q(s)\delta_H s + \sum_{j=1}^{m} e_p(t_j,0) \cdot I_j(x(t_j))$$

$$+ (e_p(\sigma(T),0)-1) \cdot \int_0^{t} e_p(s,0) \cdot q(s)\delta_H s$$

$$+(e_p(\sigma(T),0)-1)\cdot\sum_{j:t_j<t} e_p(t_j,0)\cdot I_j(x(t_j))\Big)$$

$$=\frac{e_{\ominus p}(t,0)}{e_p(\sigma(T),0)-1}\cdot\Big(e_p(\sigma(T),0)\cdot\int_0^t e_p(s,0)\cdot q(s)\delta_H s$$

$$+\int_t^{\sigma(T)} e_p(s,0)\cdot q(s)\delta_H s$$

$$+e_p(\sigma(T),0)\cdot\sum_{j:t_j<t} e_p(t_j,0)\cdot I_j(x(t_j))$$

$$+\sum_{j:t_j\in[t,T)} e_p(t_j,0)\cdot I_j(x(t_j))\Big)$$

$$=\frac{1}{1-e_{\ominus p}(\sigma(T),0)}\cdot\Big(\int_0^t (e_{\ominus p}(t,0)e_{\ominus p}(0,s))\cdot q(s)\delta_H s$$

$$+\int_t^{\sigma(T)} (e_{\ominus p}(\sigma(T),0)e_{\ominus p}(t,0)e_{\ominus p}(0,s))\cdot q(s)\delta_H s$$

$$+\sum_{j:t_j<t} (e_{\ominus p}(t,0)e_{\ominus p}(0,t_j))\cdot I_j(x(t_j))$$

$$+\sum_{j:t_j\in[t,T)} e_{\ominus p}(\sigma(T),0)(e_{\ominus p}(t,0)e_{\ominus p}(0,t_j))\cdot I_j(x(t_j))\Big)$$

$$=\frac{1}{1-e_{\ominus p}(\sigma(T),0)}\cdot\Big(\int_0^t e_{\ominus p}(t,s)\cdot q(s)\delta_H s$$

$$+\int_t^{\sigma(T)} (e_{\ominus p}(\sigma(T),0)e_{\ominus p}(t,s))\cdot q(s)\delta_H s$$

$$
\begin{aligned}
&+ \sum_{j:t_j<t} e_{\ominus p}(t,t_j)\cdot I_j(x(t_j)) \\
&+ \sum_{j:t_j\in[t,T)} \left(e_{\ominus p}(\sigma(T),0)e_{\ominus p}(t,t_j)\right)\cdot I_j(x(t_j))\Bigg) \\
&= \int_0^{\sigma(T)} g(t,s)\cdot q(s)\delta_H s + \sum_{j=1}^{m} g(t,t_j)\cdot I_j(x(t_j)),
\end{aligned}
$$

where g is defined by (4.10). *This completes the proof.*

Lemma 4.2.4 *If $x\in PC([0,\sigma(T)])$ and x,q,I_j, $j\in\{1,\ldots,m\}$, satisfy* (4.9), *then $x\in PC^{\Delta}([0,\sigma(T)])$ is a solution of the problem* (4.7), (4.8).

Proof 4.2.5 *We have*

$$
\begin{aligned}
\delta_H x(t) &= \delta_H\left(\int_0^{\sigma(T)} g(t,s)\cdot q(s)\delta_H s\right)_t + \sum_{j=1}^{m} g^{\Delta}(t,t_j)\cdot I_j(x(t_j)) \\
&= \delta_H\left(\int_0^{t} g(t,s)\cdot q(s)\delta_H s + \int_t^{\sigma(T)} g(t,s)\cdot q(s)\delta_H s\right) \\
&\quad + \sum_{j=1}^{m} g^{\Delta}(t,t_j)\cdot I_j(x(t_j)) \\
&= \frac{e_{\ominus p}(\sigma(t),t)}{1-e_{\ominus p}(\sigma(T),0)}\cdot q(t) + \int_0^{t} \frac{\ominus p(t)e_{\ominus p}(t,s)}{1-e_{\ominus p}(\sigma(T),0)}\cdot q(s)\delta_H s \\
&\quad - \frac{e_{\ominus p}(\sigma(T),0)e_{\ominus p}(\sigma(t),t)}{1-e_{\ominus p}(\sigma(T),0)}\cdot q(t) \\
&\quad + \int_t^{\sigma(T)} \frac{\ominus p(t)e_{\ominus p}(\sigma(T),0)e_{\ominus p}(t,s)}{1-e_{\ominus p}(\sigma(T),0)}\cdot q(s)\delta_H s \\
&\quad + \sum_{j=1}^{m} g^{\Delta}(t,t_j)\cdot I_j(x(t_j))
\end{aligned}
$$

$$
\begin{aligned}
&= e_{\ominus p}(\sigma(t),t)\cdot q(t)+\int_{0}^{t}\frac{\ominus p(t)e_{\ominus p}(t,s)}{1-e_{\ominus p}(\sigma(T),0)}\cdot q(s)\delta_H s \\
&\quad +\int_{t}^{\sigma(T)}\frac{\ominus p(t)e_{\ominus p}(\sigma(T),0)e_{\ominus p}(t,s)}{1-e_{\ominus p}(\sigma(T),0)}\cdot q(s)\delta_H s \\
&\quad +\sum_{j=1}^{m} g^{\Delta}(t,t_j)\cdot I_j(x(t_j)) \\
&= (1+\mu(t)\ominus p(t))\cdot q(t)+\int_{0}^{t}\frac{\ominus p(t)e_{\ominus p}(t,s)}{1-e_{\ominus p}(\sigma(T),0)}\cdot q(s)\delta_H s \\
&\quad +\int_{t}^{\sigma(T)}\frac{\ominus p(t)e_{\ominus p}(\sigma(T),0)e_{\ominus p}(t,s)}{1-e_{\ominus p}(\sigma(T),0)}\cdot q(s)\delta_H s \\
&\quad +\sum_{j=1}^{m} g^{\Delta}(t,t_j)\cdot I_j(x(t_j)),
\end{aligned}
$$

$$
\begin{aligned}
x^{\sigma}(t) &= \left(\int_{0}^{t} g(t,s)\cdot q(s)\delta_H s+\int_{t}^{\sigma(T)} g(t,s)\cdot q(s)\delta_H s\right)^{\sigma} \\
&\quad +\sum_{j=1}^{m} g(\sigma(t),t_j)\cdot I_j(x(t_j))
\end{aligned}
$$

$$
\begin{aligned}
&= \Bigg(\int_{0}^{t}\frac{e_{\ominus p}(t,s)}{1-e_{\ominus p}(\sigma(T),0)}\cdot q(s)\delta_H s \\
&\quad +\int_{t}^{\sigma(T)}\frac{e_{\ominus p}(\sigma(T),0)e_{\ominus p}(t,s)}{1-e_{\ominus p}(\sigma(T),0)}\cdot q(s)\delta_H s\Bigg)^{\sigma}
\end{aligned}
$$

$$
\begin{aligned}
& + \sum_{j=1}^{m} g(\sigma(t),t_j) \cdot I_j(x(t_j)) \\
= & \int_{0}^{\sigma(t)} \frac{e_{\ominus p}(\sigma(t),s)}{1-e_{\ominus p}(\sigma(T),0)} \cdot q(s)\delta_H s \\
& + \int_{\sigma(t)}^{\sigma(T)} \frac{e_{\ominus p}(\sigma(T),0)e_{\ominus p}(\sigma(t),s)}{1-e_{\ominus p}(\sigma(T),0)} \cdot q(s)\delta_H s \\
& + \sum_{j=1}^{m} g(\sigma(t),t_j) \cdot I_j(x(t_j)) \\
= & \int_{0}^{t} \frac{e_{\ominus p}(\sigma(t),s)}{1-e_{\ominus p}(\sigma(T),0)} \cdot q(s)\delta_H s + \int_{t}^{\sigma(t)} \frac{e_{\ominus p}(\sigma(t),s)}{1-e_{\ominus p}(\sigma(T),0)} \cdot q(s)\delta_H s \\
& - \int_{t}^{\sigma(t)} \frac{e_{\ominus p}(\sigma(T),0)e_{\ominus p}(\sigma(t),s)}{1-e_{\ominus p}(\sigma(T),0)} \cdot q(s)\delta_H s \\
& + \sum_{j=1}^{m} g(\sigma(t),t_j) \cdot I_j(x(t_j)) \\
& + \int_{t}^{\sigma(T)} \frac{e_{\ominus p}(\sigma(T),0)e_{\ominus p}(\sigma(t),s)}{1-e_{\ominus p}(\sigma(T),0)} \cdot q(s)\delta_H s \\
= & \int_{0}^{t} \frac{(1+\mu(t)\ominus p(t))e_{\ominus p}(t,s)}{1-e_{\ominus p}(\sigma(T),0)} \cdot q(s)\delta_H s \\
& + \frac{\mu(t)e_{\ominus p}(\sigma(t),t)}{1-e_{\ominus p}(\sigma(T),0)} \cdot q(t) \\
& - \frac{\mu(t)e_{\ominus p}(\sigma(T),0)e_{\ominus p}(\sigma(t),t)}{1-e_{\ominus p}(\sigma(T),0)} \cdot q(t)
\end{aligned}
$$

$$+\sum_{j=1}^{m} g(\sigma(t),t_j)\cdot I_j(x(t_j))$$

$$+\int_{t}^{\sigma(T)} \frac{e_{\ominus p}(\sigma(T),0)e_{\ominus p}(\sigma(t),s)}{1-e_{\ominus p}(\sigma(T),0)}\cdot q(s)\delta_H s$$

$$= (\mu(t)(1+\mu(t)\ominus p(t)))\cdot q(t)$$

$$+\int_{0}^{t} \frac{(1+\mu(t)\ominus p(t))e_{\ominus p}(t,s)}{1-e_{\ominus p}(\sigma(T),0)}\cdot q(s)\delta_H s$$

$$+\int_{t}^{\sigma(T)} \frac{e_{\ominus p}(\sigma(T),0)e_{\ominus p}(\sigma(t),s)}{1-e_{\ominus p}(\sigma(T),0)}\cdot q(s)\delta_H s$$

$$+\sum_{j=1}^{m} g(\sigma(t),t_j)\cdot I_j(x(t_j)),$$

$$p(t)\cdot x^{\sigma}(t) = (p(t)\mu(t)(1+\mu(t)\ominus p(t)))\cdot q(t)$$

$$+\int_{0}^{t} \frac{p(t)(1+\mu(t)\ominus p(t))e_{\ominus p}(t,s)}{1-e_{\ominus p}(\sigma(T),0)}\cdot q(s)\delta_H s$$

$$+\int_{t}^{\sigma(T)} \frac{p(t)e_{\ominus p}(\sigma(T),0)e_{\ominus p}(\sigma(t),s)}{1-e_{\ominus p}(\sigma(T),0)}\cdot q(s)\delta_H s$$

$$+\sum_{j=1}^{m} p(t)g(\sigma(t),t_j)\cdot I_j(x(t_j)),$$

$$\delta_H x(t)+p(t)\cdot x^{\sigma}(t) = (1+\mu(t)\ominus p(t))\cdot q(t)$$

$$+\int_{0}^{t} \frac{\ominus p(t)e_{\ominus p}(t,s)}{1-e_{\ominus p}(\sigma(T),0)}\cdot q(s)\delta_H s$$

$$
\begin{aligned}
& + \int_t^{\sigma(T)} \frac{\ominus p(t) e_{\ominus p}(\sigma(T),0) e_{\ominus p}(t,s)}{1 - e_{\ominus p}(\sigma(T),0)} \cdot q(s) \delta_H s \\
& + \sum_{j=1}^{m} g^{\Delta}(t,t_j) \cdot I_j(x(t_j)) \\
& + (p(t)\mu(t)(1+\mu(t) \ominus p(t))) \cdot q(t) \\
& + \int_0^{t} \frac{p(t)(1+\mu(t) \ominus p(t)) e_{\ominus p}(t,s)}{1 - e_{\ominus p}(\sigma(T),0)} \cdot q(s) \delta_H s \\
& + \int_t^{\sigma(T)} \frac{p(t) e_{\ominus p}(\sigma(T),0) e_{\ominus p}(\sigma(t),s)}{1 - e_{\ominus p}(\sigma(T),0)} \cdot q(s) \delta_H s \\
& + \sum_{j=1}^{m} p(t) g(\sigma(t),t_j) \cdot I_j(x(t_j)) \\
= {} & q(t) + \sum_{j=1}^{m} \left(g^{\Delta}(t,t_j) + p(t) g(\sigma(t),t_j) \right) \cdot I_j(x(t_j)) \\
= {} & q(t) + \sum_{j=1}^{m} \left(g^{\Delta}(t,t_j) + p(t) g(t,t_j) + \mu(t) p(t) g^{\Delta}(t,t_j) \right) \cdot I_j(x(t_j)) \\
= {} & q(t) + \sum_{j=1}^{m} \left((1+\mu(t)p(t)) g^{\Delta}(t,t_j) + p(t) g(t,t_j) \right) \cdot I_j(x(t_j)),
\end{aligned}
$$

$t \in [0, \sigma(T)]$. *If* $0 \le t \le t_j$, *then*

$$
\begin{aligned}
& (1+\mu(t)p(t)) g^{\Delta}(t,t_j) + p(t) g(t,t_j) \\
= {} & \frac{(1+\mu(t)p(t)) \ominus p(t) e_{\ominus p}(\sigma(T),0) e_{\ominus p}(t,t_j)}{1 - e_{\ominus p}(\sigma(T),0)} \\
& + \frac{p(t) e_{\ominus p}(\sigma(T),0) e_{\ominus p}(t,t_j)}{1 - e_{\ominus p}(\sigma(T),0)} \\
= {} & 0.
\end{aligned}
$$

If $0 \le t_j < t \le \sigma(T)$, *then*

$$\begin{aligned}
(1+\mu(t)p(t))g^{\Delta}(t,t_j)+p(t)g(t,t_j) &= \frac{(1+\mu(t)p(t))\ominus p(t)e_{\ominus p}(t,t_j)}{1-e_{\ominus p}(\sigma(T),0)} \\
&\quad + \frac{p(t)e_{\ominus p}(t,t_j)}{1-e_{\ominus p}(\sigma(T),0)} \\
&= 0.
\end{aligned}$$

Therefore

$$\delta_H x(t) + p(t)\cdot x^{\sigma}(t) = q(t), \quad t \in J_0.$$

Next,

$$\begin{aligned}
x(t_k^+) &= \int_0^{\sigma(T)} g(t_k^+,s)\cdot q(s)\delta_H s + \sum_{j=1}^{m} g(t_k^+,t_j)\cdot I_j(x(t_j)) \\
&= \int_0^{\sigma(T)} g(t_k^-,s)\cdot q(s)\delta_H s + \sum_{j=1}^{m} g(t_k^-,t_j)\cdot I_j(x(t_j)) \\
&\quad + \left(g(t_k^+,t_k) - g(t_k^-,t_k)\right)\cdot I_k(x(t_k)) \\
&= x(t_k^-) + I_k(x(t_k)), \quad k \in \{1,\ldots,m\},
\end{aligned}$$

and

$$\begin{aligned}
x(0) &= \int_0^{\sigma(T)} g(0,s)\cdot q(s)\delta_H s + \sum_{j=1}^{m} g(0,t_j)\cdot I_j(x(t_j)) \\
&= \int_0^{\sigma(T)} g(\sigma(T),s)\cdot q(s)\delta_H s + \sum_{j=1}^{m} g(\sigma(T),t_j)\cdot I_j(x(t_j)) \\
&= x(\sigma(T)).
\end{aligned}$$

This completes the proof.

For p, q, I_k, $k \in \{1, \ldots, m\}$, which satisfy $(H1)$ and $(H2)$, define the operator $\Phi : PC([0, \sigma(T)]) \to PC([0, \sigma(T)])$ by

$$\Phi x(t) = \int_0^{\sigma(T)} g(t,s) \cdot q(s) \delta_H s + \sum_{j=1}^{m} g(t,t_j) \cdot I_j(x(t_j)), \quad t \in [0, \sigma(T)],$$

where the function g is defined by (4.10). Note that by Lemma 4.2.2 and Lemma 4.2.4, it follows that every fixed point $x \in PC([0, \sigma(T)]$ of the operator Φ is a solution of the problem (4.7), (4.8). Let

$$M = \max_{j \in \{1, \ldots, m\}} \left\{ \sup_{t \in [0, \sigma(T)]} |g(t, t_j)| \right\}.$$

Theorem 4.2.6 *Let p, q satisfy $(H1)$, I_k, $k \in \{1, \ldots, m\}$, satisfy $(H2)$ and*

$$D(I_k(x), I_k(y)) \leq l_k D(x,y), \quad x, y \in F(\mathbb{R}), \quad M \sum_{k=1}^{m} l_k < 1.$$

Then problem (4.7), (4.8) *has a unique solution in $PC([0, \sigma(T)])$.*

Proof 4.2.7 *For $x, y \in PC([0, \sigma(T)])$, we have*

$$\begin{aligned}
D(\Phi x(t), \Phi y(t)) &= D\Bigg(\int_0^{\sigma(T)} g(t,s) q(s) \delta_H s + \sum_{j=1}^{m} g(t,t_j) I_j(x(t_j)), \\
&\qquad \int_0^{\sigma(T)} g(t,s) q(s) \delta_H s + \sum_{j=1}^{m} g(t,t_j) I_j(y(t_j)) \Bigg) \\
&= D\left(\sum_{j=1}^{m} g(t,t_j) I_j(x(t_j)), \sum_{j=1}^{m} g(t,t_j) I_j(y(t_j))) \right) \\
&\leq \sum_{j=1}^{m} |g(t,t_j)| D(I_j(x(t_j)), I_j(y(t_j))))
\end{aligned}$$

$$\leq M\sum_{j=1}^{m} l_j D(x(t_j), y(t_j))$$

$$\leq M\sum_{j=1}^{m} l_j \|x-y\|,$$

whereupon

$$\|\Phi x - \Phi y\| \leq M\sum_{j=1}^{m} l_j \|x-y\|.$$

Therefore, $\Phi : PC([0,\sigma(T)]) \to PC([0,\sigma(T)])$ *is a contraction and it has a unique fixed point in* $PC([0,\sigma(T)])$*. This completes the proof.*

Exercise 4.2.8 *Let* $\mathbb{T} = [0,1]\bigcup[2,3]\bigcup[4,5]$*, where* $[0,1]$*,* $[2,3]$ *and* $[4,5]$ *are the real-valued intervals. Consider the periodic boundary value problem*

$$\begin{aligned} \delta_H x(t) &= \ominus_H x^{\sigma}(t) + \left(\frac{1}{t^2+1}, 1, 0\right), \quad t \in \mathbb{T}\backslash\{2,4\}, \\ x(2+) &= \frac{1}{7}\cdot x(2), \\ x(3+) &= \frac{2}{3}\cdot x(3), \\ x(0) &= x(5). \end{aligned}$$

Find constants l_1 *and* l_2 *as in Theorem 4.2.6 so that the considered PBVP has a unique solution.*

4.3 PERIODIC BOUNDARY VALUE PROBLEMS FOR FIRST-ORDER NON-LINEAR FUZZY DYNAMIC EQUATIONS WITH IMPULSES-I

Let $\mathbb{T}$, $\{t_k\}_{k=0}^{m+1}$, $PC([0,\sigma(T)])$, $PC^{\Delta}([0,\sigma(T)])$, J, J_0 and M be as in Section 4.2. Consider the following first-order non-linear impulsive dynamic equation

$$\begin{aligned} \delta_H x(t) + p(t)\cdot x(\sigma(t)) &= f(t, x(\sigma(t))), \quad t \in J_0, \\ x(t_k^+) &= x(t_k) + I_k(x(t_k)), \quad k \in \{1, \ldots, m\}, \end{aligned} \tag{4.14}$$

$$x(0) = x(\sigma(T)), \tag{4.15}$$

where

(H1) $p \in \mathscr{C}([0, \sigma(T)])$, p is regressive on $[0, \sigma(T)]$,

(H2) $I_k \in \mathscr{C}(F(\mathbb{R}))$, $k \in \{1, \ldots, m\}$,

(H3) $f \in \mathscr{C}(J \times F(\mathbb{R}))$.

By Lemma 4.2.2 and Lemma 4.2.4, it follows that $x \in PC^{\Delta}([0, \sigma(T)])$ is a solution of the problem (4.14), (4.15) if and only if it satisfies the integral equation

$$x(t) = \int_0^{\sigma(T)} g(t,s)\cdot f(s, x(\sigma(s)))\delta_H s + \sum_{j=1}^{m} g(t, t_j)\cdot I_j(x(t_j)), \quad t \in [0, \sigma(T)],$$

where the function g is defined by (4.10). Define the operator Ψ : $PC([0, \sigma(T)]) \to PC([0, \sigma(T)])$ by

$$\Psi x(t) = \int_0^{\sigma(T)} g(t,s)\cdot f(s, x(\sigma(s)))\delta_H s + \sum_{j=1}^{m} g(t, t_j)\cdot I_j(x(t_j)), \quad t \in [0, \sigma(T)].$$

Theorem 4.3.1 *Let p, I_k, $k \in \{1, \ldots, m\}$, f satisfy (H1), (H2), (H3), respectively, and there exist non-negative constants α, K, β_j and γ_j, $j \in \{1, \ldots, m\}$, such that*

$$\begin{aligned} D(f(t,x), \widetilde{0}) &\leq \alpha D(x, \widetilde{0}) + K, \quad t \in J, \quad x \in F(\mathbb{R}), \\ D(I_j(x), \widetilde{0}) &\leq \beta_j D(x, \widetilde{0}) + \gamma_j, \quad x \in F(\mathbb{R}), \end{aligned}$$

$$M\left(\sum_{j=1}^{m} \beta_j + \alpha\sigma(T)\right) < 1.$$

Then problem (4.7), (4.8) has at least one solution in $PC([0, \sigma(T)])$.

Proof 4.3.2 *1. $\Psi : PC([0,\sigma(T)]) \to PC([0,\sigma(T)])$ is continuous. Let $\{x_n\}_{n\in\mathbb{N}} \subset PC([0,\sigma(T)])$, $x \in PC([0,\sigma(T)])$, be such that $x_n \to x$ in $PC([0,\sigma(T)])$, as $n \to \infty$. Then*

$$
\begin{aligned}
& D(\Psi x_n(t), \Psi x(t)) \\
&= D\Bigg(\int_0^{\sigma(T)} g(t,s)\cdot f(s,x_n(\sigma(s)))\delta_H s + \sum_{j=1}^m g(t,t_j)\cdot I_j(x_n(t_j)), \\
&\quad \int_0^{\sigma(T)} g(t,s)\cdot f(s,x(\sigma(s)))\delta_H s + \sum_{j=1}^m g(t,t_j)\cdot I_j(x(t_j))\Bigg) \\
&= D\Bigg(\int_0^{\sigma(T)} g(t,s)\cdot (f(s,x_n(\sigma(s)))\ominus_H f(s,x(\sigma(s))))\delta_H s \\
&\quad + \sum_{j=1}^m g(t,t_j)\cdot (I_j(x(t_j))\ominus_H I_j(x(t_j))), \widetilde{0}\Bigg) \\
&\le \int_0^{\sigma(T)} |g(t,s)| D(f(s,x_n(\sigma(s))), f(s,x(\sigma(s))))\Delta s \\
&\quad + \sum_{j=1}^m |g(t,t_j)| D(I_j(x_n(t_j)), I_j(x(t_j))) \\
&\le M\Bigg(\int_0^{\sigma(T)} D(f(s,x_n(\sigma(s))), f(s,x(\sigma(s))))\Delta s \\
&\quad + \sum_{j=1}^m D(I_j(x_n(t_j)), I_j(x(t_j)))\Bigg), \quad t \in [0,\sigma(T)].
\end{aligned}
$$

Hence and Lebesgue's dominated convergence theorem, it follows that

$$\|\Psi x_n - \Psi x\| \to 0, \quad as \quad n \to \infty.$$

2. *Ψ maps a bounded set on $PC([0,\sigma(T)])$ into a bounded set on $PC([0,\sigma(T)])$. Let $D \subset PC([0,\sigma(T)])$ be a bounded set such that for any $x \in D$ we have*

$$\|x\| \le L,$$

for some positive constant L. Note that there exist positive constants F_1, Q_j, $j \in \{1,\ldots,m\}$, such that

$$\begin{aligned} D(f(t,x(t)),\widetilde{0}) &\le F_1, \\ D(I_j(x(t_j)),\widetilde{0}) &\le Q_j, \quad t \in [0,\sigma(T)], \quad x \in D. \end{aligned}$$

Hence,

$$D(\Psi x(t),\widetilde{0})$$

$$= D\left(\int_0^{\sigma(T)} g(t,s)\cdot f(s,x(\sigma(s)))\delta_H s + \sum_{j=1}^m g(t,t_j)\cdot I_j(x(t_j)),\widetilde{0}\right)$$

$$\le \int_0^{\sigma(T)} |g(t,s)|D(f(s,x(\sigma(s))),\widetilde{0})\Delta s + \sum_{j=1}^m |g(t,t_j)|D(I_j(x(t_j)),\widetilde{0})$$

$$\le MF_1\sigma(T) + M\sum_{j=1}^m Q_j, \quad t \in [0,\sigma(T)].$$

3. *Let $t^1, t^2 \in [0,\sigma(T)]$. Then*

$$D(\Psi x(t^1),\Psi x(t^2))$$

$$= D\Big(\int_0^{\sigma(T)} g(t^1,s)\cdot f(s,x(\sigma(s)))\delta_H s + \sum_{j=1}^m g(t^1,t_j)\cdot I_j(x(t_j)),$$

$$\int_0^{\sigma(T)} g(t^2,s)\cdot f(s,x(\sigma(s)))\delta_H s + \sum_{j=1}^{m} g(t^2,t_j)\cdot I_j(x(t_j))\Big)$$

$$\begin{aligned}
&= D\Big(\int_0^{\sigma(T)} (g(t^1,s)-g(t^2,s))\cdot f(s,x(\sigma(s)))\delta_H s \\
&\quad + \sum_{j=1}^{m} (g(t^1,t_j)-g(t^2,t_j)\cdot I_j(x(t_j)),\widetilde{0}\Big) \\
&\leq \int_0^{\sigma(T)} |g(t^1,s)-g(t^2,s)|D(f(s,x(\sigma(s))),\widetilde{0})\Delta s \\
&\quad + \sum_{j=1}^{m} |g(t^1,t_j)-g(t^2,t_j)|D(I_j(x(t_j)),\widetilde{0}) \\
&\to 0, \quad as \quad t^1 \to t^2.
\end{aligned}$$

From here and from the previous two points, using the Arzela-Ascoli theorem, we conclude that $\Psi : PC([0,\sigma(T)]) \to PC([0,\sigma(T)])$ is a compact operator.

4. *Now we consider the equation*

$$x(t) = \lambda\Psi x(t), \quad t\in[0,\sigma(T)], \quad \lambda\in(0,1),$$

where $x \in PC([0,\sigma(T)])$. We have

$$D(x(t),\widetilde{0})$$

$$\begin{aligned}
&= D\Big(\lambda\int_0^{\sigma(T)} g(t,s)\cdot f(s,x(\sigma(s)))\delta_H s + \lambda\sum_{j=1}^{m} g(t,t_j)\cdot I_j(x(t_j)),\widetilde{0}\Big) \\
&\leq \lambda\int_0^{\sigma(T)} |g(t,s)|D(f(s,x(\sigma(s))),\widetilde{0})\Delta s + \lambda\sum_{j=1}^{m} |g(t,t_j)|D(I_j(x(t_j)),\widetilde{0})
\end{aligned}$$

$$
\begin{aligned}
&\leq \lambda M \int_0^{\sigma(T)} D(f(s,x(\sigma(s))),\widetilde{0})\Delta s + \lambda M \sum_{j=1}^{m} D(I_j(x(t_j)),\widetilde{0}) \\
&\leq \alpha\lambda M \int_0^{\sigma(T)} D(x(\sigma(s)),\widetilde{0})\Delta s + K\lambda M\sigma(T) \\
&\quad + \lambda M \sum_{j=1}^{m} (\beta_j D(x(t_j),\widetilde{0}) + \gamma_j) \\
&= \alpha\lambda M\|x\| + K\lambda M\sigma(T) + \lambda M \sum_{j=1}^{m} (\beta_j\|x\| + \gamma_j) \\
&= K\lambda M\sigma(T) + \lambda M \left(\sum_{j=1}^{m} \beta_j + \alpha\sigma(T)\right) \|x\| + \gamma_j), \quad t \in [0,\sigma(T)].
\end{aligned}
$$

Hence,

$$
\|x\| \leq K\lambda M\sigma(T) + \lambda M \left(\sum_{j=1}^{m} \beta_j + \alpha\sigma(T)\right) \|x\| + \lambda M \sum_{j=1}^{m} \gamma_j
$$

and

$$
\|x\| \leq \frac{K\lambda M\sigma(T) + \lambda M \sum\limits_{j=1}^{m} \gamma_j}{1 - \lambda M \left(\sum\limits_{j=1}^{m} \beta_j + \alpha\sigma(T)\right)}.
$$

From this and Schaefer's fixed-point theorem,[1] *it follows that the operator Ψ has a fixed point in $PC([0,\sigma(T)])$. This completes the proof.*

[1] **Theorem 4.3.3 (Schaefer's fixed-point Theorem)** *Assume that S is a normed linear space, and the operator $F : S \to S$ is compact. Define*

$$
H(F) = \{x \in S : x = \mu F x, \quad \mu \in (0,1)\}.
$$

Then, either

1. *the set $H(F)$ is unbounded, or*
2. *F has a fixed point in S.*

Now suppose that

(H4) $p : [0,T] \to (0,\infty)$ is continuous in a right-dense direction,

$t_{m+1} = \sigma(T)$, $J_1 = [0,t_1]$.

Lemma 4.3.4 *Suppose* (*H4*). *For the function g, defined by* (4.10), *we have the following estimate*

$$A = \frac{1}{e_p(\sigma(T),0)-1} \le g(t,s) \le \frac{e_p(\sigma(T),0)}{e_p(\sigma(T),0)-1} = B, \quad t,s \in [0,\sigma(T)].$$

Proof 4.3.5 *1. Let $0 \le s < t \le \sigma(T)$. Then*

$$\begin{aligned}
e_{\ominus p}(t,s) &= e^{\int_s^t \frac{1}{\mu(\tau)}\log(1+\mu(\tau)\ominus p(\tau))\Delta\tau} \\
&= e^{\int_s^t \frac{1}{\mu(\tau)}\log\left(1-\frac{\mu(\tau)p(\tau)}{1+\mu(\tau)p(\tau)}\right)\Delta\tau} \\
&= e^{\int_s^t \frac{1}{\mu(\tau)}\log\frac{1}{1+\mu(\tau)p(\tau)}\Delta\tau} \\
&= e^{-\int_s^t \frac{1}{\mu(\tau)}\log(1+\mu(\tau)p(\tau))\Delta\tau} \\
&= \frac{1}{e^{\int_s^t \frac{1}{\mu(\tau)}\log(1+\mu(\tau)p(\tau))\Delta\tau}} \\
&\ge \frac{1}{e^{\int_0^{\sigma(T)} \frac{1}{\mu(\tau)}\log(1+\mu(\tau)p(\tau))\Delta\tau}} \\
&= \frac{1}{e_p(\sigma(T),0)}, \\
g(t,s) &= \frac{e_{\ominus p}(t,s)}{1-e_{\ominus p}(\sigma(T),0)}
\end{aligned}$$

$$
\begin{aligned}
&\geq \frac{1}{1-e_{\ominus p}(\sigma(T),0)} \frac{1}{e_p(\sigma(T),0)} \\
&= \frac{1}{e_p(\sigma(T),0)-1} \\
&= A,
\end{aligned}
$$

$$
\begin{aligned}
e_{\ominus p}(t,s) &= e^{-\int_s^t \frac{1}{\mu(\tau)} \log(1+\mu(\tau)p(\tau))\Delta\tau} \\
&\leq 1,
\end{aligned}
$$

$$
\begin{aligned}
g(t,s) &= \frac{e_{\ominus p}(t,s)}{1-e_{\ominus p}(\sigma(T),0)} \\
&\leq \frac{1}{1-e_{\ominus p}(\sigma(T),0)} \\
&= \frac{e_p(\sigma(T),0)}{e_p(\sigma(T),0)-1} \\
&= B.
\end{aligned}
$$

2. *Let* $0 \leq t \leq s \leq \sigma(T)$. *Then*

$$
\begin{aligned}
e_{\ominus p}(t,s) &= e^{-\int_s^t \frac{1}{\mu(\tau)} \log(1+\mu(\tau)p(\tau))\Delta\tau} \\
&= e^{\int_t^s \frac{1}{\mu(\tau)} \log(1+\mu(\tau)p(\tau))\Delta\tau} \\
&\geq 1,
\end{aligned}
$$

$$
g(t,s) = \frac{e_{\ominus p}(\sigma(T),0)e_{\ominus p}(t,s)}{1-e_{\ominus p}(\sigma(T),0)}
$$

$$\begin{aligned} &\geq \frac{e_{\ominus p}(\sigma(T),0)}{1-e_{\ominus p}(\sigma(T),0)} \\ &= \frac{1}{e_p(\sigma(T),0)-1} \\ &= A, \end{aligned}$$

$$\begin{aligned} e_{\ominus p}(t,s) &= e^{\int\limits_t^s \frac{1}{\mu(\tau)}\log(1+\mu(\tau)p(\tau))\Delta\tau} \\ &\leq e^{\int\limits_t^{\sigma(T)} \frac{1}{\mu(\tau)}\log(1+\mu(\tau)p(\tau))\Delta\tau} \\ &\leq e^{\int\limits_0^{\sigma(T)} \frac{1}{\mu(\tau)}\log(1+\mu(\tau)p(\tau))\Delta\tau} \\ &= e_p(\sigma(T),0), \end{aligned}$$

$$\begin{aligned} g(t,s) &= \frac{e_{\ominus p}(\sigma(T),0)e_{\ominus p}(t,s)}{1-e_{\ominus p}(\sigma(T),0)} \\ &\leq \frac{e_{\ominus p}(t,s)}{e_p(\sigma(T),0)-1} \\ &\leq \frac{e_p(\sigma(T),0)}{e_p(\sigma(T),0)-1} \\ &= B. \end{aligned}$$

This completes the proof.

Lemma 4.3.6 *Let p satisfy* $(H1)$-$(H4)$. *We have*

$$\int\limits_0^{\sigma(T)} g(t,s)p(s)\Delta s = 1.$$

Proof 4.3.7 *We have*

$$\begin{aligned}
\int_0^{\sigma(T)} g(t,s)p(s)\Delta s &= \int_0^t g(t,s)p(s)\Delta s + \int_t^{\sigma(T)} g(t,s)p(s)\Delta s \\
&= \frac{1}{1-e_{\ominus p}(\sigma(T),0)}\Bigg(\int_0^t e_p(s,t)p(s)\Delta s \\
&\quad + e_{\ominus p}(\sigma(T),0)\int_t^{\sigma(T)} e_p(s,t)p(s)\Delta s\Bigg) \\
&= \frac{1}{1-e_{\ominus p}(\sigma(T),0)}\Bigg(e_p(t,t)-e_p(0,t) \\
&\quad + e_p(0,\sigma(T))(e_p(\sigma(T),t)-e_p(t,t))\Bigg) \\
&= \frac{1}{1-e_{\ominus p}(\sigma(T),0)}(1-e_p(0,t)+e_p(0,t)-e_p(0,\sigma(T))) \\
&= \frac{1-e_{\ominus p}(\sigma(T),0)}{1-e_{\ominus p}(\sigma(T),0)} \\
&= 1.
\end{aligned}$$

This completes the proof.

Denote

$$\begin{aligned}
\delta &= \frac{A}{B}, \\
f^0 &= \limsup_{t\to 0+}\max_{t\in[0,T]}\frac{D(f(t,x),\widetilde{0})}{p(t)D(x,\widetilde{0})}, \\
f_0 &= \liminf_{t\to 0+}\min_{t\in[0,T]}\frac{D(f(t,x),\widetilde{0})}{p(t)D(x,\widetilde{0})},
\end{aligned}$$

$$\begin{aligned}
f^{\infty} &= \limsup_{t\to\infty} \max_{t\in[0,T]} \frac{D(f(t,x),\widetilde{0})}{p(t)D(x,\widetilde{0})}, \\
f_{\infty} &= \liminf_{t\to\infty} \min_{t\in[0,T]} \frac{D(f(t,x),\widetilde{0})}{p(t)D(x,\widetilde{0})}, \\
f^{u} &= \sup_{t\in[\delta u,u]} \max_{t\in[0,T]} \frac{D(f(t,x),\widetilde{0})}{p(t)D(x,\widetilde{0})}, \\
f_{u} &= \inf_{t\in[\delta u,u]} \min_{t\in[0,T]} \frac{D(f(t,x),\widetilde{0})}{p(t)D(x,\widetilde{0})}, \\
I^{0}(k) &= \limsup_{x\to\widetilde{0}+} \frac{D(I_k(x),\widetilde{0})}{D(x,\widetilde{0})}, \\
I_{0}(k) &= \liminf_{x\to\widetilde{0}+} \frac{D(I_k(x),\widetilde{0})}{D(x,\widetilde{0})}, \\
I^{\infty}(k) &= \limsup_{x\to\infty} \frac{D(I_k(x),\widetilde{0})}{D(x,\widetilde{0})}, \\
I_{\infty}(k) &= \liminf_{x\to\infty} \frac{D(I_k(x),\widetilde{0})}{D(x,\widetilde{0})}, \\
I^{u}(k) &= \sup_{D(x,\widetilde{0})\in[\delta u,u]} \frac{D(I_k(x),\widetilde{0})}{D(x,\widetilde{0})}, \\
I_{u}(k) &= \inf_{D(x,\widetilde{0})\in[\delta u,u]} \frac{D(I_k(x),\widetilde{0})}{D(x,\widetilde{0})}, \quad k\in\{1,\ldots,m\}.
\end{aligned}$$

For our next results, we will use the following fixed-point theorem.

Theorem 4.3.8 *Let X be a Banach space and K be a cone in X. Assume that Ω_1, Ω_2 are open subsets of X with $0\in\Omega_1$, $\overline{\Omega_1}\subset\Omega_2$. Let*

$$\Phi : K\bigcap\left(\overline{\Omega_2}\backslash\Omega_1\right)\to K,$$

be a continuous and completely continuous operator such that

1. $\|\Phi x\| \leq \|x\|$, $x \in K\bigcap \partial\Omega_1$,
2. *there exists an* $e \in K\backslash\{0\}$ *such that* $x \neq \Phi x + \lambda e$ *for* $x \in K\bigcap \partial\Omega_2$ *and* $\lambda > 0$.

Then Φ *has a fixed point in* $K\bigcap\left(\overline{\Omega_2}\backslash\Omega_1\right)$.

Define

$$X = \{x : x \in PC([0,\sigma(T)]), \quad x(0) = x(\sigma(T))\}$$

with the norm

$$\|x\| = \sup_{t\in[0,\sigma(T)]} D(x(t),\widetilde{0}).$$

Note that X is a Banach space. Let

$$K = \{x \in X : D(x(t),\widetilde{0}) \geq \delta\|x\|, \quad t \in [0,\sigma(T)]\}.$$

Theorem 4.3.9 *Suppose* $(H2)$-$(H4)$ *and*

$$D\left(\int_0^{\sigma(T)} g(t,s)\cdot f(s,x(\sigma(s)))\delta_H s + \sum_{k=1}^m g(t,t_k)\cdot I_k(x(t_k)),\widetilde{0}\right)$$

$$\geq s\sum_{k=1}^m g(t,t_k)D\left(I_k(x(t_k)),\widetilde{0}\right), \tag{4.16}$$

for any $x \in X$, $x \geq \widetilde{0}$, *for some constant* $s \in (0,1)$. *In addition, assume and there exist* $0 < \alpha < \beta$ *such that*

$$f(t,x) \geq \widetilde{0}, \quad I_k(x) \geq \widetilde{0}, \quad t \in [0,\sigma(T)], \quad D(x,\widetilde{0}) \in [\delta\alpha,\beta],$$

$$sA\sum_{k=1}^m I_\beta(k) > 1, \quad f^\alpha + B\sum_{k=1}^m I^\alpha(k) \leq 1. \tag{4.17}$$

Then problem (4.14), (4.15) *has at least one positive solution.*

Proof 4.3.10 *As in the proof of Theorem 4.3.1, we have that* $\Psi : K \to X$ *is continuous and completely continuous. Define the sets*

$$\Omega_1 = \{x \in X : \|x\| < \alpha\},$$

$$\Omega_2 = \{x \in X : \|x\| < \beta\}.$$

1. *We will prove that* $\Psi : K \bigcap (\overline{\Omega_2} \backslash \Omega_1) \to K$. *For* $x \in K \bigcap (\overline{\Omega_2} \backslash \Omega_1)$, *we have*

$$\delta\alpha \leq D(x(t), \widetilde{0}) \leq \beta, \quad t \in [0, \sigma(T)],$$

and, using Lemma 4.3.6, we get

$$D(\Psi x(t), \widetilde{0})$$

$$= D\left(\int_0^{\sigma(T)} g(t,s) \cdot f(s, x(\sigma(s)))\delta_H s + \sum_{k=1}^{m} g(t, t_k) \cdot I_k(x(t_k)), \widetilde{0} \right)$$

$$\leq BD\left(\int_0^{\sigma(T)} f(s, x(\sigma(s)))\delta_H s + \sum_{k=1}^{m} I_k(x(t_k)), \widetilde{0} \right), \quad t \in [0, \sigma(T)],$$

$$\|\Psi x\| \leq BD\left(\int_0^{\sigma(T)} f(s, x(\sigma(s)))\delta_H s + \sum_{k=1}^{m} I_k(x(t_k)), \widetilde{0} \right),$$

and

$$\begin{aligned} D(\Psi x(t), \widetilde{0}) &\geq AD\left(\int_0^{\sigma(T)} f(s, x(\sigma(s)))\delta_H s + \sum_{k=1}^{m} I_k(x(t_k)), \widetilde{0} \right) \\ &= \frac{A}{B} BD\left(\int_0^{\sigma(T)} f(s, x(\sigma(s)))\delta_H s + \sum_{k=1}^{m} I_k(x(t_k)), \widetilde{0} \right) \\ &\geq \frac{A}{B} \|\Psi x\|, \quad t \in [0, \sigma(T)]. \end{aligned}$$

Thus, $\Psi : K \bigcap (\overline{\Omega_2} \backslash \Omega_1) \to K$.

2. *We will prove*

$$\|\Psi x\| \leq \|x\|, \quad x \in K\bigcap \partial\Omega_1.$$

We have

$$D(f(t,x),\widetilde{0}) \ \leq \ f^{\alpha}p(t)D(x,\widetilde{0}),$$

$$D(I_k(x),\widetilde{0}) \ \leq \ I^{\alpha}(k)D(x,\widetilde{0}), \quad k \in \{1,\ldots,m\},$$

$$D(x,\widetilde{0}) \ \in \ [\delta\alpha,\alpha], \quad t \in [0,\sigma(T)].$$

Hence, and the second inequality of (4.17)*, we obtain*

$$\begin{aligned} f^{\alpha} &\leq 1 - B\sum_{k=1}^{m} D(I^{\alpha}(k),\widetilde{0}), \\ D(f(t,x),\widetilde{0}) &\leq f^{\alpha}p(t)D(x,\widetilde{0}) \\ &\leq \left(1 - B\sum_{k=1}^{m} D(I^{\alpha}(k),\widetilde{0})\right)p(t)D(x,\widetilde{0}), \quad t \in [0,\sigma(T)], \end{aligned}$$

$D(x,\widetilde{0}) \in [\delta\alpha,\alpha]$. *For* $x \in K\bigcap\Omega_1$*, we get*

$$\begin{aligned} \delta\alpha &= \delta\|x\| \\ &\leq D(x(t),\widetilde{0}) \\ &\leq \alpha, \quad t \in [0,\sigma(T)], \end{aligned}$$

and, using Lemma 4.3.6, we find

$$\begin{aligned} & D(\Psi x(t),\widetilde{0}) \\ &= D\left(\int_0^{\sigma(T)} g(t,s)\cdot f(s,x(\sigma(s)))\delta_H s + \sum_{k=1}^{m} g(t,t_k)\cdot I_k(x(t_k)),\widetilde{0}\right) \\ &\leq D\left(\int_0^{\sigma(T)} g(t,s)\cdot f(s,x(\sigma(s)))\delta_H s,\widetilde{0}\right) \end{aligned}$$

$$
\begin{aligned}
& +D\left(\sum_{k=1}^{m} g(t,t_k)\cdot I_k(x(t_k)),\widetilde{0}\right) \\
\leq\ & D\left(\int_0^{\sigma(T)} g(t,s)\cdot f(s,x(\sigma(s)))\delta_H s,\widetilde{0}\right) \\
& +\sum_{k=1}^{m} g(t,t_k)D(I_k(x(t_k)),\widetilde{0}) \\
\leq\ & \left(1-B\sum_{k=1}^{m} I^{\alpha}(k)\right)\int_0^{\sigma(T)} g(t,s)p(s)D(x(\sigma(s)),\widetilde{0})\Delta s \\
& +B\sum_{k=1}^{m} I^{\alpha}(k)D(x(t_k),\widetilde{0}) \\
\leq\ & \left(\left(1-B\sum_{k=1}^{m} I^{\alpha}(k)\right)\int_0^{\sigma(T)} g(t,s)p(s)\Delta s+B\sum_{k=1}^{m} I^{\alpha}(k)\right)\|x\| \\
=\ & \left(1-B\sum_{k=1}^{m} I^{\alpha}(k)+B\sum_{k=1}^{m} I^{\alpha}(k)\right)\|x\| \\
=\ & \|x\|,\quad t\in[0,\sigma(T)],
\end{aligned}
$$

whereupon

$$\|\Psi x\| \leq \|x\|.$$

3. *Let e be the unit fuzzy number. Then $e \in K\backslash\{0\}$. Assume that there exist $x_0 \in K\bigcap\partial\Omega_2$ and a $\lambda_0 > 0$ such that*

$$x_0 = \Psi x_0 + \lambda_0 \cdot e.$$

Since $x_0 \in K\bigcap\partial\Omega_2$, we obtain

$$\delta\beta \;=\; \delta\|x_0\|$$

$$\leq \|x_0\|$$

$$\leq \beta.$$

Also,

$$D(I_k(x),\widetilde{0}) \geq I_\beta(k)D(x,\widetilde{0}), \quad D(x,\widetilde{0}) \in [\delta\beta,\beta], \quad k \in \{1,\ldots,m\}.$$

Let

$$\mu = \min_{t\in[0,\sigma(T)]} D(x_0(t),\widetilde{0}).$$

Then, using Lemma 4.3.6, we obtain

$$\begin{aligned}
& D(x_0(t),\widetilde{0}) \\
= \;& D(\Psi x_0(t) + \lambda_0 \cdot e,\widetilde{0}) \\
= \;& D\Bigg(\int_0^{\sigma(T)} g(t,s)\cdot f(s,x_0(\sigma(s)))\delta_H s \\
& + \sum_{k=1}^{m} g(t,t_k)\cdot I_k(x_0(t_k)) + \lambda_0 \cdot e,\widetilde{0}\Bigg) \\
\geq \;& D\left(\int_0^{\sigma(T)} g(t,s)\cdot f(s,x(\sigma(s)))\delta_H s + \sum_{k=1}^{m} g(t,t_k)\cdot I_k(x(t_k)),\widetilde{0}\right) \\
\geq \;& s\sum_{k=1}^{m} g(t,t_k) D(I_k(x(t_k)),\widetilde{0}) \\
\geq \;& s\sum_{k=1}^{m} g(t,t_k) I_\beta(k) D(x(t_k),\widetilde{0}) \\
\geq \;& s\mu A \sum_{k=1}^{m} I_\beta(k) \\
> \;& \mu, \quad t \in [0,\sigma(T)],
\end{aligned}$$

which is a contradiction.

From here and from Theorem 4.3.8, it follows that the problem (4.14), (4.15) *has at least one positive solution. This completes the proof.*

Corollary 4.3.11 *Suppose* $(H1)$–$(H4)$, (4.16), *and there exist constants* α, β *such that* $0 < \alpha < \beta$ *and*

$$f(t,x) \geq \widetilde{0}, \quad I_k(x) \geq \widetilde{0}, \quad t \in [0, \sigma(T)], \quad D(x, \widetilde{0}) \in [\delta\alpha, \beta],$$

and

$$sA \sum_{k=1}^{m} I_\infty(k) > 1, \quad f^0 + B \sum_{k=1}^{m} I^0(k) \leq 1.$$

Then problem (4.14), (4.15) *has at least one positive solution.*

Proof 4.3.12 *By the definitions of* f^0 *and* I^0, *it follows that there exists an* $\alpha > 0$ *such that*

$$\begin{aligned} I^0(k) &\geq \limsup_{x \in [\delta\alpha, \alpha]} \frac{D(I_k(x), \widetilde{0})}{D(x, \widetilde{0})}, \\ f^0 &\geq \limsup_{x \in [\delta\alpha, \alpha]} \max_{t \in [0,T]} \frac{D(f(t,x), \widetilde{0})}{p(t) D(x, \widetilde{0})}. \end{aligned}$$

Hence,

$$I^\alpha(k) \leq I^0(k), \quad f^0 \geq f^\alpha.$$

Next, by the definitions of f_∞ *and* I_∞, *it follows that there exists a* $\beta > \alpha$ *such that*

$$\begin{aligned} I_\infty(k) &\leq \liminf_{x \in [\delta\beta, \beta]} \frac{D(I_k(x), \widetilde{0})}{D(x, \widetilde{0})}, \\ f_\infty &\leq \liminf_{x \in [\delta\beta, \beta]} \min_{t \in [0,T]} \frac{D(f(t,x), \widetilde{0})}{p(t) D(x, \widetilde{0})}. \end{aligned}$$

From here,

$$I_\beta(k) \geq I_\infty(k),$$

$$f_\beta \geq f_\infty.$$

Then

$$sA\sum_{k=1}^{m} I_\beta(k) > 1,$$

$$f^\alpha + B\sum_{k=1}^{m} I^\alpha(k) \leq 1.$$

Now, using Theorem 4.3.9, it follows that problems (4.14), (4.15) *has at least one positive solution. This completes the proof.*

Theorem 4.3.13 *Suppose* $(H1)$–$(H4)$, (4.16) *and there exist* $0 < \alpha < \beta$ *such that*

$$f(t,x) \geq \widetilde{0}, \quad I_k(x) \geq \widetilde{0}, \quad t \in [0,\sigma(T)], \quad D(x,\widetilde{0}) \in [\delta\alpha,\beta],$$

$$f^\beta + B\sum_{k=1}^{m} I^\beta(k) \leq 1, \quad sA\sum_{k=1}^{m} I_\alpha(k) > 1. \tag{4.18}$$

Then problem (4.14), (4.15) *has at least one positive solution.*

Proof 4.3.14 *Let the operator* $\Psi : K \to X$ *and the sets* Ω_1, Ω_2 *be as in Theorem 4.3.9. As in the proof of Theorem 4.3.1, we have that* $\Psi : K \to X$ *is continuous and completely continuous.*

1. *As in the proof of Theorem 4.3.9, we have* $\Psi : K\bigcap\left(\overline{\Omega_2}\backslash\Omega_1\right) \to K$.

2. *We will prove*

   $$\|\Psi x\| \leq \|x\|, \quad x \in K\bigcap\partial\Omega_2.$$

   We have

   $$D(f(t,x),\widetilde{0}) \leq f^\beta p(t)D(x,\widetilde{0}), \quad D(I_k(x),\widetilde{0}) \leq I^\beta(k)D(x,\widetilde{0}),$$

   $k \in \{1,\ldots,m\}$, $D(x,\widetilde{0}) \in [\delta\beta,\beta]$, $t \in [0,\sigma(T)]$. *Hence and the first inequality of* (4.18), *we obtain*

   $$f^\beta \leq 1 - B\sum_{k=1}^{m} I^\beta(k),$$

$$\begin{aligned} D(f(t,x),\widetilde{0}) &\leq f_0^{\beta} p(t) D(x,\widetilde{0}) \\ &\leq \left(1 - B\sum_{k=1}^{m} I^{\beta}(k)\right) p(t) D(x,\widetilde{0}), \quad t \in [0,\sigma(T)], \end{aligned}$$

$D(x,\widetilde{0}) \in [\delta\beta,\beta]$. For $x \in K\bigcap\Omega_2$, we get

$$\begin{aligned} \delta\beta &= \delta\|x\| \\ &\leq \beta, \quad t \in [0,\sigma(T)], \end{aligned}$$

and, using Lemma 4.3.6, we find

$$\begin{aligned} & D(\Psi x(t),\widetilde{0}) \\ = \;& D\left(\int_0^{\sigma(T)} g(t,s)\cdot f(s,x(\sigma(s)))\delta_H s + \sum_{k=1}^{m} g(t,t_k)\cdot I_k(x(t_k)),\widetilde{0}\right) \\ \leq \;& D\left(\int_0^{\sigma(T)} g(t,s)\cdot f(s,x(\sigma(s)))\delta_H s,\widetilde{0}\right) \\ & + D\left(\sum_{k=1}^{m} g(t,t_k)\cdot I_k(x(t_k)),\widetilde{0}\right) \\ \leq \;& \int_0^{\sigma(T)} g(t,s) D(f(s,x(\sigma(s))),\widetilde{0})\Delta s + \sum_{k=1}^{m} g(t,t_k) D(I_k(x(t_k)),\widetilde{0}) \\ \leq \;& \left(1 - B\sum_{k=1}^{m} I^{\beta}(k)\right) \int_0^{\sigma(T)} g(t,s) p(s) D(x(s),\widetilde{0})\Delta s \\ & + \sum_{k=1}^{m} g(t,t_k) I^{\beta}(k) D(x(t_k),\widetilde{0}) \end{aligned}$$

$$\leq \left(\left(1-B\sum_{k=1}^{m} I^{\beta}(k)\right)\int_{0}^{\sigma(T)} g(t,s)p(s)\Delta s + B\sum_{k=1}^{m} I^{\beta}(k)\right)\|x\|$$

$$= \|x\|, \quad t\in[0,\sigma(T)],$$

whereupon

$$\|\Psi x\| \leq \|x\|.$$

3. *Let e be the unit fuzzy number. Then* $e\in K\backslash\{0\}$. *Assume that there exist and* $x_0\in K\bigcap\partial\Omega_1$ *and a* $\lambda_0>0$ *such that*

$$x_0 = \Psi x_0 + \lambda_0\cdot e.$$

Since $x_0\in K\bigcap\partial\Omega_1$, *we obtain*

$$\begin{aligned} \delta\alpha &= \delta\|x_0\| \\ &\leq \|x_0\| \\ &\leq \alpha. \end{aligned}$$

Also,

$$D(f(t,x),\widetilde{0}) \geq f_{\alpha}p(t)D(x,\widetilde{0}), \ t\in[0,\sigma(T)], \ D(x,\widetilde{0})\in[\delta\alpha,\alpha],$$

and

$$D(I_k(x),\widetilde{0}) \geq I_{\alpha}(k)D(x,\widetilde{0}), \quad D(x,\widetilde{0})\in[\delta\alpha,\alpha], \quad k\in\{1,\ldots,m\}.$$

Let

$$\mu = \min_{t\in[0,\sigma(T)]} D(x_0(t),\widetilde{0}).$$

Then, using Lemma 4.3.6, we obtain

$$\begin{aligned} \mu &= D(x_0(t),\widetilde{0}) \\ &= D(\Psi x_0(t)+\lambda_0\cdot e,\widetilde{0}) \end{aligned}$$

$$
\begin{aligned}
&= D\Big(\int_0^{\sigma(T)} g(t,s)\cdot f(s,x_0(\sigma(s)))\delta_H s \\
&+ \sum_{k=1}^{m} g(t,t_k)\cdot I_k(x_0(t_k)) + \lambda_0, \widetilde{0}\Big) \\
&\geq D\left(\int_0^{\sigma(T)} g(t,s)\cdot f(s,x(\sigma(s)))\delta_H s + \sum_{k=1}^{m} g(t,t_k)\cdot I_k(x(t_k)), \widetilde{0}\right) \\
&\geq s\sum_{k=1}^{m} g(t,t_k) D(I_k(x_0(t_k)), \widetilde{0}) \\
&\geq s\sum_{k=1}^{m} g(t,t_k) I_\alpha(k) D(x_0(t_k), \widetilde{0}) \\
&\geq sA\mu \sum_{k=1}^{m} I_\alpha(k) \\
&> \mu.
\end{aligned}
$$

This is a contradiction.

From here and from Theorem 4.3.8, it follows that the problem (4.14), (4.15) *has at least one positive solution. This completes the proof.*

Corollary 4.3.15 *Suppose* $(H1)$–$(H4)$, (4.16) *and*

$$f(t,x) \geq \widetilde{0}, \quad I_k(x) \geq \widetilde{0}, \quad t \in [0, \sigma(T)], \quad D(x, \widetilde{0}) \in [\delta\alpha, \beta],$$

$$sA\sum_{k=1}^{m} I_0(k) > 1, \quad f^\infty + B\sum_{k=1}^{m} I^\infty(k) \leq 1.$$

Then problem (4.14), (4.15) *has at least one positive solution.*

Proof 4.3.16 *By the definitions of* f_0 *and* I_0*, it follows that there exists an* $\alpha > 0$ *such that*

$$I_0(k) \leq \liminf_{x\in[\delta\alpha,\alpha]} \frac{D(I_k(x), \widetilde{0})}{D(x, \widetilde{0})}, \quad f_0 \geq \liminf_{x\in[\delta\alpha,\alpha]} \min_{t\in[0,T]} \frac{D(f(t,x), \widetilde{0})}{p(t)D(x, \widetilde{0})}.$$

Hence,

$$I_\alpha(k) \geq I_0(k), \quad f_0 \leq f_\alpha.$$

Next, by the definitions of f^∞ *and* I^∞, *it follows that there exists an* $\beta > \alpha$ *such that*

$$I^\infty(k) \geq \limsup_{x \in [\delta\beta, \beta]} \frac{D(I_k(x), \widetilde{0})}{D(x, \widetilde{0})}, \quad f^\infty \geq \limsup_{x \in [\delta\beta, \beta]} \max_{t \in [0,T]} \frac{D(f(t,x), \widetilde{0})}{p(t) D(x, \widetilde{0})}.$$

From here,

$$I^\beta(k) \leq I^\infty(k), \quad f^\infty \geq f^\beta.$$

Then

$$sA \sum_{k=1}^{m} I_\alpha(k) > 1, \quad f^\beta + B \sum_{k=1}^{m} I^\beta(k) \leq 1.$$

Now, using Theorem 4.3.13, it follows that problem (4.14), (4.15) *has at least one positive solution. This completes the proof.*

Exercise 4.3.17 *Find conditions for the function g so that the problem*

$$\begin{aligned} \delta_H x(t) + p(t) \cdot x(\sigma(t)) &= g(t), \quad t \in J_0, \\ x(t_k^+) &= x(t_k^-), \quad k \in \{1, \ldots, m\}, \\ x(0) &= x(\sigma(T)), \end{aligned}$$

where p satisfies $(H1)$ *and* $(H4)$, *has at least one positive solution.*

Theorem 4.3.18 *Suppose* $(H1)$–$(H4)$, (4.16) *and there exist* $0 < \alpha < \rho < \beta$ *such that*

$$f(t,x) \geq \widetilde{0}, \quad I_k(x) \geq \widetilde{0}, \quad t \in [0, \sigma(T)], \quad D(x, \widetilde{0}) \in [\delta\alpha, \beta],$$

$$f^\alpha + B \sum_{k=1}^{m} I^\alpha(k) \leq 1, \quad sA \sum_{k=1}^{m} I_\rho(k) > 1,$$

$$f^{\beta}+B\sum_{k=1}^{m}I^{\beta}(k)\ \leq\ 1.$$

Then problem (4.14), (4.15) *has at least two positive solutions.*

Proof 4.3.19 *Let* Ψ*,* Ω_1*,* Ω_2 *be as in the proof of Theorem 4.3.9. Define*

$$\Omega_3=\{x\in X: \|x\|<\rho\}.$$

By Theorem 4.3.9, it follows that the problem (4.14), (4.15) *has at least one positive solution in* $\overline{\Omega_3}\backslash\Omega_1$*. By Theorem 4.3.13, it follows that the problem* (4.14), (4.15) *has at least one positive solution in* $\overline{\Omega_2}\backslash\Omega_3$*. This completes the proof.*

Theorem 4.3.20 *Suppose* $(H1)$*-*$(H4)$, (4.16) *and there exist* $0<\alpha<\rho<\beta$ *such that*

$$f(t,x)\geq\widetilde{0},\quad I_k(x)\geq\widetilde{0},\quad t\in[0,\sigma(T)],\quad D(x,\widetilde{0})\in[\delta\alpha,\beta],$$

$$sA\sum_{k=1}^{m}I^{\alpha}(k)\ >\ 1,\quad f_{\rho}+B\sum_{k=1}^{m}I_{\rho}(k)\leq 1,$$

$$sA\sum_{k=1}^{m}I^{\beta}(k)\ >\ 1.$$

Then problem (4.14), (4.15) *has at least two positive solutions.*

Proof 4.3.21 *Let* Ψ*,* Ω_1*,* Ω_2 *and* Ω_3 *be as in the proof of Theorem 4.3.18. By Theorem 4.3.13, it follows that the problem* (4.14), (4.15) *has at least one positive solution in* $\overline{\Omega_3}\backslash\Omega_1$*. By Theorem 4.3.9, it follows that the problem* (4.14), (4.15) *has at least one positive solution in* $\overline{\Omega_2}\backslash\Omega_3$*. This completes the proof.*

Theorem 4.3.22 *Suppose* $(H1)$*-*$(H4)$, (4.16) *and there exist positive constants*

$$0<\alpha_1<\beta_1<\ldots<\alpha_r<\beta_r$$

such that

$$f(t,x) \geq \widetilde{0}, \quad I_k(x) \geq \widetilde{0}, \quad t \in [0,T], \quad D(x,\widetilde{0}) \in [\delta\alpha_1, \beta_r].$$

Then problem (4.14), (4.15) *has at least r positive solutions x_j, $j \in \{1,\ldots,r\}$, satisfying*

$$\alpha_j \leq \|x_j\| \leq \beta_j, \quad j \in \{1,\ldots,r\},$$

if one of the following conditions hold.

1.

$$sA\sum_{k=1}^{m} I_{\alpha_j}(k) > 1, \quad f^{\beta_j} + B\sum_{k=1}^{m} I^{\beta_j}(k) \leq 1, \quad j \in \{1,\ldots,r\}.$$

2.

$$f^{\alpha_j} + B\sum_{k=1}^{m} I^{\alpha_j}(k) \leq 1, \quad sA\sum_{k=1}^{m} I_{\beta_j}(k) > 1, \quad j \in \{1,\ldots,r\}.$$

Proof 4.3.23 *Let* Ψ *be as in the proof of Theorem 4.3.9. Define*

$$\begin{aligned} \Omega_{\alpha_j} &= \{x \in X : \|x\| < \alpha_j\}, \\ \Omega_{\beta_j} &= \{x \in X : \|x\| < \beta_j\}, \quad j \in \{1,\ldots,r\}. \end{aligned}$$

1. *By Theorem 4.3.13, it follows that the problem* (4.14), (4.15) *has at least one positive solution $x_1 \in K\bigcap(\overline{\Omega_{\beta_1}}\backslash\Omega_{\alpha_1})$. By Theorem 4.3.9, it follows that the problem* (4.14), (4.15) *has at least one positive solution $x_2 \in K\bigcap(\overline{\Omega_{\alpha_2}}\backslash\Omega_{\beta_1})$. And so on.*

2. *By Theorem 4.3.9, it follows that the problem* (4.14), (4.15) *has at least one positive solution $x_1 \in K\bigcap(\overline{\Omega_{\beta_1}}\backslash\Omega_{\alpha_1})$. By Theorem 4.3.13, it follows that the problem* (4.14), (4.15) *has at least one positive solution $x_2 \in K\bigcap(\overline{\Omega_{\alpha_2}}\backslash\Omega_{\beta_1})$. And so on. This completes the proof.*

Suppose $(H4)$ and

(H5) $f \in \mathscr{C}(J_0 \times F([0,\infty)))$, $f \geq \widetilde{0}$ on $J_0 \times F([0,\infty))$,

(H6) $I_k \in \mathscr{C}(F([0,\infty)))$, $I_k \geq \widetilde{0}$ on $F([0,\infty))$, $k \in \{1,\ldots,m\}$.

In the following, we will use the following fixed-point theorem.

Theorem 4.3.24 (Guo-Krasnosel'skii fixed-point Theorem) *Let X be a Banach space with a cone K. Let Ω_1, Ω_2 be bounded open subsets of X with*

$$0 \in \Omega_1 \subset \overline{\Omega_1} \subset \Omega_2,$$

$\Phi : K\bigcap(\overline{\Omega_2}\backslash\Omega_1) \to K$ is a completely continuous operator such that either

1.

$$\|\Phi x\| \leq \|x\|, \quad x \in K\bigcap\partial\Omega_1, \quad \|\Phi x\| \geq \|x\|, \quad x \in K\bigcap\partial\Omega_2,$$

or

2.

$$\|\Phi x\| \geq \|x\|, \quad x \in K\bigcap\partial\Omega_1, \quad \|\Phi x\| \leq \|x\|, \quad x \in K\bigcap\partial\Omega_2.$$

Then Φ has at least one fixed point in $K\bigcap(\overline{\Omega_2}\backslash\Omega_1)$.

Denote

$$\max f_0 = \lim_{x\to\widetilde{0}} \max_{t\in[0,T]} \frac{D(f(t,x),\widetilde{0})}{D(x,\widetilde{0})}, \quad \min f_0 = \lim_{x\to\widetilde{0}} \min_{t\in[0,T]} \frac{D(f(t,x),\widetilde{0})}{D(x,\widetilde{0})},$$

$$\max f_\infty = \lim_{x\to\infty} \max_{t\in[0,T]} \frac{D(f(t,x),\widetilde{0})}{D(x,\widetilde{0})}, \quad \min f_\infty = \lim_{x\to\infty} \min_{t\in[0,T]} \frac{D(f(t,x),\widetilde{0})}{D(x,\widetilde{0})}.$$

Theorem 4.3.25 *Suppose (H4)-(H6) and*

$$D\left(\int_0^{\sigma(T)} g(t,s)\cdot f(s,x(\sigma(s)))\delta_H s + \sum_{k=1}^{m} g(t,t_k)\cdot I_k(x(t_k)),\widetilde{0}\right)$$

$$\geq sA \int_0^{\sigma(T)} D(f(s,x(\sigma(s))),\widetilde{0})\Delta s, \tag{4.19}$$

for any $x \in X$, $x \geq \widetilde{0}$ *and for some* $s \in (0,1)$. *In addition, assume that*

$$\max f_0 = 0, \quad \min f_\infty = \infty, \quad \lim_{x \to 0} \frac{D(I_k(x), \widetilde{0})}{D(x, \widetilde{0})} = 0, \quad k \in \{1, \ldots, m\}.$$

Then problem (4.14), (4.15) *has at least one positive solution.*

Proof 4.3.26 *Let* Ψ *be as in the proof of Theorem 4.3.9. As in the proof of Theorem 4.3.1, we have* $\Psi : X \to X$ *is completely continuous. Also, as in the proof of Theorem 4.3.9, it follows that* $\Psi : K \to K$. *Since*

$$\max f_0 = 0, \quad \textit{and} \quad \lim_{x \to 0} \frac{D(I_k(x), \widetilde{0})}{D(x, \widetilde{0})} = 0,$$

for $\varepsilon < (B(\sigma(T) + m))^{-1}$, *there exists a positive constant* r *such that*

$$D(f(t,x), \widetilde{0}) \leq \varepsilon \cdot D(x, \widetilde{0}), \quad D(I_k(x), \widetilde{0}) \leq \varepsilon D(x, \widetilde{0}), \quad k \in \{1, \ldots, m\},$$

$0 < D(x, \widetilde{0}) \leq r$. *Define*

$$\Omega_r = \{x \in X : \|x\| < r\}.$$

Then, for $x \in \Omega_r$, *we get*

$$\begin{aligned}
& D(\Psi x(t), \widetilde{0}) \\
= \; & D\left(\int_0^{\sigma(T)} g(t,s) \cdot f(s, x(\sigma(s))) \delta_H s + \sum_{k=1}^{m} g(t, t_k) \cdot I_k(x(t_k)), \widetilde{0} \right) \\
\leq \; & BD\left(\int_0^{\sigma(T)} f(s, x(\sigma(s))) \delta_H s + \sum_{k=1}^{m} I_k(x(t_k)), \widetilde{0} \right) \\
\leq \; & B\left(D\left(\int_0^{\sigma(T)} f(s, x(\sigma(s))) \delta_H s, \widetilde{0} \right) + D\left(\sum_{k=1}^{m} I_k(x(t_k)), \widetilde{0} \right) \right)
\end{aligned}$$

$$\leq B\left(D\left(\int_0^{\sigma(T)} f(s,x(\sigma(s)))\delta_H s,\widetilde{0}\right)+\sum_{k=1}^{m} D(I_k(x(t_k)),\widetilde{0})\right)$$

$$\leq B\varepsilon\left(\int_0^{\sigma(T)} D(x(\sigma(s)),\widetilde{0})\Delta s+\sum_{k=1}^{m} D(x(t_k),\widetilde{0})\right)$$

$$\leq B\varepsilon(\sigma(T)+m)\|x\|, \quad t\in[0,\sigma(T)],$$

whereupon

$$\begin{aligned}\|\Psi x\| &\leq B\varepsilon(\sigma(T)+m)\|x\| \\ &\leq B\frac{\sigma(T)+m}{B(\sigma(T)+m)}\|x\| \\ &= \|x\|.\end{aligned}$$

Since $\min f_\infty=\infty$, *for* $N>(sA\sigma(T)\delta)^{-1}$, *we get that there exists a constant* $R>\frac{r}{\delta}$ *such that*

$$D(f(t,x),\widetilde{0})\geq ND(x,\widetilde{0}) \quad \textit{for} \quad D(x,\widetilde{0})\geq\delta R.$$

Define

$$\Omega_R=\{x\in X:\|x\|<R\}.$$

Then, for $x\in K\bigcap\partial\Omega_R$, *we have*

$$\begin{aligned}D(x(t),\widetilde{0}) &\geq \delta\|x\| \\ &= \delta R, \quad t\in[0,\sigma(T)],\end{aligned}$$

and

$$D(\Psi x(t),\widetilde{0})$$

$$=D\left(\int_0^{\sigma(T)} g(t,s)\cdot f(s,x(\sigma(s)))\delta_H s+\sum_{k=1}^{m} g(t,t_k)\cdot I_k(x(t_k)),\widetilde{0}\right)$$

$$\begin{aligned}
&\geq sA \int\limits_0^{\sigma(T)} D(f(s,x(\sigma(s))),\widetilde{0})\Delta s \\
&\geq sAN \int\limits_0^{\sigma(T)} D(x(\sigma(s)),\widetilde{0})\Delta s \\
&\geq sAN\delta R\sigma(T) \\
&\geq R \\
&= \|x\|, \quad t \in [0,\sigma(T)],
\end{aligned}$$

whereupon

$$\|\Psi x\| \geq \|x\|.$$

Hence and with Theorem 4.3.24, it follows that problem (4.14), (4.15) *has at least one positive solution. This completes the proof.*

Theorem 4.3.27 *Suppose* $(H4)$-$(H6)$, (4.19) *and*

$$\max f_\infty = 0, \quad \min f_0 = \infty, \quad \lim_{x\to\infty} \frac{D(I_k(x),\widetilde{0})}{D(x,\widetilde{0})} = 0, \quad k \in \{1,\ldots,m\}.$$

Then problem (4.14), (4.15) *has at least one positive solution.*

Proof 4.3.28 *Let* Ψ *be as in the proof of Theorem 4.3.9. As in the proof of Theorem 4.3.1, we have* $\Psi : X \to X$ *is completely continuous. Also, as in the proof of Theorem 4.3.9, it follows that* $\Psi : K \to K$*. Since*

$$\max f_\infty = 0 \quad \textit{and} \quad \lim_{x\to\infty} \frac{D(I_k(x),\widetilde{0})}{D(x,\widetilde{0})} = 0,$$

for $\varepsilon \leq (B(\sigma(T)+m))^{-1}$*, there exists a positive constant r such that*

$$D(f(t,x),\widetilde{0}) \leq \varepsilon D(x,\widetilde{0}), \quad D(I_k(x),\widetilde{0}) \leq \varepsilon D(x,\widetilde{0}), \quad k \in \{1,\ldots,m\},$$

$D(x,\widetilde{0}) \geq \delta r$. Let

$$\Omega_r = \{x \in X : \|x\| < r\}.$$

Then, for $x \in K \bigcap \partial\Omega_r$, we have $\|x\| = r$ and

$$\begin{aligned}
& D(\Psi x(t),\widetilde{0}) \\
= \; & D\left(\int_0^{\sigma(T)} g(t,s) \cdot f(s,x(\sigma(s)))\delta_H s + \sum_{k=1}^{m} g(t,t_k) \cdot I_k(x(t_k)),\widetilde{0} \right) \\
\leq \; & BD\left(\int_0^{\sigma(T)} f(s,x(\sigma(s)))\delta_H s + \sum_{k=1}^{m} I_k(x(t_k)),\widetilde{0} \right) \\
\leq \; & B\left(D\left(\int_0^{\sigma(T)} f(s,x(\sigma(s)))\delta_H s,\widetilde{0} \right) + D\left(\sum_{k=1}^{m} I_k(x(t_k)),\widetilde{0} \right) \right) \\
\leq \; & B\left(\int_0^{\sigma(T)} D(f(s,x(\sigma(s))),\widetilde{0})\Delta s + \sum_{k=1}^{m} D(I_k(x(t_k)),\widetilde{0}) \right) \\
\leq \; & B\varepsilon\left(\int_0^{\sigma(T)} D(x(\sigma(s)),\widetilde{0})\delta s + \sum_{k=1}^{m} D(x(t_k),\widetilde{0}) \right) \\
\leq \; & B\varepsilon(\sigma(T)+m)\|x\| \\
\leq \; & \|x\|, \quad t \in [0,\sigma(T)],
\end{aligned}$$

whereupon

$$\|\Psi x\| \leq \|x\|.$$

Because $\min f_0 = \infty$, *for $N > (sA\sigma(T)\delta)^{-1}$, there exists a constant $R < \delta r$ such that*

$$D(f(t,x),\widetilde{0}) \geq ND(x,\widetilde{0}) \quad \textit{for} \quad 0 < D(x,\widetilde{0}) \leq R.$$

Define

$$\Omega_R = \{x \in X : \|x\| < R\}.$$

Then, for $x \in K \bigcap \partial\Omega_R$, we have

$$\begin{aligned} D(x(t),\widetilde{0}) &\geq \delta\|x\| \\ &= \delta R, \quad t \in [0,\sigma(T)], \end{aligned}$$

and

$$D(\Psi x(t),\widetilde{0})$$

$$\begin{aligned} &= D\left(\int_0^{\sigma(T)} g(t,s)\cdot f(s,x(\sigma(s)))\delta_H s + \sum_{k=1}^{m} g(t,t_k)\cdot I_k(x(t_k)),\widetilde{0}\right) \\ &\geq sA\int_0^{\sigma(T)} D(f(s,x(\sigma(s))),\widetilde{0})\Delta s \\ &\geq sAN\int_0^{\sigma(T)} D(x(\sigma(s)),\widetilde{0})\Delta s \\ &\geq sAN\sigma(T)\|x\| \\ &\geq \|x\|, \quad t \in [0,\sigma(T)], \end{aligned}$$

whereupon

$$\|\Psi x\| \geq \|x\|.$$

Hence and with Theorem 4.3.24, it follows that problem (4.14), (4.15) *has at least one positive solution. This completes the proof.*

Theorem 4.3.29 *Suppose* $(H4)$-$(H6)$, (4.19) *and*

$$\min f_0 = \infty,$$

there exists a positive constant R such that

$$D(f(t,x),\widetilde{0}) \le \frac{D(x,\widetilde{0})}{2B\sigma(T)}, \quad 0 < D(x,\widetilde{0}) \le R, \quad t \in [0,\sigma(T)],$$

and

$$D(I_k(x),\widetilde{0}) \le \frac{D(x,\widetilde{0})}{2mB}, \quad k \in \{1,\dots,m\}, \quad x \in \mathbb{R}.$$

Then problem (4.14), (4.15) *has at least one positive solution.*

Proof 4.3.30 *Let Ψ be as in the proof of Theorem 4.3.9. As in the proof of Theorem 4.3.1, we have $\Psi : X \to X$ is completely continuous. Also, as in the proof of Theorem 4.3.9, it follows that $\Psi : K \to K$. Since* $\min f_0 = \infty$, *for $N > (sA\sigma(T)\delta)^{-1}$, there exists a positive constant $r < \dfrac{R}{2BN\sigma(T)}$ such that*

$$D(f(t,x),\widetilde{0}) \ge ND(x,\widetilde{0}), \quad 0 < D(x,\widetilde{0}) \le r, \quad t \in [0,\sigma(T)].$$

Define

$$\Omega_r = \{x \in X : \|x\| < r\}.$$

Then, for any $x \in K \bigcap \partial\Omega_r$, we have

$$\begin{aligned} D(x(t),\widetilde{0}) &\ge \delta\|x\| \\ &= \delta r, \quad t \in [0,\sigma(T)], \end{aligned}$$

and

$$\begin{aligned} &D(\Psi x(t),\widetilde{0}) \\ &= D\left(\int_0^{\sigma(T)} g(t,s)\cdot f(s,x(\sigma(s)))\delta_H s + \sum_{k=1}^{m} g(t,t_k)\cdot I_k(x(t_k)),\widetilde{0}\right) \\ &\ge sA\int_0^{\sigma(T)} D(f(s,x(\sigma(s))),\widetilde{0})\Delta s \end{aligned}$$

$$
\begin{aligned}
&\geq sAN \int_0^{\sigma(T)} D(x(\sigma(s)),\widetilde{0})\Delta s \\
&\geq sAN\delta r\sigma(T) \\
&\geq r \\
&= \|x\|, \quad t \in [0,\sigma(T)],
\end{aligned}
$$

whereupon

$$
\|\Psi x\| \geq \|x\|.
$$

Let

$$
\Omega_R = \{x \in X : \|x\| < R\}.
$$

Then, for $x \in K \bigcap \partial\Omega_R$, *we obtain*

$$
D(\Psi x(t),\widetilde{0}) \tag{4.20}
$$

$$
\begin{aligned}
&= D\left(\int_0^{\sigma(T)} g(t,s)\cdot f(s,x(\sigma(s)))\delta_H s + \sum_{k=1}^{m} g(t,t_k)\cdot I_k(x(t_k)),\widetilde{0}\right) \\
&\leq BD\left(\int_0^{\sigma(T)} f(s,x(\sigma(s)))\delta_H s + \sum_{k=1}^{m} I_k(x(t_k)),\widetilde{0}\right) \\
&\leq B\left(D\left(\int_0^{\sigma(T)} f(s,x(\sigma(s)))\delta_H s,\widetilde{0}\right) + D\left(\sum_{k=1}^{m} I_k(x(t_k)),\widetilde{0}\right)\right) \\
&\leq B\left(\int_0^{\sigma(T)} D(f(s,x(\sigma(s))),\widetilde{0})\Delta s + \sum_{k=1}^{m} g(t,t_k)\cdot D(I_k(x(t_k)),\widetilde{0})\right)
\end{aligned}
$$

$$\begin{aligned}
&\leq B\left(\frac{R}{2B\sigma(T)}\sigma(T)+\frac{1}{2mB}\sum_{k=1}^{m}D(x(t_k),\widetilde{0})\right)\\
&\leq B\left(\frac{R}{2B}+\frac{mR}{2mB}\right)\\
&= R\\
&= \|x\|, \quad t\in[0,\sigma(T)],
\end{aligned}$$

whereupon

$$\|\Psi x\| \leq \|x\|.$$

Hence and with Theorem 4.3.24, it follows that problem (4.14), (4.15) *has at least one positive solution. This completes the proof.*

Theorem 4.3.31 *Suppose* $(H4)$-$(H6)$, (4.19) *and*

$$\min f_\infty = \infty,$$

and there exists a positive constant R such that

$$D(f(t,s),\widetilde{0}) \leq \frac{R}{2B\sigma(T)}, \quad 0 < D(x,\widetilde{0}) \leq R, \quad t\in[0,\sigma(T)],$$

$$D(I_k(x),\widetilde{0}) \leq \frac{D(x,\widetilde{0})}{2mB}, \quad k\in\{1,\ldots,m\}, \quad x\in F(\mathbb{R}).$$

Then problem (4.14), (4.15) *has at least one positive solution.*

Proof 4.3.32 *Let* Ψ *be as in the proof of Theorem 4.3.9. As in the proof of Theorem 4.3.1, we have* $\Psi : X \to X$ *is completely continuous.Also, as in the proof of Theorem 4.3.9, it follows that* $\Psi : K \to K$. *Since* $\min f_\infty = \infty$, *for* $M^* \geq (sA\sigma(T)\delta)^{-1}$, *there exists a positive constant* $R^* > \frac{R}{\delta}$ *such that*

$$D(f(t,x),\widetilde{0}) \geq M^* D(x,\widetilde{0}), \quad D(x,\widetilde{0}) \geq \delta R^*, \quad t\in[0,\sigma(T)].$$

Let

$$\Omega_{R^*} = \{x \in X : \|x\| < R^*\}.$$

Then, for any $x \in K \bigcap \partial\Omega_{R^*}$, *we have*

$$\begin{aligned} D(x(t),\widetilde{0}) &\geq \delta\|x\| \\ &= \delta R^*, \quad t \in [0,\sigma(T)], \end{aligned}$$

and

$$\begin{aligned} & D(\Psi x(t),\widetilde{0}) \\ &= D\left(\int_0^{\sigma(T)} g(t,s)\cdot f(s,x(\sigma(s)))\delta_H s + \sum_{k=1}^{m} g(t,t_k)\cdot I_k(x(t_k)),\widetilde{0}\right) \\ &\geq sA\int_0^{\sigma(T)} D(f(s,x(\sigma(s))),\widetilde{0})\Delta s \\ &\geq sAM^*\int_0^{\sigma(T)} D(x(\sigma(s)),\widetilde{0})\Delta s \\ &\geq sAM^*\delta R^*\sigma(T) \\ &\geq R^* \\ &= \|x\|, \quad t \in [0,\sigma(T)], \end{aligned}$$

whereupon

$$\|\Psi x\| \geq \|x\|.$$

Hence, (4.20) *and Theorem 4.3.24, it follows that problem* (4.14), (4.15) *has at least one positive solution. This completes the proof.*

4.4 PERIODIC BOUNDARY VALUE PROBLEMS FOR FIRST-ORDER NON-LINEAR FUZZY DYNAMIC EQUATIONS WITH IMPULSES-II

Definition 4.4.1 *We say that a time scale $\mathbb{T}$ is periodic if there exists a $p > 0$ such that if $t \in \mathbb{T}$, then $t \pm p \in \mathbb{T}$. The smallest p is called the period of the time scale.*

Example 4.4.2 *Let $\mathbb{T} = h\mathbb{Z}$, where $h > 0$. Then $\mathbb{T}$ is a periodic time scale with period h.*

Remark 4.4.3 *If $\mathbb{T}$ is a periodic time scale with period p, then*

$$\sigma(t \pm np) = \sigma(t) \pm np, \quad t \in \mathbb{T}.$$

Hence,

$$\begin{aligned} \mu(t \pm np) &= \sigma(t \pm np) - (t \pm np) \\ &= \sigma(t) \pm np - t \mp np \\ &= \sigma(t) - t \\ &= \mu(t), \quad t \in \mathbb{T}. \end{aligned}$$

Definition 4.4.4 *Let $\mathbb{T} \neq \mathbb{R}$ be a periodic time scale with period p. We say that a function $f : \mathbb{T} \to F(\mathbb{R})$ is periodic with period T if there exists an $n \in \mathbb{N}$ such that $T = np$ and $f(t \pm T) = f(t)$ for any $t \in \mathbb{T}$ and T is the smallest number such that $f(t \pm T) = f(t)$.*

Below we suppose that $\mathbb{T}$ is a periodic time scale with period p such that $0, t_j \in \mathbb{T}$, $j \in \{1, \ldots, m\}$, where $T = lp$ for some $l \in \mathbb{N}$, $0 < t_j < t_{j+1}$ and each t_j is dense in $\mathbb{T}$, $j \in \{1, \ldots, m\}$. In this section, we will investigate the following BVP

$$\begin{aligned} \delta_H y(t) + a(t) \cdot y^{\sigma}(t) &= f(t, y(t)), \quad t \in \mathbb{T}, \\ y(t_j^+) \ominus_H y(t_j^-) &= I(t_j, y(t_j)), \quad j \in \{1, \ldots, m\}, \\ y(0) &= y(T), \end{aligned} \tag{4.21}$$

where

(H1) $a \in \mathscr{R}$, $a(t+T) = a(t)$, $t \in \mathbb{T}$, $\eta = e_{\ominus a}(T,0) < 1$,

(H2) $f \in \mathscr{C}(\mathbb{T} \times F(\mathbb{R}))$, $f(t+T, y(t+T)) = f(t, y(t))$, $t \in \mathbb{T}$.

Define $t_{m+1} = T$, $J_0 = [0, t_1]$, $J_j = (t_j, t_{j+1}]$, $j \in \{1, \ldots, m\}$,

$$\begin{aligned} PC_p &= \Big\{ y : \mathbb{T} \to F(\mathbb{R}), \quad y(t \pm T) = y(t), \quad y \in \mathscr{C}(J_j), \\ &\qquad \exists y(t_j^{\pm}), \quad y(t_j^-) = y(t_j), \quad j \in \{1, \ldots, m\} \Big\}, \\ PC_p^1 &= \{ y : \mathbb{T} \to F(\mathbb{R}), \quad y(t \pm T) = y(t), \quad y \in \mathscr{C}^1(J_j), \end{aligned}$$

$j \in \{1, \ldots, m\}\}$. The set PC_p is a Banach space endowed with the norm

$$\|y\| = \max_{j \in \{1, \ldots, m\}} \{ \sum_{t \in J_j} D(y(t), \widetilde{0}) \}.$$

Lemma 4.4.5 *Let $h \in PC_p$. Then the function $y \in PC_p^1$ is a solution of the BVP*

$$\begin{aligned} \delta_H y(t) + a(t) \cdot y^{\sigma}(t) &= h(t), \quad t \in \mathbb{T}, \\ y(t_j^+) \ominus_H y(t_j^-) &= I(t_j, y(t_j)), \quad j \in \{1, \ldots, m\}, \qquad (4.22) \\ y(0) &= y(T), \end{aligned}$$

if and only if $y \in PC_p$ and y satisfies the integral equation

$$y(t) = \int_0^T G(t,s) \cdot h(s) \delta_H s + \sum_{j=1}^{m} G(t, t_j) \cdot I(t_j, y(t_j)), \qquad (4.23)$$

where

$$G(t,s) = \frac{1}{1-\eta} \begin{cases} e_{\ominus a}(t,s) & 0 \le s \le t \le T \\ \eta e_{\ominus a}(t,s) & 0 \le t < s \le T. \end{cases}$$

Proof 4.4.6 *By Theorem 3.2.1, it follows that $y \in PC_p^1$ is a solution of the fuzzy impulsive dynamic equation*

$$\begin{aligned} \delta_H y(t) + a(t) \cdot y^\sigma(t) &= h(t), \quad t \in \mathbb{T}, \\ y(t_j^+) \ominus_H y(t_j^-) &= I(t_j, y(t_j)), \quad j \in \{1, \ldots, m\}, \end{aligned} \tag{4.24}$$

if and only if $y \in PC_p$ and it satisfies the integral equation

$$\begin{aligned} y(t) &= \int_0^t e_{\ominus a}(t,s) \cdot h(s) \delta_H s + e_{\ominus a}(t,0) \cdot y(0) \\ &+ \sum_{\{j: t_j < t\}} e_{\ominus a}(t, t_j) \cdot I(t_j, y(t_j)), \end{aligned} \tag{4.25}$$

$t \in [0, T]$.

1. *Let $y \in PC_p^1$ be a solution of the BVP* (4.22)*. Then y satisfies* (4.24) *and hence, y satisfies* (4.25)*. From here,*

$$\begin{aligned} y(0) &= y(T) \\ &= \int_0^T e_{\ominus a}(T,s) \cdot h(s) \delta_H s + \eta \cdot y(0) \\ &+ \sum_{j=1}^m e_{\ominus a}(T, t_j) \cdot I(t_j, y(t_j)), \end{aligned}$$

whereupon

$$(1-\eta) \cdot y(0) = \int_0^T e_{\ominus a}(T,s) \cdot h(s) \delta_H s + \sum_{j=1}^m e_{\ominus a}(T, t_j) \cdot I(t_j, y(t_j)),$$

or

$$y(0) = \frac{1}{1-\eta} \cdot \left(\int_0^T e_{\ominus a}(T,s) \cdot h(s) \delta_H s + \sum_{j=1}^m e_{\ominus a}(T, t_j) \cdot I(t_j, y(t_j)) \right).$$

From the last expression and from (4.25), we get

$$\begin{aligned}
y(t) &= \int_0^t e_{\ominus a}(t,s)\cdot h(s)\delta_H s \\
&+\frac{e_{\ominus a}(t,0)}{1-\eta}\left(\int_0^T e_{\ominus a}(T,s)\cdot h(s)\delta_H s \right.\\
&\left.+\sum_{j=1}^{m} e_{\ominus a}(T,t_j)\cdot I(t_j,y(t_j))\right)\\
&+\sum_{\{j:t_j<t\}} e_{\ominus a}(t,t_j)\cdot I(t_j,y(t_j))\\
&= \int_0^t e_{\ominus a}(t,s)\cdot h(s)\delta_H s\\
&+\frac{e_{\ominus a}(t,0)}{1-\eta}\cdot\left(\int_0^t e_{\ominus a}(T,s)\cdot h(s)\delta_H s\right.\\
&+\sum_{\{j:t_j<t\}} e_{\ominus a}(T,t_j)\cdot I(t_j,y(t_j))\\
&\left.+\int_t^T e_{\ominus a}(T,s)\cdot h(s)\delta_H s+\sum_{\{j:t_j\geq t\}} e_{\ominus a}(T,t_j)\cdot I(t_j,y(t_j))\right)\\
&+\sum_{\{j:t_j<t\}} e_{\ominus a}(t,t_j)\cdot I(t_j,y(t_j))\\
&= \int_0^t e_{\ominus a}(t,s)\cdot h(s)\delta_H s+\frac{\eta}{1-\eta}\cdot\int_0^t e_{\ominus a}(t,s)\cdot h(s)\delta_H s\\
&+\frac{\eta}{1-\eta}\sum_{\{j:t_j<t\}} e_{\ominus a}(t,t_j)\cdot I(t_j,y(t_j))
\end{aligned}$$

$$
\begin{aligned}
&+\frac{e_{\ominus a}(t,0)}{1-\eta}\cdot\int_t^T e_{\ominus a}(T,s)\cdot h(s)\delta_H s \\
&+\frac{e_{\ominus a}(t,0)}{1-\eta}\sum_{\{j:t_j\geq t\}} e_{\ominus a}(T,t_j)\cdot I(t_j,y(t_j)) \\
&+\sum_{\{j:t_j<t\}} e_{\ominus a}(t,t_j)\cdot I(t_j,y(t_j)) \\
= {} & \frac{1}{1-\eta}\cdot\int_0^t e_{\ominus a}(t,s)\cdot h(s)\delta_H s \\
&+\frac{1}{1-\eta}\cdot\sum_{\{j:t_j<t\}} e_{\ominus a}(t,t_j)\cdot I(t_j,y(t_j)) \\
&+\frac{\eta}{1-\eta}\int_t^T e_{\ominus a}(t,s)\cdot h(s)\delta_H s \\
&+\frac{\eta}{1-\eta}\cdot\sum_{\{j:t_j\geq t\}} e_{\ominus a}(T,t_j)\cdot I(t_j,y(t_j)) \\
= {} & \int_0^T G(t,s)\cdot h(s)\delta_H s+\sum_{j=1}^m G(t,t_j)\cdot I(t_j,y(t_j)), \quad t\in[0,T],
\end{aligned}
$$

i.e., we get (4.23).

2. *Let* $y\in PC_p$ *satisfies* (4.23). *Then* $y\in PC_p^1$ *and* y *satisfies* (4.24). *Next,*

$$
\begin{aligned}
y(t) = {} & \int_0^t G(t,s)\cdot h(s)\delta_H s+\int_t^T G(t,s)\cdot h(s)\delta_H s \\
&+\sum_{\{j:t_j<t\}} G(t,t_j)\cdot I(t_j,y(t_j))+\sum_{\{j:t_j\geq t\}} G(t,t_j)\cdot I(t_j,y(t_j)) \\
= {} & \frac{1}{1-\eta}\int_0^t e_{\ominus a}(t,s)\cdot h(s)\delta_H s+\frac{\eta}{1-\eta}\int_t^T e_{\ominus a}(t,s)\cdot h(s)\delta_H s
\end{aligned}
$$

$$+\frac{1}{1-\eta}\sum_{\{j:t_j\leq t\}} e_{\ominus a}(t,t_j)\cdot I(t_j,y(t_j))$$

$$+\frac{\eta}{1-\eta}\sum_{\{j:t_j>t\}} e_{\ominus a}(t,t_j)\cdot I(t_j,y(t_j)),\quad t\in[0,T],$$

whereupon

$$\begin{aligned}
y(T) &= \frac{1}{1-\eta}\cdot\int_0^T e_{\ominus a}(T,s)\cdot h(s)\delta_H s \\
&\quad +\frac{1}{1-\eta}\cdot\sum_{j=1}^{m} e_{\ominus a}(T,t_j)\cdot I(t_j,y(t_j)), \\
y(0) &= \frac{\eta}{1-\eta}\cdot\int_0^T e_{\ominus a}(0,s)\cdot h(s)\delta_H s \\
&\quad +\frac{\eta}{1-\eta}\cdot\sum_{j=1}^{m} e_{\ominus a}(0,t_j)\cdot I(t_j,y(t_j)) \\
&= \frac{1}{1-\eta}\cdot\int_0^T e_{\ominus a}(T,s)\cdot h(s)\delta_H s \\
&\quad +\frac{1}{1-\eta}\cdot\sum_{j=1}^{m} e_{\ominus a}(T,t_j)\cdot I(t_j,y(t_j)) \\
&= y(T).
\end{aligned}$$

This completes the proof.

Let A be a positive constant such that

$$\sup_{s,t\in[0,T]} G(t,s)\leq A.$$

We have

$$0\leq G(t,s)\leq A,\quad t,s\in[0,T].$$

For $y \in PC_p$, define the operator

$$\Phi y(t) = \int_0^T G(t,s)\cdot f(s,y(s))\delta_H s + \sum_{j=1}^m G(t,t_j)\cdot I(t_j,y(t_j)), \quad t \in [0,T].$$

Theorem 4.4.7 *Let $(H1)$ and $(H2)$ hold. Suppose that there exist positive constants L, E_j, $j \in \{1,\dots,m\}$, such that*

$$\begin{aligned} D(f(t,y),f(t,x)) &\leq LD(y,x), \quad t \in \mathbb{T}, \quad x,y \in \mathbb{R}, \\ D(I(t_j,y(t_j)),I(t_j,x(t_j))) &\leq E_j D(y(t_j),x(t_j)), \quad y,x \in PC_p, \end{aligned}$$

$j \in \{1,\dots,m\}$, and

$$A(LT + \sum_{j=1}^m E_j) < 1.$$

The problem (4.21) *has a unique solution in PC_p.*

Proof 4.4.8 *Note that $\Phi : PC_p \to PC_p$. For $x,y \in PC_p$, we get*

$$\begin{aligned} & D(\Phi y(t), \Phi x(t)) \\ = {} & D\Big(\int_0^T G(t,s)\cdot f(s,y(s))\delta_H s + \sum_{j=1}^m G(t,t_j)\cdot I(t_j,y(t_j)), \\ & \int_0^T G(t,s)\cdot f(s,x(s))\delta_H s + \sum_{j=1}^m G(t,t_j)\cdot I(t_j,x(t_j))\Big) \\ \leq {} & D\Big(\int_0^T G(t,s)\cdot f(s,y(s))\delta_H s, \int_0^T G(t,s)\cdot f(s,x(s))\delta_H s\Big) \\ & + D\Big(\sum_{j=1}^m G(t,t_j)\cdot I(t_j,x(t_j)), \sum_{j=1}^m G(t,t_j)\cdot I(t_j,y(t_j))\Big) \\ \leq {} & D\Big(\int_0^T G(t,s)\cdot f(s,y(s))\delta_H s, \int_0^T G(t,s)\cdot f(s,x(s))\delta_H s\Big) \end{aligned}$$

$$
\begin{aligned}
&+\sum_{j=1}^{m} D\Big(G(t,t_j)\cdot I(t_j,x(t_j)),G(t,t_j)\cdot I(t_j,y(t_j))\Big)\\
\leq\ & \int_0^T G(t,s)D(f(s,y(s)),f(s,x(s)))\Delta s\\
&+\sum_{j=1}^{m} G(t,t_j)D(I(t_j,y(t_j)),I(t_j,x(t_j)))\\
\leq\ & A\Bigg(\int_0^T D(f(s,y(s)),f(s,x(s)))\Delta s\\
&+\sum_{j=1}^{m} D(I(t_j,y(t_j)),I(t_j,x(t_j)))\Bigg)\\
\leq\ & A\left(L\int_0^T D(y(s),x(s))\Delta s+\sum_{j=1}^{m} E_j D(y(t_j),x(t_j))\right)\\
\leq\ & A\left(LT+\sum_{j=1}^{m} E_j\right)\|y-x\|, \quad t\in[0,T],
\end{aligned}
$$

whereupon

$$
\begin{aligned}
\|\Phi y-\Phi x\| &\leq A\left(LT+\sum_{j=1}^{m} E_j\right)\|y-x\|\\
&< \|y-x\|.
\end{aligned}
$$

Hence and the contraction mapping principle, it follows that the operator Φ has a unique fixed point in PC_p. This completes the proof.

Theorem 4.4.9 *Let $(H1)$ and $(H2)$ hold. Assume that there exist functions g_1, g_2, g_3, $g_4 : PC_p \to PC_p$ with*

$$
\alpha_1 = \int_0^T g_1(s)\Delta s < \infty,
$$

$$\beta_1 = \int_0^T g_2(s)\Delta s < \infty,$$

$$\alpha_2 = \sum_{j=1}^{m} g_3(t_j) < \infty,$$

$$\beta_2 = \sum_{j=1}^{m} g_4(t_j) < \infty,$$

such that

$$D(f(t,y),\widetilde{0}) \leq g_1(t) + g_2(t)D(y,\widetilde{0}), \quad t \in \mathbb{T}, \quad y \in F(\mathbb{R}),$$

$$D(I(t,y),\widetilde{0}) \leq g_3(t) + g_4(t)D(y,\widetilde{0}), \quad t \in \mathbb{T}, \quad y \in F(\mathbb{R}).$$

Suppose that

$$A(\beta_1 + \beta_2) < 1.$$

The problem (4.21) *has at least one solution in* PC_p.

Proof 4.4.10 *Define*

$$H(\Phi) = \{y \in PC_p : y(t) = \mu \cdot \Phi y(t), \quad t \in \mathbb{T}, \quad \mu \in (0,1)\}.$$

For $y \in H(\Phi)$, *we have*

$$D(y(t),\widetilde{0})$$

$$= D\left(\mu \cdot \int_0^T G(t,s)\cdot f(s,y(s))\delta_H s + \mu \sum_{j=1}^{m} G(t,t_j)\cdot I(t_j,y(t_j)),\widetilde{0}\right)$$

$$\leq \mu\left(D\left(\int_0^T G(t,s)\cdot f(s,y(s))\delta_H s,\widetilde{0}\right) + D\left(\sum_{j=1}^{m} G(t,t_j)\cdot I(t_j,y(t_j)),\widetilde{0}\right)\right)$$

$$\leq \mu\left(D\left(\int_0^T G(t,s)\cdot f(s,y(s))\delta_H s,\widetilde{0}\right) + \sum_{j=1}^{m} D\left(G(t,t_j)\cdot I(t_j,y(t_j)),\widetilde{0}\right)\right)$$

$$\begin{aligned}
&\leq \mu \int_0^T G(t,s)D(f(s,y(s)),\widetilde{0})\Delta s + \mu \sum_{j=1}^{m} G(t,t_j)D(I(t_j,y(t_j)),\widetilde{0}) \\
&\leq \mu\Bigg(\int_0^T G(t,s)g_1(s)\Delta s + \int_0^T G(t,s)g_2(s)D(y(s),\widetilde{0})\Delta s \\
&\quad + \sum_{j=1}^{m} G(t,t_j)g_3(t_j) + \sum_{j=1}^{m} G(t,t_j)g_4(t_j)D(y(t_j),\widetilde{0})\Bigg) \\
&\leq \mu A(\alpha_1 + \beta_1\|y\| + \alpha_2 + \beta_2\|y\|) \\
&= \mu A(\alpha_1+\alpha_2) + \mu A(\beta_1+\beta_2)\|y\|, \quad t \in [0,T],
\end{aligned}$$

whereupon

$$\|y\| \leq \mu A(\alpha_1+\alpha_2) + \mu A(\beta_1+\beta_2)\|y\|,$$

and

$$(1-\mu A(\beta_1+\beta_2))\|y\| \leq \mu A(\alpha_1+\alpha_2),$$

or

$$\|y\| \leq \frac{\mu A(\alpha_1+\alpha_2)}{1-\mu A(\beta_1+\beta_2)}.$$

Therefore, $H(\Phi)$ is bounded. Hence and Schaefer's fixed-point theorem, it follows that problem (4.21) *has at least one fixed point in PC_p. This completes the proof.*

Theorem 4.4.11 *Let* $(H1)$ *and* $(H2)$ *hold and*

$$\lim_{D(y,\widetilde{0})\to\infty} \frac{D(f(t,y),\widetilde{0})}{D(y,\widetilde{0})} = 0 \quad \textit{uniformly},$$

$$\lim_{D(y,\widetilde{0})\to\infty} \frac{D(I(t,y),\widetilde{0})}{D(y,\widetilde{0})} = 0 \quad \textit{uniformly}.$$

Then problem (4.21) *has at least one solution in PC_p.*

Proof 4.4.12 *Suppose that the set*

$$H(\Phi) = \{y \in PC_p : y(t) = \mu \cdot \Phi y(t), \quad t \in \mathbb{T}, \quad \mu \in (0,1)\},$$

is unbounded. Then there exist sequences $\{y_k\}_{k\in\mathbb{N}}$ *and* $\{\mu_k\}_{k\in\mathbb{N}}$ *with* $\|y_k\| > k$ *and* $\mu_k \in (0,1)$*, such that*

$$\begin{aligned} y_k(t) &= \mu_k \cdot \int_0^T G(t,s) \cdot f(s, y_k(s))\delta_H s \\ &\quad + \mu_k \cdot \sum_{j=1}^m G(t,t_j) \cdot I(t_j, y_k(t_j)), \quad t \in [0,T], \quad k \in \mathbb{N}. \end{aligned}$$

Define

$$v_k(t) = \frac{1}{\|y_k\|} \cdot y_k(t), \quad k \in \mathbb{N}, \quad t \in [0,T].$$

Then $\|v_k\| = 1$, $k \in \mathbb{N}$, *and*

$$\begin{aligned} v_k(t) &= \mu_k \cdot \int_0^T \left(G(t,s)\frac{1}{\|y_k\|} \right) \cdot f(s, y_k(s))\delta_H s \\ &\quad + \mu_k \cdot \sum_{j=1}^m \left(G(t,t_j)\frac{1}{\|y_k\|} \right) \cdot I(t_j, y_k(t_j)) \\ &\to 0, \quad as \quad k \to \infty, \quad t \in [0,T]. \end{aligned}$$

This contradicts with $\|v_k\| = 1$, $k \in \mathbb{N}$. *Therefore,* $H(\Phi)$ *is bounded. From here and from Schaefer's fixed-point theorem, it follows that the problem* (4.21) *has at least one solution in* PC_p*. This completes the proof.*

Exercise 4.4.13 *Assume* $(H1)$, $(H2)$ *and* f *and* I *are bounded. Prove that the problem* (4.21) *has at least one solution in* PC_p*.*

4.5 FUZZY GRONWALL-TYPE INEQUALITIES

Suppose that $\mathbb{T}$ is a time scale with forward jump operator and delta differentiation operators σ and Δ, respectively. Let $I \subset \mathbb{T}$ and $t_0 \in I$.

In this section, we will deduce some fuzzy Gronwall-type inequalities.

Theorem 4.5.1 *(The Fuzzy Gronwall-Type Inequality) Suppose that* $p : I \to \mathbb{R}$, $y, f : I \to F(\mathbb{R})$, $y \in \mathscr{C}^1(I)$ *and* $f \in \mathscr{C}(I)$, $p \in \mathscr{R}^+(I)$. *If*

$$\delta_H y(t) \leq p(t) \cdot y(t) + f(t), \quad t \in I, \tag{4.26}$$

then

$$\begin{aligned} y(t) &\leq e_p(t,t_0) \cdot \left(y(t_0) + \int_{t_0}^{t} e_{\ominus p}(\sigma(\tau), t_0) \cdot f(\tau) \delta_H \tau \right) \\ &= e_p(t,t_0) \cdot y(t_0) + \int_{t_0}^{t} e_p(t, \sigma(\tau)) \cdot f(\tau) \delta_H \tau, \quad t \geq t_0, \quad t \in I. \end{aligned}$$

Proof 4.5.2 *By inequality* (4.26), *we find*

$$\delta_H y(t) - p(t) \cdot y(t) \leq f(t), \quad t \in I.$$

Hence,

$$\begin{aligned} \delta_H \left(e_{\ominus p}(t,t_0) \cdot y(t) \right) &= e_{\ominus p}(\sigma(t), t_0) \cdot \delta_H y(t) + ((\ominus p)(t) e_{\ominus p}(t,t_0)) \cdot y(t) \\ &= ((1 + \mu(t)(\ominus p)(t))\, e_{\ominus p}(t,t_0)) \cdot \delta_H y(t) + ((\ominus p)(t) e_{\ominus p}(t,t_0)) \cdot y(t) \\ &= \left(\left(1 - \frac{\mu(t)p(t)}{1+\mu(t)p(t)} \right) e_{\ominus p}(t,t_0) \right) \cdot \delta_H y(t) \\ &\quad \ominus_H \left(\frac{p(t)}{1+\mu(t)p(t)} e_{\ominus p}(t,t_0) \right) \cdot y(t) \\ &= \left(\frac{1}{1+\mu(t)p(t)} e_{\ominus p}(t) \right) \cdot \delta_H y(t) \ominus_H \left(\frac{p(t)}{1+\mu(t)p(t)} e_{\ominus p}(t) \right) \cdot y(t) \\ &= \left(\frac{1}{1+\mu(t)p(t)} e_{\ominus p}(t,t_0) \right) \cdot (\delta_H y(t) \ominus_H p(t) \cdot y(t)) \end{aligned}$$

$$= e_{\ominus p}(\sigma(t),t_0)\cdot(\delta_H y(t)\ominus_H p(t)\cdot y(t))$$

$$\leq e_{\ominus p}(\sigma(t),t_0)\cdot f(t), \quad t\in I,$$

We fuzzy integrate the last inequality from t_0 to $t\in I$, $t\geq t_0$, and we arrive at

$$(e_{\ominus p}(s,t_0)\cdot y(s))\Big|_{s=t_0}^{s=t} \leq \int_{t_0}^{t} e_{\ominus p}(\sigma(\tau),t_0)\cdot f(\tau)\delta_H\tau, \quad t\geq t_0, \quad t\in I,$$

or

$$e_{\ominus p}(t,t_0)\cdot y(t)\ominus_H y(t_0)\leq \int_{t_0}^{t} e_{\ominus p}(\sigma(\tau),t_0)\cdot f(\tau)\delta_H\tau, \quad t\geq t_0, \quad t\in I,$$

or

$$e_{\ominus p}(t,t_0)\cdot y(t)\leq y(t_0)+\int_{t_0}^{t} e_{\ominus p}(\sigma(\tau),t_0)\cdot f(\tau)\delta_H\tau, \quad t\geq t_0, \quad t\in I,$$

or

$$y(t)\leq e_p(t,t_0)\cdot\left(y(t_0)+\int_{t_0}^{t} e_{\ominus p}(\sigma(\tau),t_0)\cdot f(\tau)\delta_H\tau\right), \ t\geq t_0, \quad t\in I.$$

This completes the proof.

Theorem 4.5.3 *(The Fuzzy Gronwall-Type Inequality) Suppose that $y,f:I\to F(\mathbb{R})$, $p:I\to\mathbb{R}$, $y\in\mathscr{C}^1(I)$ and $f\in\mathscr{C}(I)$, $p\in\mathscr{R}^+(I)$. If*

$$\delta_H y(t)\leq(-p(t))\cdot y(\sigma(t))+f(t), \quad t\in I, \tag{4.27}$$

then

$$\begin{aligned} y(t) &\leq e_{\ominus p}(t,t_0)\cdot\left(y(t_0)+\int_{t_0}^{t} e_p(\tau,t_0)\cdot f(\tau)\delta_H\tau\right) \\ &= e_{\ominus p}(t,t_0)\cdot y(t_0)+\int_{t_0}^{t} e_{\ominus p}(t,\tau)\cdot f(\tau)\delta_H\tau, \quad t\geq t_0, \quad t\in I. \end{aligned}$$

Proof 4.5.4 *By inequality* (4.27), *we find*

$$\delta_H y(t) + p(t)\cdot y(\sigma(t)) \leq f(t), \quad t \geq t_0.$$

Then

$$\begin{aligned} \delta_H(y(t)e_p(t,t_0)) &= e_p(t,t_0)\cdot\delta_H y(t) + (p(t)e_p(t,t_0))\cdot y(\sigma(t)) \\ &= e_p(t,t_0)\cdot(\delta_H y(t) + p(t)\cdot y(\sigma(t))) \\ &\leq e_p(t,t_0)\cdot f(t), \quad t \in I. \end{aligned}$$

Now, we fuzzy integrate the last inequality from t_0 to $t \in I$, $t \geq t_0$, and we obtain

$$(e_p(s,t_0)\cdot y(s))\Big|_{s=t_0}^{s=t} \leq \int_{t_0}^{t} e_p(\tau,t_0)\cdot f(\tau)\delta_H\tau, \quad t \geq t_0, \quad t \in I,$$

or

$$e_p(t,t_0)\cdot y(t) \ominus_H y(t_0) \leq \int_{t_0}^{t} e_p(\tau,t_0)\cdot f(\tau)\delta_H\tau, \quad t \geq t_0, \quad t \in I,$$

or

$$e_p(t,t_0)\cdot y(t) \leq y(t_0) + \int_{t_0}^{t} e_p(\tau,t_0)\cdot f(\tau)\delta_H\tau, \quad t \geq t_0, \quad t \in I,$$

or

$$y(t) \leq e_{\ominus p}(t,t_0)\cdot\left(y(t_0) + \int_{t_0}^{t} e_p(\tau,t_0)\cdot f(\tau)\delta_H\tau\right), \quad t \geq t_0, \quad t \in I.$$

This completes the proof.

Remark 4.5.5 *Suppose that all conditions of Theorem 4.5.1 hold. Since*

$$y(t) = y(\sigma(t)) \ominus_H \mu(t)\cdot\delta_H y(t), \quad t \in I,$$

the inequality (4.26) *takes the form*

$$\delta_H y(t) \leq p(t) \cdot y(\sigma(t)) \ominus_H (p(t)\mu(t)) \cdot \delta_H y(t) + f(t), \quad t \in I,$$

or

$$(1 + p(t)\mu(t)) \cdot \delta_H y(t) \leq p(t) \cdot y(\sigma(t)) + f(t), \quad t \in I,$$

or

$$\delta_H y(t) \leq \frac{p(t)}{1 + p(t)\mu(t)} \cdot y(\sigma(t)) + \frac{1}{1 + p(t)\mu(t)} \cdot f(t), \quad t \in I,$$

or

$$\delta_H y(t) \leq (-(\ominus p)(t)) \cdot y(\sigma(t)) + \frac{1}{1 + p(t)\mu(t)} \cdot f(t), \quad t \in I.$$

Now, we apply Theorem 4.5.3 and we get

$$\begin{aligned} y(t) &\leq e_p(t,t_0) \cdot \left(y(t_0) + \int_{t_0}^{t} \left(\frac{1}{1 + p(\tau)\mu(\tau)} e_{\ominus p}(\tau, t_0) \right) \cdot f(\tau)\delta_H \tau \right) \\ &= e_p(t,t_0) \cdot \left(y(t_0) + \int_{t_0}^{t} e_{\ominus p}(\sigma(\tau), t_0) \cdot f(\tau)\delta_H \tau \right), \quad t \geq t_0, \ t \in I. \end{aligned}$$

Thus, we obtain the conclusion of Theorem 4.5.1.

Remark 4.5.6 *Assume that all conditions of Theorem 4.5.3 hold. Since*

$$y(\sigma(t)) = y(t) + \mu(t) \cdot \delta_H y(t), \quad t \in I,$$

the inequality (4.27) *takes the form*

$$\delta_H y(t) \leq (-p(t)) \cdot y(t) \ominus_H (p(t)\mu(t)) \cdot \delta_H y(t) + f(t), \quad t \in I,$$

or

$$(1 + p(t)\mu(t)) \cdot \delta_H y(t) \leq (-p(t)) \cdot y(t) + f(t), \quad t \in I,$$

or

$$\delta_H y(t) \leq \left(-\frac{p(t)}{1 + p(t)\mu(t)} \right) \cdot y(t) + \frac{1}{1 + p(t)\mu(t)} \cdot f(t), \quad t \in I,$$

or

$$\delta_H y(t) \leq ((\ominus p)(t)) \cdot y(t) + \frac{1}{1+p(t)\mu(t)} \cdot f(t), \quad t \in I.$$

Now, we apply Theorem 4.5.1 and we obtain

$$\begin{aligned} y(t) &\leq e_{\ominus p}(t,t_0) \cdot \left(y(t_0) + \int_{t_0}^{t} \frac{e_p(\sigma(\tau))}{1+p(\tau)\mu(\tau)} \cdot f(\tau)\delta_H \tau \right) \\ &= e_{\ominus p}(t,t_0) \cdot \left(y(t_0) + \int_{t_0}^{t} e_p(\tau,t_0) \cdot f(\tau)\delta_H \tau \right), \quad t \geq t_0, \quad t \in I. \end{aligned}$$

Thus, we find the conclusion of Theorem 4.5.3.

As above, one can prove the following results.

Theorem 4.5.7 *(The Fuzzy Gronwall-Type Inequality) Suppose that $p: I \to \mathbb{R}$, $y, f: I \to F(\mathbb{R})$, $y \in \mathscr{C}^1(I)$ and $f \in \mathscr{C}(I)$, $p \in \mathscr{R}^+(I)$. If*

$$\delta_H y(t) \geq p(t) \cdot y(t) + f(t), \quad t \in I,$$

then

$$\begin{aligned} y(t) &\geq e_p(t,t_0) \cdot \left(y(t_0) + \int_{t_0}^{t} e_{\ominus p}(\sigma(\tau),t_0) \cdot f(\tau)\delta_H \tau \right) \\ &= e_p(t,t_0) \cdot y(t_0) + \int_{t_0}^{t} e_p(t,\sigma(\tau)) \cdot f(\tau)\delta_H \tau, \quad t \geq t_0, \quad t \in I. \end{aligned}$$

Theorem 4.5.8 *(The Fuzzy Gronwall-Type Inequality) Suppose that $y, f: I \to F(\mathbb{R})$, $p: I \to \mathbb{R}$, $y \in \mathscr{C}^1(I)$ and $f \in \mathscr{C}(I)$, $p \in \mathscr{R}^+(I)$. If*

$$\delta_H y(t) \geq (-p(t)) \cdot y(\sigma(t)) + f(t), \quad t \in I,$$

then

$$y(t) \geq e_{\ominus p,\beta}(t) \left(y(t_0) + \int_{t_0}^{t} e_{p,\beta}(\tau) f(\tau)\delta_H \tau \right), \quad t \geq t_0, \quad t \in I.$$

4.6 EXTREMAL SOLUTIONS OF PERIODIC BOUNDARY VALUE PROBLEMS FOR FIRST-ORDER FUZZY IMPULSIVE INTEGRO-DYNAMIC EQUATIONS OF MIXED TYPE

Suppose that $\mathbb{T}$ is a time scale with the forward jump operator and the delta differentiation operators σ and Δ, respectively. Let $0, t_j, T \in \mathbb{T}$, $j \in \{1, \ldots, m\}$. In this section we will investigate the following PBVP

$$\begin{aligned} \delta_H y(t) &= f(t, y(t), Ty(t), Sy(t)), \quad t \in J \backslash \{t_j\}_{j=1}^m, \\ y(t_j^+) &= y(t_j^-) + I_j(y(t_j^-)), \quad j \in \{1, \ldots, m\}, \\ y(0) &= y(T), \end{aligned} \tag{4.28}$$

where

$$0 < t_1 < \ldots < t_m < t_{m+1} = T,$$

$J = [0, T]$, t_j, $j \in \{1, \ldots, m\}$, are right-dense,

$$\begin{aligned} Ty(t) &= \int_0^t k(t,s) \cdot y(s) \delta_H s, \\ Sy(t) &= \int_0^t h(t,s) \cdot y(s) \delta_H s, \quad t \in J, \end{aligned}$$

(H1) $k \in \mathscr{C}(D)$, $k \geq 0$ on D,

$$D = \{(t,s) \in J \times J : t \geq s\}, \quad k_0 = \max_{(t,s) \in D} k(t,s),$$

(H2) $h \in \mathscr{C}(J \times J)$, $h \geq 0$ on $J \times J$, $h_0 = \max_{(t,s) \in J \times J} h(t,s)$,

(H3) $f \in \mathscr{C}(J \times F(\mathbb{R}) \times F(\mathbb{R}) \times F(\mathbb{R}))$,

(H4) $I_j \in \mathscr{C}(F(\mathbb{R}))$, $j \in \{1, \ldots, m\}$.

We will start with the following useful fuzzy impulsive dynamic inequalities.

Theorem 4.6.1 *Assume that the sequence* $\{t_k\}_{k\in\mathbb{N}}$ *satisfies*

$$0 \le t_0 < t_1 < t_2 < \ldots < t_k < \ldots, \qquad \lim_{k\to\infty} t_k = \infty,$$

$m \in PC^1(\mathbb{T})$ *is right-dense continuous at* t_k, $k \in \mathbb{N}$,

$$\inf_{t\in J}(\mu(t)p(t)) > -1,$$

and

$$\begin{aligned} \delta_H m(t) &\le p(t)\cdot m(t) + q(t), \quad t \in J\backslash\{t_k\}_{k\in\mathbb{N}}, \\ m(t_k^+) &\le d_k \cdot m(t_k^-) + b_k, \quad k \in \mathbb{N}, \end{aligned} \tag{4.29}$$

where $p, q \in \mathscr{C}(\mathbb{T})$, $d_k \ge 0$ *are real constants, and* b_k *are fuzzy real constants,* $k \in \mathbb{N}$. *Then*

$$\begin{aligned} m(t) \le & \prod_{t_0<t_k<t} (d_k e_p(t,t_0)) \cdot m(t_0) + \sum_{t_0<t_k<t} \left(\prod_{t_k<t_j<t} d_j e_p(t,t_k) \right) \cdot b_k \\ & + \int_{t_0}^{t} \prod_{s<t_k<t} (d_k e_p(t,\sigma(s))) \cdot q(s)\delta_H s, \end{aligned} \tag{4.30}$$

for any $t \ge t_0$.

Proof 4.6.2 *For* $t \in [t_0, t_1]$, *by the fuzzy Gronwall inequality, we get*

$$m(t) \le e_p(t,t_0)\cdot m(t_0) + \int_{t_0}^{t} e_p(t,\sigma(s))\cdot q(s)\delta_H s.$$

Therefore, the inequality (4.30) *is true for any* $t \in [t_0, t_1]$. *Now assume that the inequality* (4.30) *holds for* $t \in [t_0, t_n]$ *for some* $n \in \mathbb{N}$. *Let* $t \in (t_n, t_{n+1}]$. *Then, using the fuzzy Gronwall inequality, we get*

$$m(t) \le e_p(t,t_n)\cdot m(t_n) + \int_{t_n}^{t} e_p(t,\sigma(s))\cdot q(s)\delta_H s.$$

Next, by (4.29)*, using* (4.30)*, we have*

$$\begin{aligned}
m(t) &\leq e_p(t,t_n)\cdot(d_n\cdot m(t_n^-)+b_n)+\int_{t_n}^{t} e_p(t,\sigma(s))\cdot q(s)\delta_H s \\
&= (d_n e_p(t,t_n))\cdot m(t_n^-) \\
&\quad +e_p(t,t_n)\cdot b_n+\int_{t_n}^{t} e_p(t,\sigma(s))\cdot q(s)\delta_H s \\
&\leq (d_n e_p(t,t_n))\cdot\Bigg(\prod_{t_0<t_k<t_n}(d_k e_p(t_n,t_0))\cdot m(t_0) \\
&\quad +\sum_{t_0<t_k<t_n}\Bigg(\prod_{t_k<t_j<t_n} d_j e_p(t,t_k)\Bigg)\cdot b_k \\
&\quad +\int_{t_0}^{t_n}\prod_{s<t_k<t_n}(d_k e_p(t,\sigma(s)))\cdot q(s)\delta_H s\Bigg) \\
&\quad +e_p(t,t_n)\cdot b_n+\int_{t_n}^{t} e_p(t,\sigma(s))\cdot q(s)\delta_H s \\
&\leq \prod_{t_0<t_k<t}(d_k e_p(t,t_0))\cdot m(t_0) \\
&\quad +\sum_{t_0<t_k<t}\Bigg(\prod_{t_k<t_j<t} d_j e_p(t,t_k)\Bigg)\cdot b_k \\
&\quad +\int_{t_0}^{t}\prod_{s<t_k<t}(d_k e_p(t,\sigma(s)))\cdot q(s)\delta_H s.
\end{aligned}$$

Then (4.30) *is true for* $t\in[t_0,t_{n+1}]$*. By induction, we obtain the result that the inequality* (4.30) *is true for any* $t\geq t_0$*. This completes the proof.*

As in above, one can prove the following impulsive dynamic inequality.

Theorem 4.6.3 *Assume that the sequence* $\{t_k\}_{k\in\mathbb{N}}$ *satisfies*

$$0 \le t_0 < t_1 < t_2 < \ldots < t_k < \ldots, \qquad \lim_{k\to\infty} t_k = \infty,$$

$m \in PC^1(\mathbb{T})$ *is right-dense continuous at* t_k, $k \in \mathbb{N}$,

$$\inf_{t\in J}(\mu(t)p(t)) > -1,$$

and

$$\begin{aligned} \delta_H m(t) &\ge p(t)\cdot m(t) + q(t), \quad t \in J\backslash\{t_k\}_{k\in\mathbb{N}}, \\ m(t_k^+) &\ge d_k \cdot m(t_k^-) + b_k, \quad k \in \mathbb{N}, \end{aligned}$$

where $p, q \in \mathscr{C}(\mathbb{T})$, $d_k \ge 0$ *are real constants, and* b_k *are fuzzy real constants,* $k \in \mathbb{N}$. *Then*

$$\begin{aligned} m(t) \ \ge\ & \prod_{t_0<t_k<t} (d_k e_p(t,t_0)) \cdot m(t_0) + \sum_{t_0<t_k<t} \left(\prod_{t_k<t_j<t} d_j e_p(t,t_k) \right) \cdot b_k \\ & + \int_{t_0}^{t} \prod_{s<t_k<t} (d_k e_p(t,\sigma(s))) \cdot q(s)\delta_H s, \end{aligned}$$

for any $t \ge t_0$.

Let

$$\Omega = PC(J) \bigcap \mathscr{C}^1(J\backslash\{t_j\}_{j=1}^m).$$

Theorem 4.6.4 *Let* $t_0 = 0$ *and* $y \in \Omega$ *satisfies*

$$\delta_H y(t) \ge \ominus_H a(t)\cdot y(t) \ominus_H b(t)\cdot Ty(t) \ominus_H c(t)\cdot Sy(t), \ \ t \in J\backslash\{t_j\}_{j=1}^m,$$

$$y(t_j^+) \ge y(t_j) \ominus_H L_j \cdot y(t_j), \quad j \in \{1, \ldots, m\},$$

$$y(0) \ge y(T),$$

where $a,b,c \in \mathscr{C}(J)$, $a,b,c \geq 0$ on J, a is not identically vanishing, $\sup_{t\in J}(\mu(t)a(t)) < 1$, $0 \leq L_j \leq 1$, $j \in \{1,\ldots,m\}$. If

$$(Bk_0 + Ch_0)e_{\ominus(-a)}(T,0) \leq \frac{\left(\prod_{\{0<t_k<T\}}(1-L_k)\right)^2}{\int\limits_0^T \prod_{\{s<t_k<T\}}(1-L_k)\Delta s}, \tag{4.31}$$

with

$$\begin{aligned} B &= \sup_{t\in J}\left(b(t)\int\limits_0^t e_{\ominus(-a)}(\sigma(t),s)\Delta s\right), \\ C &= \sup_{t\in J}\left(c(t)\int\limits_0^t e_{\ominus(-a)}(\sigma(t),s)\Delta s\right), \end{aligned}$$

then $y(t) \geq \widetilde{0}$, $t \in J$.

Proof 4.6.5 *Note that*

$$\ominus(-a(t)) = \frac{a(t)}{1-\mu(t)a(t)}, \quad t \in J,$$

and

$$\begin{aligned} e_{\ominus(-a)}(t,0) &= e^{\int\limits_0^t \frac{1}{\mu(\tau)}\log\frac{1}{1-\mu(\tau)a(\tau)}\Delta\tau} \\ &\geq 1, \quad t \in J. \end{aligned}$$

Set

$$z(t) = e_{\ominus(-a)}(t,0)\cdot y(t), \quad t \in J.$$

Then $z \in \Omega$ and

$$\begin{aligned} \delta_H z(t) &= e_{\ominus(-a)}(\sigma(t),0)\cdot\delta_H y(t) \\ &+ \left((\ominus(-a))(t)e_{\ominus(-a)}(t,0)\right)\cdot y(t) \end{aligned}$$

$$
\begin{aligned}
&= e_{\ominus(-a)}(\sigma(t),0)\cdot\delta_H y(t)\\
&\quad + \left(\frac{a(t)}{1-\mu(t)a(t)}e_{\ominus(-a)}(t,0)\right)\cdot y(t)\\
&= e_{\ominus(-a)}(\sigma(t),0)\cdot\delta_H y(t) + \left(a(t)e_{\ominus(-a)}(\sigma(t),0)\right)\cdot y(t)\\
&= e_{\ominus(-a)}(\sigma(t),0)\cdot(\delta_H y(t) + a(t)\cdot y(t))\\
&\geq \ominus_H \left(b(t)e_{\ominus(-a)}(\sigma(t),0)\right)\cdot\int_0^t k(t,s)\cdot y(s)\delta_H s\\
&\quad \ominus_H \left(c(t)e_{\ominus(-a)}(\sigma(t),0)\right)\cdot\int_0^t h(t,s)\cdot y(s)\delta_H s\\
&= \ominus_H \left(b(t)e_{\ominus(-a)}(\sigma(t),0)\right)\cdot\int_0^t k(t,s)e_{\ominus(-a)}(0,s)\cdot z(s)\delta_H s\\
&\quad \ominus_H \left(c(t)e_{\ominus(-a)}(\sigma(t),0)\right)\cdot\int_0^t \left(h(t,s)e_{\ominus(-a)}(0,s)\right)\cdot z(s)\delta_H s\\
&= \ominus_H b(t)\cdot\int_0^t \left(k(t,s)e_{\ominus(-a)}(\sigma(t),s)\right)\cdot z(s)\delta_H s\\
&\quad \ominus_H c(t)\cdot\int_0^t \left(h(t,s)e_{\ominus(-a)}(\sigma(t),s)\right)\cdot z(s)\delta_H s, \quad t\in J\backslash\{t_j\}_{j=1}^m,
\end{aligned}
$$

$$
\begin{aligned}
z(t_j^+)\ominus_H z(t_j) &= e_{\ominus(-a)}(t_j^+,0)\cdot y(t_j^+)\ominus_H e_{\ominus(-a)}(t_j,0)\cdot y(t_j)\\
&= e_{\ominus(-a)}(t_j,0)\cdot\left(y(t_j^+)\ominus_H y(t_j)\right)
\end{aligned}
$$

$$
\begin{aligned}
&\geq \ominus_H L_j e_{\ominus(-a)}(t_j,0)\cdot y(t_j)\\
&= \ominus_H L_j \cdot z(t_j), \quad j\in\{1,\ldots,m\},\\
z(0) &= y(0)\\
&\geq \left(e_{\ominus(-a)}(T,0)e_{-a}(T,0)\right)\cdot y(T)\\
&= e_{-a}(T,0)\cdot z(T).
\end{aligned}
$$

Thus, z satisfies the following impulsive dynamic inequality

$$
\begin{aligned}
\delta_H z(t) &\geq \ominus_H b(t)\cdot \int_0^t \left(k(t,s)e_{\ominus(-a)}(\sigma(t),s)\right)\cdot z(s)\delta_H s \qquad (4.32)\\
&\quad \ominus_H c(t)\cdot \int_0^t \left(h(t,s)e_{\ominus(-a)}(\sigma(t),s)\right)\cdot z(s)\delta_H s,\\
&\quad t\in J\backslash\{t_j\}_{j=1}^m,\\
z(t_j^+)\ominus_H z(t_j) &\geq \ominus_H L_j\cdot z(t_j), \quad j\in\{1,\ldots,m\},\\
z(0) &\geq e_{-a}(T,0)\cdot z(T).
\end{aligned}
$$

Assume that

$$z(t)\geq\widetilde{0}, \quad t\in J,$$

is not true. Then we have the following two cases.

1. *There exists a $t_1^*\in J$ such that $z(t_1^*)<0$ and $z(t)\leq\widetilde{0}$, $t\in J$. Hence,*

$$\delta_H z(t) \geq \widetilde{0}, \quad t\in J\backslash\{t_j\}_{j=1}^m,$$

$$z(t_j^+) \ominus_H z(t_j) \;\geq\; \widetilde{0}, \quad j \in \{1, \ldots, m\}.$$

Therefore, z is non-decreasing on J. Thus,

$$\begin{aligned} z(0) &\leq z(t_1^*) \\ &< \widetilde{0}, \end{aligned}$$

$$\begin{aligned} z(0) &\leq z(T) \\ &\leq \widetilde{0}. \end{aligned}$$

Hence,

$$\begin{aligned} z(0) &\leq z(T) \\ &\leq e_{\ominus(-a)}(T, 0) \cdot z(0), \end{aligned}$$

or

$$\left(1 - e_{\ominus(-a)}(T, 0)\right) \leq 0,$$

which is a contradiction.

2. *There exist $t_1^*, t_2^* \in J$ such that*

$$z(t_1^*) < \widetilde{0} \quad \textit{and} \quad z(t_2^*) > \widetilde{0}.$$

Let

$$\lambda = \sup_{t \in J} z(t).$$

Then $\lambda > \widetilde{0}$ and there exists some $j \in \{1, \ldots, m\}$ such that $t_j < t_0^* \leq t_{j+1}$ and

$$z(t_0^*) = \lambda \quad \textit{or} \quad z(t_j^+) = \lambda.$$

Without loss of generality, assume that $z(t_0^*) = \lambda$. *By* (4.32), *we get*

$$\begin{aligned}\delta_H z(t) &\geq \ominus_H \left(k_0 b(t) \int\limits_0^t e_{\ominus(-a)}(\sigma(t), s)\Delta s \right) \cdot \lambda \\ &\quad \ominus_H \left(h_0 c(t) \int\limits_0^t e_{\ominus(-a)}(\sigma(t), s)\Delta s \right) \cdot \lambda \\ &\geq \ominus_H (Bk_0 + Ch_0) \cdot \lambda, \quad t \in J \backslash \{t_k\}_{k=1}^m.\end{aligned}$$

For $t \in [t_0^*, T]$, $k \in \{j+1, \ldots, m\}$, *we have*

$$\begin{aligned}\delta_H z(t) &\geq \ominus_H (Bk_0 + Ch_0) \cdot \lambda, \quad t \neq t_k, \\ z(t_k^+) &\geq (1 - L_k) \cdot z(t_k).\end{aligned}$$

Hence and with Theorem 4.6.3, it follows that

$$z(t) \geq \prod_{t_0^* < t_k < t} (1 - L_k) \cdot \lambda + \int\limits_{t_0^*}^t \prod_{s < t_k < t} ((1 - L_k)(-(Bk_0 + Ch_0))) \cdot \lambda \delta_H s, \tag{4.33}$$

$t \in [t_0^*, T]$. *In particular,*

$$z(T) \geq \prod_{t_0^* < t_k < T} (1 - L_k) \cdot \lambda \ominus_H \left((Bk_0 + Ch_0) \int\limits_{t_0^*}^T \prod_{s < t_k < T} (1 - L_k) \Delta s \right) \cdot \lambda. \tag{4.34}$$

If $z(T) < \widetilde{0}$, *we get*

$$\widetilde{0} > \prod_{t_0^* < t_k < T} (1 - L_k) \cdot \lambda - \left((Bk_0 + Ch_0) \int\limits_{t_0^*}^T \prod_{s < t_k < T} (1 - L_k) \Delta s \right) \cdot \lambda,$$

or

$$\prod_{t_0^*<t_k<T}(1-L_k)\cdot\lambda < \left((Bk_0+Ch_0)\int_{t_0^*}^{T}\prod_{s<t_k<T}(1-L_k)\Delta s\right)\cdot\lambda,$$

or

$$\begin{aligned} Bk_0+Ch_0 &> \frac{\prod_{t_0^*<t_k<T}(1-L_k)}{\int_{t_0^*}^{T}\prod_{s<t_k<t}(1-L_k)\Delta s} \\ &\geq \frac{\prod_{0<t_k<T}(1-L_k)}{\int_{0}^{T}\prod_{s<t_k<T}(1-L_k)\Delta s}, \end{aligned}$$

whereupon

$$\begin{aligned} (Bk_0+Ch_0)e_{\ominus(-a)}(T,0) &\geq Bk_0+Ch_0 \\ &> \frac{\prod_{0<t_k<T}(1-L_k)}{\int_{0}^{T}\prod_{s<t_k<T}(1-L_k)\Delta s}. \end{aligned}$$

This is a contradiction. Therefore $z(T)\geq\widetilde{0}$ *and from here,*

$$\begin{aligned} z(0) &\geq e_{-a}(T,0)\cdot z(T) \\ &\geq \widetilde{0}. \end{aligned}$$

Consequently, $0<t_1^*<T$. *Let* $t_r<t_1^*\leq t_{r+1}$ *for some* $r\in\{1,\ldots,m\}$. *Assume that* $t_0^*<t_1^*$. *So,* $j\geq r$. *By (4.33), we obtain*

$$\begin{aligned} \widetilde{0} &> z(t_1^*) \\ &\geq \prod_{\{t_0^*<t_k<t_1^*\}}(1-L_k)\cdot\lambda \end{aligned}$$

$$+\int_{t_0^*}^{t_1^*} \prod_{\{s<t_k<t_1^*\}} (1-L_k)\cdot(\ominus_H(Bk_0+Ch_0)\cdot\lambda)\,\delta_H s,$$

whereupon

$$\left((Bk_0+Ch_0)\int_{t_0^*}^{t_1^*} \prod_{\{s<t_k<t_1^*\}} (1-L_k)\Delta s\right)\cdot\lambda \geq \prod_{\{t_0^*<t_k<t_1^*\}} (1-L_k)\cdot\lambda,$$

or

$$\begin{aligned} Bk_0+Ch_0 &> \frac{\prod_{\{t_0^*<t_k<t_1^*\}}(1-L_k)}{\int_{t_0^*}^{t_1^*}\prod_{\{s<t_k<t_1^*\}}(1-L_k)\Delta s} \\ &\geq \frac{\prod_{\{0<t_k<T\}}(1-L_k)}{\int_0^T\prod_{\{s<t_k<T\}}(1-L_k)\Delta s}, \end{aligned}$$

or

$$\begin{aligned} (Bk_0+Ch_0)e_{\ominus(-a)}(T,0) &\geq Bk_0+Ch_0 \\ &\geq \frac{\prod_{\{0<t_k<T\}}(1-L_k)}{\int_0^T\prod_{\{s<t_k<T\}}(1-L_k)\Delta s}. \end{aligned}$$

This is a contradiction. Now, assume that $t_1^*<t_0^*$. So, $r\leq j$. Thus,

$$\begin{aligned} \delta_H z(t) &\geq \ominus_H(Bk_0+Ch_0)\cdot\lambda, \quad t\in J\backslash\{t_j\}_{j=1}^m, \\ z(t_j^+) &\geq (1-L_j)\cdot z(t_j), \quad j\in\{1,\ldots,m\}. \end{aligned}$$

Hence and with Theorem 4.6.3, it follows that

$$z(t) \geq \prod_{0<t_k<t}(1-L_k)\cdot z(0)$$

$$+\int_0^t \prod_{s<t_k<t}(1-L_k)\cdot(\ominus_H(Bk_0+Ch_0)\cdot\lambda)\,\delta_H s,$$

$t\in J$. From here,

$$\begin{aligned}\widetilde{0} &> z(t_1^*)\\ &\geq \prod_{0<t_k<t}(1-L_k)\cdot z(0)\\ &\quad+\int_0^t \prod_{s<t_k<t}(1-L_k)\cdot(\ominus_H(Bk_0+Ch_0)\cdot\lambda)\,\delta_H s,\end{aligned}$$

which implies

$$\prod_{0<t_k<t}(1-L_k)\cdot z(0)<\left((Bk_0+Ch_0)\int_0^{t_1^*}\prod_{s<t_k<t_1^*}(1-L_k)\Delta s\right)\cdot\lambda,$$

and

$$\begin{aligned}&\left(e_{-a}(T,0)\prod_{0<t_k<t}(1-L_k)\right)\cdot z(T)\\ &<\left((Bk_0+Ch_0)\int_0^{t_1^*}\prod_{s<t_k<t_1^*}(1-L_k)\Delta s\right)\cdot\lambda.\end{aligned}$$

Hence and (4.34), *we obtain*

$$\begin{aligned}&\left((Bk_0+Ch_0)\int_0^{t_1^*}\prod_{s<t_k<t_1^*}(1-L_k)\Delta s\right)\cdot\lambda\\ &>\left(e_{-a}(T,0)\prod_{0<t_k<t_1^*}(1-L_k)\right)\cdot\left(\prod_{t_0^*<t_k<T}(1-L_k)\cdot\lambda\right.\end{aligned}$$

$$\ominus_H \left(\left(Bk_0 + Ch_0 \right) \int_{t_0^*}^{T} \prod_{s<t_k<T} (1 - L_k) \Delta s \right) \cdot \lambda \Big),$$

or

$$\prod_{0<t_k<t_1^*} (1 - L_k) \Big(\prod_{t_0^*<t_k<T} (1 - L_k) \cdot \lambda$$

$$\ominus_H \left(\left(Bk_0 + Ch_0 \right) \int_{t_0^*}^{T} \prod_{s<t_k<T} (1 - L_k) \delta_H s \right) \cdot \lambda \Big)$$

$$\leq \left((Bk_0 + Ch_0) e_{\ominus(-a)}(T,0) \int_{0}^{t_1^*} \prod_{s<t_k<t_1^*} (1 - L_k) \Delta s \right) \cdot \lambda,$$

or

$$\prod_{0<t_k<t_1^*} (1 - L_k) \Big(\prod_{t_0^*<t_k<T} (1 - L_k) \Big)$$

$$\leq (Bk_0 + Ch_0) \int_{t_0^*}^{T} \prod_{s<t_k<T} (1 - L_k) \Delta s$$

$$\leq (Bk_0 + Ch_0) e_{\ominus(-a)}(T,0) \int_{0}^{t_1^*} \prod_{s<t_k<t_1^*} (1 - L_k) \Delta s.$$

From the last inequality, we get

$$\left(\prod_{\{0<t_k<T\}} (1 - L_k) \right)^2$$

$$= \left(\prod_{\{0<t_k<T\}} (1 - L_k) \right) \left(\prod_{\{0<t_k<T\}} (1 - L_k) \right)$$

$$
\begin{aligned}
&\le \prod_{\{0<t_k<t_1^*\}} (1-L_k) \prod_{\{t_0^*<t_k<T\}} (1-L_k) \prod_{\{0<t_k<T\}} (1-L_k) \\
&\le (Bk_0+Ch_0) \\
&\quad \times \prod_{\{0<t_k<t_1^*\}} (1-L_k) \prod_{\{0<t_k<T\}} (1-L_k) \int_{t_0^*}^{T} \prod_{\{s<t_k<T\}} (1-L_k)\Delta s \\
&\quad +(Bk_0+Ch_0)e_{\ominus(-a)}(T,0) \\
&\quad \times \prod_{\{0<t_k<T\}} (1-L_k) \int_{0}^{t_1^*} \prod_{\{s<t_k<t_1^*\}} (1-L_k)\Delta s \\
&< (Bk_0+Ch_0)e_{\ominus(-a)}(T,0) \int_0^T \prod_{s<t_k<T} (1-L_k)\Delta s.
\end{aligned}
$$

This is a contradiction.

Therefore $z(t) \ge \widetilde{0}$, $t \in J$. Hence, $y(t) \ge \widetilde{0}$, $t \in J$. This completes the proof.

Suppose that

(H5) $\delta \in PC(J)$, $\eta \in \Omega$,

(H6) $a,b,c \in \mathscr{C}(J)$, $a,b,c \ge 0$ on J, a is not identically vanishing,

(H7) $0 \le L_j < 1$, $j \in \{1,\dots,m\}$,

(H8) $\sup_{t\in J}(\mu(t)a(t)) < 1$.

Consider the following linear PBVP

$$
\begin{aligned}
\delta_H y(t) + a(t)\cdot y(t) &= \ominus_H b(t)\cdot Ty(t) \ominus_H c(t)\cdot Sy(t) + \delta(t), \\
&\qquad t \in J\backslash\{t_j\}_{j=1}^m, \\
y(t_j^+) \ominus_H y(t_j) &= \ominus_H L_j \cdot y(t_j) + I_j(\eta(t_j)) + L_j\cdot \eta(t_j), \\
&\qquad j \in \{1,\dots,m\}, \\
y(0) &= y(T).
\end{aligned}
\tag{4.35}
$$

Lemma 4.6.6 *Suppose (H4)-(H6). Then $y \in \Omega$ is a solutions of the PBVP* (4.35) *if and only if $y \in PC(J)$ and it satisfies the integral equation*

$$\begin{aligned} y(t) &= \int_0^T G(t,s)\cdot(\delta(s) \ominus_H b(s)\cdot Ty(s) \ominus_H c(s)\cdot Sy(s))\,\delta_H s \\ &\quad + \sum_{\{j:0<t_j<T\}} \left(G(t,t_j)e_{-a}(\sigma(t_j),t_j)\right) \\ &\quad \cdot\left(\ominus_H L_j\cdot y(t_j) + I_j(\eta(t_j)) + L_j\cdot\eta(t_j)\right), \end{aligned} \tag{4.36}$$

$t \in J$, where

$$G(t,s) = \frac{1}{1-e_{-a}(T,0)}\begin{cases} e_{-a}(t,\sigma(s)) & 0 \le s < t \le T \\ e_{-a}(T,0)e_{-a}(t,\sigma(s)) & 0 \le t \le s \le T. \end{cases}$$

Proof 4.6.7 *By Theorem 3.1.1, it follows that $y \in \Omega$ is a solution of the impulsive dynamic equation*

$$\begin{aligned} \delta_H y(t) + a(t)\cdot y(t) &= \ominus_H b(t)\cdot Ty(t) \ominus_H c(t)\cdot Sy(t) + \delta(t), \\ &\quad t \in J\backslash\{t_j\}_{j=1}^m, \\ y(t_j^+) \ominus_H y(t_j) &= \ominus_H L_j\cdot y(t_j) + I_j(\eta(t_j)) + L_j\cdot\eta(t_j), \\ &\quad j \in \{1,\dots,m\}, \end{aligned} \tag{4.37}$$

if and only if $y \in PC(J)$ and it satisfies the integral equation

$$\begin{aligned} y(t) &= \int_0^t e_{-a}(t,\sigma(s))\cdot(\ominus_H b(s)\cdot Ty(s) \ominus_H c(s)\cdot Sy(s) + \delta(s))\,\delta_H s \\ &\quad + e_{-a}(t,0)\cdot y(0) \\ &\quad + \sum_{\{j:0<t_j<t\}} e_{-a}(t,t_j)\cdot\left(\ominus_H L_j\cdot y(t_j) + I_j(\eta(t_j)) + L_j\cdot\eta(t_j)\right), \\ &\quad t \in J. \end{aligned} \tag{4.38}$$

1. *Let $y \in \Omega$ satisfies* (4.35). *Then y satisfies* (4.38). *Hence,*

$$y(0) = y(T)$$

$$= \int_0^T e_{-a}(T,\sigma(s)) \cdot (\ominus_H b(s) \cdot Ty(s) \ominus_H c(s) \cdot Sy(s) + \delta(s))\, \delta_H s$$

$$+ e_{-a}(T,0) \cdot y(0)$$

$$+ \sum_{\{j:0<t_j<T\}} e_{-a}(T,t_j) \cdot \left(\ominus_H L_j \cdot y(t_j) + I_j(\eta(t_j)) + L_j \cdot \eta(t_j)\right),$$

whereupon

$$(1 - e_{-a}(T,0)) \cdot y(0)$$
$$= \int_0^T e_{-a}(T,\sigma(s)) \cdot (\ominus_H b(s) \cdot Ty(s) \ominus_H c(s) \cdot Sy(s) + \delta(s))\, \delta_H s$$

$$+ \sum_{\{j:0<t_j<T\}} e_{-a}(T,t_j) \cdot \left(\ominus_H L_j \cdot y(t_j) + I_j(\eta(t_j)) + L_j \cdot \eta(t_j)\right),$$

and

$$y(0) = \frac{1}{1 - e_{-a}(T,0)} \cdot$$
$$\times \int_0^T e_{-a}(T,\sigma(s)) \cdot (\ominus_H b(s) \cdot Ty(s) \ominus_H c(s) \cdot Sy(s) + \delta(s))\, \delta_H s$$

$$+ \frac{1}{1 - e_{-a}(T,0)} \ominus_H \sum_{\{j:0<t_j<T\}} e_{-a}(T,t_j) \cdot$$
$$\times \left(\ominus_H L_j \cdot y(t_j) + I_j(\eta(t_j)) + L_j \cdot \eta(t_j)\right).$$

Then

$$
\begin{aligned}
y(t) &= \int_0^t e_{-a}(t,\sigma(s)) \cdot (\ominus_H b(s) \cdot Ty(s) \ominus_H c(s) \cdot Sy(s) + \delta(s))\, \delta_H s \\
&+ \frac{e_{-a}(t,0)}{1-e_{-a}(T,0)} \\
&\cdot \int_0^t e_{-a}(T,\sigma(s)) \cdot (\ominus_H b(s) \cdot Ty(s) \ominus_H c(s) \cdot Sy(s) + \delta(s))\, \delta_H s \\
&+ \frac{e_{-a}(t,0)}{1-e_{-a}(T,0)} \\
&\cdot \int_t^T e_{-a}(T,\sigma(s)) \cdot (\ominus_H b(s) \cdot Ty(s) \ominus_H c(s) \cdot Sy(s) + \delta(s))\, \delta_H s \\
&+ \frac{e_{-a}(t,0)}{1-e_{-a}(T,0)} \\
&\cdot \sum_{\{j:0<t_j<t\}} e_{-a}(T,t_j) \cdot \left(\ominus_H L_j \cdot y(t_j) \ominus_H I_j(\eta(t_j)) + L_j \cdot \eta(t_j)\right) \\
&+ \frac{e_{-a}(t,0)}{1-e_{-a}(T,0)} \\
&\cdot \sum_{\{j:t\le t_j<T\}} e_{-a}(t,t_j) \cdot \left(\ominus_H L_j \cdot y(t_j) \ominus_H I_j(\eta(t_j)) + L_j \cdot \eta(t_j)\right)
\end{aligned}
$$

$$+\sum_{\{j:0<t_j<t\}} e_{-a}(t,t_j)\cdot\left(\ominus_H L_j\cdot y(t_j)+I_j(\eta(t_j))+L_j\cdot\eta(t_j)\right)$$

$$=\frac{1}{1-e_{-a}(T,0)}$$

$$\cdot\int_0^t e_{-a}(t,\sigma(s))\cdot(\ominus_H b(s)\cdot Ty(s)\ominus_H c(s)\cdot Sy(s)+\delta(s))\,\delta_H s$$

$$+\frac{e_{-a}(T,0)}{1-e_{-a}(T,0)}$$

$$\cdot\int_t^T e_{-a}(t,\sigma(s))\cdot(\ominus_H b(s)\cdot Ty(s)\ominus_H c(s)\cdot Sy(s)+\delta(s))\,\delta_H s$$

$$+\frac{1}{1-e_{-a}(T,0)}\cdot$$

$$\sum_{\{j:0<t_j<t\}} e_{-a}(t,t_j)\cdot\left(\ominus_H L_j\cdot y(t_j)+I_j(\eta(t_j))+L_j\cdot\eta(t_j)\right)$$

$$+\frac{e_{-a}(T,0)}{1-e_{-a}(T,0)}$$

$$\cdot\sum_{\{j:t\le t_j<T\}} e_{-a}(t,t_j)\cdot\left(\ominus_H L_j\cdot y(t_j)+I_j(\eta(t_j))+L_j\cdot\eta(t_j)\right)$$

$$= \int_0^T G(t,s) \cdot (\ominus_H b(s) \cdot Ty(s) \ominus_H c(s) \cdot Sy(s) + \delta(s))\, \delta_H s$$

$$+ \sum_{\{j:0<t_j<T\}} \left(G(t,t_j) e_{-a}(\sigma(t_j),t_j) \right)$$

$$\cdot \left(\ominus_H L_j \cdot y(t_j) + I_j(\eta(t_j)) + L_j \cdot \eta(t_j) \right),$$

$t \in J$, i.e., y satisfies (4.36).

2. *Let $y \in PC(J)$ satisfies* (4.36). *Then $y \in \Omega$ and y satisfies* (4.37). *We have*

$$y(0)$$

$$= \frac{e_{-a}(T,0)}{1 - e_{-a}(T,0)}$$

$$\cdot \int_0^T e_{-a}(0,\sigma(s)) \cdot (\ominus_H b(s) \cdot Ty(s) \ominus_H c(s) \cdot Sy(s) + \delta(s))\, \delta_H s$$

$$+ \frac{e_{-a}(T,0)}{1 - e_{-a}(T,0)}$$

$$\cdot \sum_{\{j:0<t_j<T\}} e_{-a}(0,t_j) \cdot \left(\ominus_H L_j \cdot y(t_j) + I_j(\eta(t_j)) + L_j \cdot \eta(t_j) \right),$$

$$y(T)$$

$$= \frac{1}{1 - e_{-a}(T,0)}$$

$$\cdot \int_0^T e_{-a}(T,\sigma(s)) \cdot (\ominus_H b(s) \cdot Ty(s) \ominus_H c(s) \cdot Sy(s) + \delta(s))\, \delta_H s$$

$$+ \frac{1}{1 - e_{-a}(T,0)}$$

$$\cdot \sum_{\{j:0<t_j<T\}} e_{-a}(T,\sigma(s)) \cdot (\ominus_H b(s) \cdot Ty(s) \ominus_H c(s) \cdot Sy(s) + \delta(s))\, \delta_H s$$

$$= y(0).$$

Then y satisfies (4.35). *This completes the proof.*

Let $(H8)$ hold and

$$\sup_{t,s \in J} G(t,s) = A.$$

Then, we have

$$0 \le G(t,s) \le A, \quad t,s \in J.$$

Theorem 4.6.8 *Assume* $(H1)$, $(H2)$, $(H4) - (H8)$ *and*

$$A \left(\int_0^T (k_0 s b(s) + h_0 s c(s))\, \Delta s + \sum_{j=1}^{m} L_j \right) < 1. \tag{4.39}$$

Then PBVP (4.35) *has a unique solution in* Ω.

Proof 4.6.9 *For* $y \in \Omega$ *define the operator*

$$\begin{aligned} \Phi y(t) &= \int_0^T G(t,s) \cdot (\delta(s) \ominus_H b(s) \cdot Ty(s) \ominus_H c(s) \cdot Sy(s))\, \delta_H s \\ &+ \sum_{\{j:0<t_j<t\}} \left(G(t,t_j) e_{-a}(\sigma(t_j),t_j)\right) \\ &\cdot \left(\ominus_H L_j \cdot y(t_j) + I_j(\eta(t_j)) + L_j \cdot \eta(t_j)\right), \end{aligned}$$

$t \in J$. *Note that* $\Phi : \Omega \to \Omega$. *For* $x, y \in \Omega$, *we get*

$$
\begin{aligned}
& D(\Phi y(t), \Phi x(t)) \\
= \; & D\Bigg(\int_0^T G(t,s)\cdot(\delta(s)\ominus_H b(s)\cdot Ty(s)\ominus_H c(s)\cdot Sy(s))\,\delta_H s \\
& + \sum_{\{j:0<t_j<t\}} G(t,t_j)e_{-a}(\sigma(t_j),t_j) \\
& \cdot(\ominus_H L_j\cdot y(t_j)+I_j(\eta(t_j))+L_j\cdot\eta(t_j)), \\
& \int_0^T G(t,s)\cdot(\delta(s)\ominus_H b(s)\cdot Tx(s)\ominus_H c(s)\cdot Sx(s))\,\delta_H s \\
& + \sum_{\{j:0<t_j<t\}} G(t,t_j)e_{-a}(\sigma(t_j),t_j) \\
& \cdot(\ominus_H L_j\cdot x(t_j)+I_j(\eta(t_j))+L_j\cdot\eta(t_j))\Bigg) \\
= \; & D\Bigg(\int_0^T G(t,s)\,(\ominus_H b(s)T(y\ominus_H x)(s)\ominus_H c(s)S(y\ominus_H x)(s))\,\delta_H s \\
& + \sum_{\{j:0<t_j<t\}} G(t,t_j)e_{-a}(\sigma(t_j),t_j)\cdot(\ominus_H L_j\cdot D(y,x)(t_j)),\widetilde{0}\Bigg) \\
\le \; & D\Bigg(\int_0^T G(t,s)\cdot(\ominus_H b(s)\cdot T(y\ominus_H x)(s)\ominus_H c(s)\cdot S(y\ominus_H x)(s))\,\delta_H s,\widetilde{0}\Bigg) \\
& + D\Bigg(\sum_{\{j:0<t_j<t\}} G(t,t_j)e_{-a}(\sigma(t_j),t_j)\cdot(\ominus_H L_j\cdot(y\ominus_H x)(t_j)),\widetilde{0}\Bigg) \\
\le \; & \int_0^T G(t,s)\,(b(s)TD(y,x)(s)+c(s)SD(y,x)(s))\,\delta_H s \\
& + \sum_{\{j:0<t_j<t\}} G(t,t_j)e_{-a}(\sigma(t_j),t_j)\,(L_j D(y,x)(t_j)) \\
\le \; & A\left(\int_0^T (k_0 s b(s)+h_0 s c(s))\Delta s+\sum_{j=1}^m L_j\right)\|y-x\|, \quad t\in J,
\end{aligned}
$$

whereupon

$$
\|\Phi y-\Phi x\| \;\le\; A\left(\int_0^T (k_0 s b(s)+h_0 s c(s))\Delta s+\sum_{j=1}^m L_j\right)\|y-x\|
$$

$$< \ \|y-x\|.$$

Hence and on the principle of contraction mapping, we conclude that the operator Φ has a unique fixed point in Ω. This completes the proof.

Theorem 4.6.10 *Assume that the following conditions are met.*

1. *There exist functions $u_0, v_0 \in \Omega$, $u_0(t) \leq v_0(t)$, $t \in J$, such that*

$$\begin{aligned}
\delta_H u_0(t) &\leq f(t,u_0(t),Tu_0(t),Su_0(t)), \quad t \in J\backslash\{t_j\}_{j=1}^m,\\
u_0(t_j^+) \ominus_H u_0(t_j^-) &\leq I_j(u_0(t_j)), \quad j \in \{1,\ldots,m\},\\
u_0(0) &\leq u_0(T),\\
\delta_H v_0(t) &\geq f(t,v_0(t),Tv_0(t),Sv_0(t)), \quad t \in J\backslash\{t_j\}_{j=1}^m,\\
v_0(t_j^+) \ominus_H v_0(t_j^-) &\geq I_j(v_0(t_j)), \quad j \in \{1,\ldots,m\},\\
v_0(0) &\geq v_0(T).
\end{aligned} \tag{4.40}$$

2. *The function $f \in \mathscr{C}(J \times F(\mathbb{R}) \times F(\mathbb{R}))$ satisfies*

$$\begin{aligned}
& f(t,u_2,v_2,w_2) \ominus_H f(t,u_1,v_1,w_1) \\
\geq \ & \ominus_H a(t) \cdot (u_2 \ominus_H u_1) \ominus_H b(t) \cdot (v_2 \ominus_H v_1) \ominus_H c(t) \cdot (w_2 \ominus_H w_1),
\end{aligned} \tag{4.41}$$

whenever

$$u_0(t) \ \leq \ u_1 \leq u_2 \leq v_0(t),$$

$$Tu_0(t) \ \leq \ v_1 \leq v_2 \leq Tv_0(t),$$

$$Su_0(t) \ \leq \ w_1 \leq w_2 \leq Sv_0(t), \quad t \in J,$$

where a, b, c satisfy (H6)–(H8).

3. *The functions $I_j \in \mathscr{C}(\mathbb{R})$, $j \in \{1,\ldots,m\}$, satisfy*

$$I_j(x) \ominus_H I_j(y) \geq \ominus_H L_j \cdot (x \ominus_H y), \tag{4.42}$$

whenever

$$u_0(t_j) \leq y \leq x \leq v_0(t_j),$$

$$0 \leq L_j < 1, \quad j \in \{1,\ldots,m\}.$$

Furthermore, assume (4.31) *and* (4.39). *Then PBVP* (4.28) *has minimal and maximal solutions u^* and v^*, respectively. Moreover, there exist monotone iteration sequences*

$$\{u_n(t)\}_{n\in\mathbb{N}}, \quad \{v_n(t)\}_{n\in\mathbb{N}} \subset [u_0, v_0]$$

such that

$$u_n(t) \to u^*(t),$$

$$v_n(t) \to v^*(t), \quad as \quad n \to \infty \quad uniformly \quad on \quad t \in J,$$

where $\{u_n(t)\}_{n\in\mathbb{N}}$ and $\{v_n(t)\}_{n\in\mathbb{N}}$ satisfy

$$\delta_H u_n(t) = f(t,u_{n-1}(t),Tu_{n-1}(t),Su_{n-1}(t)) \ominus_H a(t)\cdot(u_n \ominus_H u_{n-1})(t) \tag{4.43}$$

$$\ominus_H b(t)\cdot T(u_n \ominus_H u_{n-1})(t) \ominus_H c(t)\cdot S(u_n \ominus_H u_{n-1})(t), \quad t \in J\backslash\{t_j\}_{j=1}^m,$$

$$u_n(t_j^+) = u_n(t_j) \ominus_H L_j\cdot u_n(t_j) + I_j(u_{n-1}(t_j)) + L_j\cdot u_{n-1}(t_j),$$

$$j \in \{1,\ldots,m\},$$

$$u_n(0) = u_n(T),$$

$$v_n^\Delta(t) = f(t,v_{n-1}(t),Tv_{n-1}(t),Sv_{n-1}(t)) \ominus_H a(t)(v_n \ominus_H v_{n-1})(t)$$

$$\ominus_H b(t)\cdot T(v_n \ominus_H v_{n-1})(t) \ominus_H c(t)\cdot S(v_n] \ominus_H v_{n-1})(t),$$

$$t \in J\backslash\{t_j\}_{j=1}^m,$$

$$v_n(t_j^+) = v_n(t_j) \ominus_H L_j \cdot v_n(t_j) + I_j(v_{n-1}(t_j)) + L_j \cdot v_{n-1}(t_j),$$

$$j \in \{1,\ldots,m\},$$

$$v_n(0) = v_n(T), \quad n \in \mathbb{N},$$

$$u_0 \le u_1 \le \ldots \le u_n \le \ldots \le u^* \le v^* \le \ldots \le v_n \le \ldots \le v_1 \le v_0.$$

Proof 4.6.11 *Note that for any $u_{n-1}, v_{n-1} \in \Omega$, using Theorem 4.6.8, it follows that* (4.43) *has a unique solution $u_n, v_n \in \Omega$, respectively. We have*

$$\begin{aligned}\delta_H(u_1 \ominus_H u_0)(t) &= \delta_H u_1(t) \ominus_H \delta_H u_0(t)\\ &\ge f(t,u_0(t),Tu_0(t),Su_0(t)) \ominus_H a(t)\cdot(u_1 \ominus_H u_0)(t)\\ &\ominus_H b(t)\cdot T(u_1 \ominus_H u_0)(t) \ominus_H c(t)\cdot S(u_1 \ominus_H u_0)(t)\\ &\ominus_H f(t,u_0(t),Tu_0(t),Su_0(t))\\ &= \ominus_H a(t)\cdot(u_1 \ominus_H u_0)(t) \ominus_H b(t)\cdot T(u_1 \ominus_H u_0)(t)\\ &\ominus_H c(t)\cdot S(u_1 \ominus_H u_0)(t), \quad t \in J\backslash\{t_j\}_{j=1}^m,\end{aligned}$$

and

$$\begin{aligned}&(u_1 \ominus_H u_0)(t_j^+) \ominus_H (u_1 \ominus_H u_0)(t_j)\\ =\ & u_1(t_j^+) \ominus_H u_1(t_j) \ominus_H (u_0(t_j^+) \ominus_H u_0(t_j))\\ =\ & \ominus_H L_j \cdot u_1(t_j) + I_j(u_0(t_j)) + L_j \cdot u_0(t_j) \ominus_H I_j(u_0(t_j))\\ =\ & \ominus_H L_j \cdot (u_1 \ominus_H u_0)(t_j), \quad j \in \{1,\ldots,m\},\end{aligned}$$

$$\begin{aligned}(u_1 \ominus_H u_0)(0) &= u_1(0) \ominus_H u_0(0) \\ &\geq u_1(T) \ominus_H u_0(T) \\ &= (u_1 \ominus_H u_0)(T),\end{aligned}$$

and

$$\begin{aligned}\delta_H(v_1 \ominus_H v_0)(t) &= \delta_H v_1(t) \ominus_H \delta_H v_0(t) \\ &\leq f(t, v_0(t), Tv_0(t), Sv_0(t)) \ominus_H a(t) \cdot (v_1 \ominus_H v_0)(t) \\ &\quad \ominus_H b(t) \cdot T(v_1 \ominus_H v_0)(t) \ominus_H c(t) \cdot S(v_1 \ominus_H v_0)(t) \\ &\quad \ominus_H f(t, v_0(t), Tv_0(t), Sv_0(t)) \\ &= \ominus_H a(t) \cdot (v_1 \ominus_H v_0)(t) \ominus_H b(t) \cdot T(v_1 \ominus_H v_0)(t) \\ &\quad \ominus_H c(t) \cdot S(v_1 \ominus_H v_0)(t), \quad t \in J \setminus \{t_j\}_{j=1}^m,\end{aligned}$$

and

$$\begin{aligned}&(v_1 \ominus_H v_0)(t_j^+) \ominus_H (v_1 \ominus_H v_0)(t_j) \\ &= v_1(t_j^+) \ominus_H v_1(t_j) \ominus_H (v_0(t_j^+) \ominus_H v_0(t_j)) \\ &= \ominus_H L_j \cdot v_1(t_j) + I_j(v_0(t_j)) + L_j \cdot v_0(t_j) \ominus_H I_j(v_0(t_j)) \\ &= \ominus_H L_j \cdot (v_1 \ominus_H v_0)(t_j), \quad j \in \{1, \ldots, m\},\end{aligned}$$

and

$$(v_1 \ominus_H v_0)(0) = v_1(0) \ominus_H v_0(0)$$

$$\leq v_1(T) \ominus_H v_0(T)$$

$$= (v_1 \ominus_H v_0)(T),$$

and

$$\delta_H(v_1 \ominus_h u_1)(t) = \delta_H v_1(t) \ominus_H \delta_H u_1(t)$$

$$= f(t, v_0(t), Tv_0(t), Sv_0(t)) \ominus_H a(t) \cdot (v_1 \ominus_H v_0)(t)$$

$$\ominus_H b(t) \cdot T(v_1 \ominus_H v_0)(t) \ominus_H c(t) \cdot S(v_1 \ominus_H v_0)(t)$$

$$\ominus_H f(t, u_0(t), Tu_0(t), Su_0(t)) \ominus_H a(t) \cdot (u_1 \ominus_H u_0)(t)$$

$$\ominus_H b(t) \cdot T(u_1 \ominus_H u_0)(t) \ominus_H c(t) \cdot S(u_1 \ominus_H u_0)(t)$$

$$\geq \ominus_H a(t) \cdot (v_0 \ominus_H u_0)(t) \ominus_H b(t) \cdot T(v_0 \ominus_H u_0)(t)$$

$$\ominus_H c(t) \cdot S(v_0 \ominus_H u_0)(t) \ominus_H a(t) \cdot (v_1 \ominus_H v_0)(t)$$

$$\ominus_H b(t) \cdot T(v_1 \ominus_H v_0)(t) \ominus_H c(t) \cdot S(v_1 \ominus_H v_0)(t)$$

$$\ominus_H a(t) \cdot (u_1 \ominus_H u_0)(t) \ominus_H b(t) \cdot T(u_1 \ominus_H u_0)(t)$$

$$\ominus_H c(t) \cdot S(u_1 \ominus_H u_0)(t)$$

$$= \ominus_H a(t) \cdot (v_1 \ominus_H u_1)(t) \ominus_H b(t) \cdot T(v_1 \ominus_H u_1)(t)$$

$$\ominus_H c(t) \cdot S(v_1 \ominus_H u_1)(t), \quad t \in J \backslash \{t_j\}_{j=1}^m,$$

and

$$(v_1 \ominus_H u_1)(t_j^+) \ominus_H (v_1 \ominus_H u_1)(t_j)$$

$$
\begin{aligned}
&= (v_1(t_j^+) \ominus_H v_1(t_j)) \ominus_H (u_1(t_j^+) \ominus_H u_1(t_j)) \\
&= \ominus_H L_j \cdot v_1(t_j) + I_j(v_0(t_j)) + L_j \cdot u_0(t_j) \\
&\quad + L_j \cdot u_1(t_j) \ominus_H I_j(v_0(t_j)) \ominus_H L_j \cdot v_0(t_j) \\
&\geq \ominus_H L_j \cdot (v_1 \ominus_H u_1)(t_j), \quad j \in \{1, \ldots, m\},
\end{aligned}
$$

$$
(v_1 \ominus_H u_1)(0) \geq (v_1 \ominus_H u_1)(T).
$$

Hence and with Theorem 4.6.4, it follows that

$$
u_0 \leq u_1 \leq v_1 \leq v_0 \quad on \quad J.
$$

Assume that

$$
u_{n-2} \leq u_{n-1} \leq v_{n-1} \leq v_{n-2} \quad on \quad J,
$$

for some $n \in \mathbb{N}$, $n \geq 2$. We will prove that

$$
u_{n-1} \leq u_n \leq v_n \leq v_{n-1} \quad on \quad J. \tag{4.44}
$$

Indeed, we have

$$
\begin{aligned}
\delta_H(u_n \ominus_H u_{n-1})(t) &= \delta_H u_n(t) \ominus_H \delta_H u_{n-1}(t) \\
&= f(t, u_{n-1}(t), Tu_{n-1}(t), Su_{n-1}(t)) \ominus_H a(t) \cdot (u_n \ominus_H u_{n-1})(t) \\
&\quad \ominus_H b(t) \cdot T(u_n \ominus_H u_{n-1})(t) \ominus_H c(t) \cdot S(u_n \ominus_H u_{n-1})(t) \\
&\quad \ominus_H f(t, u_{n-2}(t), Tu_{n-2}(t), Su_{n-2}(t)) \\
&\quad + a(t) \cdot (u_{n-1} \ominus_H u_{n-2})(t)
\end{aligned}
$$

$$+b(t)\cdot T(u_{n-1}\ominus_H u_{n-2})(t)$$

$$+c(t)\cdot S(u_{n-1}\ominus_H u_{n-2})(t)$$

$$= \ominus_H a(t)\cdot(u_n\ominus_H u_{n-1})(t)\ominus_H b(t)\cdot T(u_n\ominus_H u_{n-1})(t)$$

$$\ominus_H c(t)\cdot S(u_n\ominus_H u_{n-1})(t), \quad t\in J\backslash\{t_j\}_{j=1}^m,$$

and

$$\begin{aligned}
&(u_n\ominus_H u_{n-1})(t_j^+)\ominus_H(u_n\ominus_H u_{n-1})(t_j)\\
=\ & u_n(t_j^+)\ominus_H u_n(t_j)\ominus_H(u_{n-1}(t_j^+)\ominus_H u_{n-1}(t_j))\\
=\ & \ominus_H L_j\cdot u_n(t_j)+I_j(u_{n-1}(t_j))+L_j\cdot u_{n-1}(t_j)\ominus_H I_j(u_{n-2}(t_j))
\end{aligned}$$

$$\ominus_H L_j u_{n-2}(t_j)$$

$$= \ominus_H L_j\cdot(u_n\ominus_H u_{n-1})(t_j)+I_j(u_{n-1}\ominus_H u_{n-2})(t)$$

$$+L_j\cdot(u_{n-1}\ominus_H u_{n-2})(t_j)$$

$$= \ominus_H L_j\cdot(u_n\ominus_H u_{n-1})(t_j), \quad j\in\{1,\ldots,m\},$$

and

$$(u_n\ominus_H u_{n-1})(0) = u_n(0)\ominus_H u_{n-1}(0)$$

$$= u_n(T)\ominus_H u_{n-1}(T)$$

$$= (u_n\ominus_H u_{n-1})(T),$$

and

$$\delta_H(v_n\ominus_H v_{n-1})(t) = \delta_H v_n(t)\ominus_H \delta_H v_{n-1}(t)$$

$$= f(t, v_{n-1}(t), Tv_{n-1}(t), Sv_{n-1}(t)) \ominus_H a(t) \cdot (v_n \ominus_H v_{n-1})(t)$$

$$\ominus_H b(t) \cdot T(v_n \ominus_H v_{n-1})(t) \ominus_H c(t) \cdot S(v_n \ominus_H v_{n-1})(t)$$
$$\ominus_H f(t, v_{n-2}(t), Tv_{n-2}(t), Sv_{n-2}(t))$$
$$+a(t) \cdot (v_{n-1} \ominus_H v_{n-2})(t)$$
$$+b(t) \cdot T(v_{n-1} \ominus_H v_{n-2})(t)$$
$$+c(t) \cdot S(v_{n-1} \ominus_H v_{n-2})(t)$$
$$= \ominus_H a(t) \cdot (v_n \ominus_H v_{n-1})(t) \ominus_H b(t) \cdot T(v_n \ominus_H v_{n-1})(t)$$
$$\ominus_H c(t) \cdot S(v_n \ominus_H v_{n-1})(t), \quad t \in J \backslash \{t_j\}_{j=1}^{m},$$

and

$$(v_n \ominus_H v_{n-1})(t_j^+) \ominus_H (v_n \ominus_H v_{n-1})(t_j)$$

$$= v_n(t_j^+) \ominus_H v_n(t_j) \ominus_H (v_{n-1}(t_j^+) \ominus_H v_{n-1}(t_j))$$

$$= \ominus_H L_j \cdot v_n(t_j) + I_j(v_{n-1}(t_j)) + L_j \cdot v_{n-1}(t_j) \ominus_H I_j(v_{n-2}(t_j))$$
$$+L_j \cdot v_{n-1}(t_j) \ominus_H L_j \cdot v_{n-2}(t_j)$$
$$\leq \ominus_H L_j \cdot (v_n \ominus_H v_{n-1})(t_j), \quad j \in \{1, \ldots, m\},$$

and

$$\begin{aligned}(v_n \ominus_H v_{n-1})(0) &= v_n(0) \ominus_H v_{n-1}(0) \\ &= v_n(T) \ominus_H v_{n-1}(T) \\ &= (v_n \ominus_H v_{n-1})(T),\end{aligned}$$

and

$$\delta_H(v_n \ominus_H u_n)(t) = \delta_H v_n(t) \ominus_H \delta_H u_n(t)$$

$$= f(t, v_{n-1}(t), Tv_{n-1}(t), Sv_{n-1}(t)) \ominus_H a(t) \cdot (v_n \ominus_H v_{n-1})(t)$$

$$
\begin{aligned}
&\ominus_H b(t)\cdot T(v_n \ominus_H v_{n-1})(t) \ominus_H c(t)\cdot S(v_n \ominus_H v_{n-1})(t)\\
&\ominus_H f(t,u_{n-1}(t),Tu_{n-1}(t),Su_{n-1}(t)) + a(t)\cdot(u_n \ominus_H u_{n-1})(t)\\
&+b(t)\cdot T(u_n \ominus_H u_{n-1})(t) + c(t)\cdot S(u_n \ominus_H u_{n-1})(t)\\
\geq\ &\ominus_H a(t)\cdot(v_{n-1} \ominus_H u_{n-1})(t) \ominus_H b(t)\cdot T(v_{n-1} \ominus_H u_{n-1})(t)\\
&\ominus_H c(t)\cdot S(v_{n-1} \ominus_H u_{n-1})(t) \ominus_H a(t)\cdot(v_n \ominus_H v_{n-1})(t)\\
&\ominus_H b(t)\cdot T(v_n \ominus_H v_{n-1})(t) \ominus_H c(t)\cdot S(v_n \ominus_H v_{n-1})(t)\\
&\ominus_H a(t)\cdot(u_n \ominus_H u_{n-1})(t) \ominus_H b(t)\cdot T(u_n \ominus_H u_{n-1})(t)\\
&\ominus_H c(t)\cdot S(u_n \ominus_H u_{n-1})(t)\\
=\ &\ominus_H a(t)\cdot(v_n \ominus_H u_n)(t) \ominus_H b(t)\cdot T(v_n \ominus_H u_n)(t)\\
&\ominus_H c(t)\cdot S(v_n \ominus_H u_n)(t), \quad t \in J\backslash\{t_j\}_{j=1}^{m},
\end{aligned}
$$

and

$$
\begin{aligned}
&(v_n \ominus_H u_n)(t_j^+) \ominus_H (v_n \ominus_H u_n)(t_j)\\
=\ &(v_n(t_j^+) \ominus_H v_n(t_j)) \ominus_H (u_n(t_j^+) \ominus_H u_n(t_j))\\
=\ &\ominus_H L_j\cdot v_n(t_j) + I_j(v_{n-1}(t_j)) + L_j\cdot v_{n-1}(t_j)\\
&+L_j\cdot u_n(t_j) \ominus_H I_j(u_{n-1}(t_j)) \ominus_H L_j\cdot u_{n-1}(t_j)\\
\geq\ &\ominus_H L_j\cdot(v_n \ominus_H u_n)(t_j), \quad j \in \{1,\ldots,m\},
\end{aligned}
$$

and

$$(v_n \ominus_H u_n)(0) = (v_n \ominus_H u_n)(T).$$

From here and from Theorem 4.6.4, we conclude that (4.44) *holds for any* $n \in \mathbb{N}$. *By Lemma 4.6.6, it follows that*

$$\begin{aligned}
& u_n(t) \\
= & \int_0^t G(t,s) \cdot \Big(f(s, u_{n-1}(s), Tu_{n-1}(s), Su_{n-1}(s)) + a(s) \cdot u_{n-1}(s) \\
& \ominus_H b(s) \cdot T(u_n \ominus_H u_{n-1})(s) \ominus_H c(s) \cdot S(u_n \ominus_H u_{n-1})(s) \Big) \delta_H s \\
& + \sum_{\{j: 0 < t_j < T\}} (G(t,t_j) e_{-a}(\sigma(t_j), t_j)) \\
& \cdot \left(\ominus_H L_j \cdot u_n(t_j) + I_j(u_{n-1}(t_j)) + L_j \cdot u_{n-1}(t_j) \right), \\
& v_n(t) \\
= & \int_0^t G(t,s) \cdot \Big(f(s, v_{n-1}(s), Tv_{n-1}(s), Sv_{n-1}(s)) + a(s) \cdot v_{n-1}(s) \\
& \ominus_H b(s) \cdot T(v_n \ominus_H v_{n-1})(s) \ominus_H c(s) \cdot S(v_n \ominus_H v_{n-1})(s) \Big) \delta_H s \\
& + \sum_{\{j: 0 < t_j < T\}} (G(t,t_j) e_{-a}(\sigma(t_j), t_j)) \\
& \cdot \left(\ominus_H L_j \cdot v_n(t_j) + I_j(v_{n-1}(t_j)) + L_j \cdot v_{n-1}(t_j) \right),
\end{aligned} \tag{4.45}$$

$t \in J$, $n \in \mathbb{N}$. *Note that the sequences* $u_n^{\Delta}(t)\}_{n \in \mathbb{N}}$ *and* $\{v_n^{\Delta}(t)\}_{n \in \mathbb{N}}$ *are uniformly bounded on J. Then there exist* $u^*, v^* \in \Omega$ *such that*

$$u_n(t) \to u^*(t), \quad v_n(t) \to v^*(t), \quad as \quad n \to \infty,$$

uniformly on J. Hence and (4.45), *we obtain*

$$\begin{aligned} u^*(t) &= \int_0^T G(t,s)\cdot(f(s,u^*(s),Tu^*(s),Su^*(s))+a(s)\cdot u^*(s))\,\delta_H s \\ &\quad + \sum_{\{j:0<t_j<T\}} (G(t,t_j)e_{-a}(\sigma(t_j),t_j))\cdot I_j(u^*(t_j)), \\ v^*(t) &= \int_0^T G(t,s)\cdot(f(s,v^*(s),Tv^*(s),Sv^*(s))+a(s)\cdot v^*(s))\,\delta_H s \\ &\quad + \sum_{\{j:0<t_j<T\}} (G(t,t_j)e_{-a}(\sigma(t_j),t_j))\cdot I_j(v^*(t_j)), \quad t\in J. \end{aligned}$$

Let $w\in[u_0,v_0]$ *be a solution of the PBVP* (4.28), *i.e.,*

$$\begin{aligned} \delta_H w(t) &= f(t,w(t),Tw(t),Sw(t)), \quad t\in J\backslash\{t_j\}_{j=1}^m, \\ w(t_j^+)\ominus_H w(t_j) &= I_j(w(t_j)), \quad j\in\{1,\ldots,m\}, \\ w(0) &= w(T). \end{aligned}$$

We have

$$\begin{aligned} &\delta_H(w\ominus_H u_1)(t) \\ &\quad = \delta_H w(t) - \delta_H u_1(t) \\ &\quad \geq f(t,w(t),Tw(t),Sw(t))\ominus_H f(t,u_0(t),Tu_0(t),Su_0(t)) \\ &\quad \geq \ominus_H a(t)\cdot(u_1\ominus_H u_0)(t)+b_0(t)\cdot T(u_1\ominus_H u_0)(t) \\ &\quad\quad +c(t)\cdot S(u_1\ominus_H u_0)(t) \end{aligned}$$

$$\geq f(t,w(t),Tw(t),Sw(t)) \ominus_H f(t,u_0(t),Tu_0(t),Su_0(t))$$
$$\geq \ominus_H a(t)\cdot(w\ominus_H u_0)(t)\ominus_H b(t)\cdot T(w\ominus_H u_0)(t)$$

$$\ominus_H c(t)\cdot S(w\ominus_H u_0)(t),$$

and

$$\begin{aligned}(w\ominus_H u_1)(t_j^+)\ominus_H (w\ominus_H u_1)(t_j)\\ = w(t_j^+)\ominus_H w(t_j)\ominus_H (u_1(t_j^+)\ominus_H u_1(t_j))\end{aligned}$$

$$= \quad I_j(w(t_j))+L_j\cdot u_1(t_j)\ominus_H I_j(u_0(t_j))\ominus_H L_j\cdot u_0(t_j)$$

$$\geq \quad \ominus_H L_j\cdot(w(t_j)\ominus_H u_0(t_j))+L_j\cdot(u_1(t_j)\ominus_H u_0(t_j))$$

$$= \quad \ominus_H L_j\cdot(w(t_j)\ominus_H u_1(t_j)),$$

and

$$\begin{aligned}(w\ominus_H u_1)(0) &= w(0)\ominus_H u_1(0)\\ &= (w\ominus_H u_1)(T),\end{aligned}$$

and

$$\delta_H(w\ominus_H v_1)(t) \quad = \quad \delta_H w(t)\ominus_H \delta_H v_1(t)$$

$$= \quad f(t,w(t),Tw(t),Sw(t))$$

$$\ominus_H f(t,v_0(t),Tv_0(t),Sv_0(t))$$

$$+a(t)\cdot(v_1\ominus_H v_0)(t)+b(t)\cdot T(v_1\ominus_H v_0)(t)$$

$$+c(t)\cdot S(v_1\ominus_H v_0)(t)$$

$$\leq f(t,w(t),Tw(t),Sw(t)) \ominus_H f(t,v_0(t),Tv_0(t),Sv_0(t))$$

$$\leq \ominus_H a(t)\cdot(w\ominus_H v_0)(t)\ominus_H b(t)\cdot T(w\ominus_H v_0)(t)$$

$$\ominus_H c(t)\cdot S(w\ominus_H v_0)(t),$$

and

$$\begin{aligned}
&(w\ominus_H v_1)(t_j^+)\ominus_H (w-v_1)(t_j) \\
&\quad = w(t_j^+)\ominus_H w(t_j)\ominus_H (v_1(t_j^+)\ominus_H v_1(t_j)) \\
&= I_j(w(t_j))+L_j\cdot v_1(t_j)\ominus_H I_j(v_0(t_j))\ominus_H L_j\cdot v_0(t_j) \\
&\geq \ominus_H L_j\cdot(w(t_j)\ominus_H v_0(t_j))+L_j\cdot(v_1(t_j)\ominus_H v_0(t_j)) \\
&= \ominus_H L_j\cdot(w(t_j)\ominus_H v_1(t_j)),
\end{aligned}$$

and

$$\begin{aligned}
(w\ominus_H v_1)(0) &= w(0)\ominus_H v_1(0) \\
&= (w\ominus_H v_1)(T).
\end{aligned}$$

Hence and with Theorem 4.6.4, it follows that

$$u_1(t)\leq w(t)\leq v_1(t),\quad t\in J.$$

Assume that

$$u_{n-1}(t)\leq w(t)\leq v_{n-1}(t),\quad t\in J,$$

for some $n\in\mathbb{N}$, $n\geq 2$. We will prove that

$$u_n(t)\leq w(t)\leq v_n(t),\quad t\in J. \tag{4.46}$$

We have

$$\delta_H(w\ominus_H u_n)(t) = \delta_H w(t)-\delta_H u_n(t)$$

$$
\begin{aligned}
&\geq f(t,w(t),Tw(t),Sw(t)) \ominus_H f(t,u_{n-1}(t),Tu_{n-1}(t),Su_{n-1}(t)) \\
&\geq \ominus_H a(t)\cdot(w\ominus_H u_{n-1})(t)\ominus_H b(t)\cdot T(w\ominus_H u_{n-1})(t) \\
&\ominus_H c(t)\cdot S(w\ominus_H u_{n-1})(t),
\end{aligned}
$$

and

$$
\begin{aligned}
& (w\ominus_H u_n)(t_j^+)\ominus_H (w\ominus_H u_n)(t_j) \\
&= w(t_j^+)\ominus_H w(t_j)\ominus_H (u_n(t_j^+)\ominus_H u_n(t_j)) \\
&= I_j(w(t_j))+L_j\cdot u_n(t_j)\ominus_H I_j(u_{n-1}(t_j))\ominus_H L_j\cdot u_{n-1}(t_j) \\
&\geq \ominus_H L_j\cdot(w(t_j)\ominus_H u_{n-1}(t_j))+L_j\cdot(u_n(t_j)\ominus_H u_{n-1}(t_j)) \\
&= \ominus_H L_j\cdot(w(t_j)\ominus_H u_n(t_j)),
\end{aligned}
$$

and

$$
\begin{aligned}
(w\ominus_H u_n)(0) &= w(0)\ominus_H u_n(0) \\
&= (w\ominus_H u_n)(T),
\end{aligned}
$$

and

$$
\begin{aligned}
\delta_H(w\ominus_H v_n)(t) &= \delta_H w(t)-\delta_H v_n(t) \\
&\leq f(t,w(t),Tw(t),Sw(t)) \ominus_H f(t,v_{n-1}(t),Tv_{n-1}(t),Sv_{n-1}(t)) \\
&\leq \ominus_H a(t)\cdot(w\ominus_H v_{n-1})(t)\ominus_H b(t)\cdot T(w\ominus_H v_{n-1})(t) \\
&\ominus_H c(t)\cdot S(w\ominus_H v_{n-1})(t),
\end{aligned}
$$

and

$$
(w\ominus_H v_n)(t_j^+)\ominus_H (w\ominus_H v_n)(t_j)
$$

$$
\begin{aligned}
&= w(t_j^+) \ominus_H w(t_j) \ominus_H (v_n(t_j^+) \ominus_H v_n(t_j)) \\
&= I_j(w(t_j)) + L_j \cdot v_n(t_j) \ominus_H I_j(v_{n-1}(t_j)) \ominus_H L_j \cdot v_{n-1}(t_j) \\
&\geq \ominus_H L_j \cdot (w(t_j) \ominus_H v_{n-1}(t_j)) + L_j \cdot (v_n(t_j) \ominus_H v_{n-1}(t_j)) \\
&= \ominus_H L_j \cdot (w(t_j) \ominus_H v_n(t_j)),
\end{aligned}
$$

and

$$
\begin{aligned}
(w \ominus_H v_n)(0) &= w(0) \ominus_H v_n(0) \\
&= (w \ominus_H v_n)(T).
\end{aligned}
$$

Hence and with Theorem 4.6.4, it follows that (4.46) *holds for any* $n \in \mathbb{N}$. *Letting* $n \to \infty$ *in* (4.46), *we get*

$$u^*(t) \leq w(t) \leq v^*(t), \quad t \in J.$$

This completes the proof.

4.7 MULTIPLE POSITIVE SOLUTIONS FOR FIRST-ORDER FUZZY IMPULSIVE INTEGRAL BOUNDARY VALUE PROBLEMS

Let $\mathbb{T}$ be a time scale with a forward jump operator and a delta differentiation operator σ and Δ, respectively. Consider the following first-order non-linear fuzzy impulsive integral boundary value problem

$$
\begin{aligned}
\delta_H x(t) + p(t) \cdot x(\sigma(t)) &= f(t, x(\sigma(t))), \quad t \in J = [0,T] \backslash \{t_1, \ldots, t_m\}, \\
x(t_j^+) &= x(t_j^-) + I_j(x(t_j)), \quad j \in \{1, \ldots, m\}, \\
\alpha \cdot x(0) \ominus_H \beta \cdot x(\sigma(T)) &= \int_0^{\sigma(T)} g(s) \cdot x(s) \delta_H s,
\end{aligned}
\tag{4.47}
$$

where

(H1) $p \in \mathscr{C}([0,\sigma(T)])$, $p \geq 0$ on $[0,\sigma(T)]$,

(H2) $I_j \in \mathscr{C}(F(\mathbb{R}))$, $I_j \geq \widetilde{0}$ on $[0,\infty)$, $j \in \{1,\ldots,m\}$,

(H3) g is a non-negative integrable function on $[0,\sigma(T)]$,

(H4) α and β are non-negative real constants,

$$\Gamma = \alpha - \beta e_p(0,\sigma(T)) - \int_0^{\sigma(T)} g(s) e_p(0,s)\Delta s > 0,$$

(H5) $f \in \mathscr{C}([0,\sigma(T)] \times F(\mathbb{R}))$, $f \geq \widetilde{0}$ on $[0,\sigma(T)] \times F([0,\infty))$,

$$0 < t_1 < \ldots < t_m < T,$$

t_j, $j \in \{1,\ldots,m\}$, are right-dense. Let $PC[0,\sigma(T)]$ and $PC^{\Delta}([0,\sigma(T)])$ be as in Section 4.2

Lemma 4.7.1 *Suppose* $(H1)$, $(H3)$, $(H4)$, $h \in \mathscr{C}([0,\sigma(T)])$, $v_j \in F(\mathbb{R})$, $j \in \{1,\ldots,m\}$. *Then* $x \in PC([0,\sigma(T)])$ *is a solution of the integral equation*

$$x(t) = \int_0^{\sigma(T)} G(t,s) \cdot h(s) \delta_H s + \sum_{j=1}^{m} G(t,t_j) \cdot v_j, \quad t \in [0,\sigma(T)], \tag{4.48}$$

where

$$G(t,s) = \begin{cases} \dfrac{1}{\Gamma} e_{\ominus p}(t,s) \left(\alpha - \displaystyle\int_0^{\sigma(s)} g(\tau) e_{\ominus p}(\tau,0)\Delta\tau \right) \\ \text{for} \quad 0 \leq s \leq t \leq \sigma(T) \quad \text{and} \\ \dfrac{1}{\Gamma} e_{\ominus p}(t,s) \left(\beta e_{\ominus p}(\sigma(T),0) + \displaystyle\int_{\sigma(s)}^{\sigma(T)} g(r) e_{\ominus p}(r,0)\Delta r \right) \\ \text{for} \quad 0 \leq t \leq s \leq \sigma(T), \end{cases}$$

if and only if $x \in PC^{\Delta}([0,\sigma(T)])$ *is a solution of the BVP*

$$\begin{aligned} \delta_H x(t) + p(t)\cdot x(\sigma(t)) &= h(t), \quad t \in J, \\ x(t_j^+) &= x(t_j^-) + v_j, \quad j \in \{1,\dots,m\}, \\ \alpha \cdot x(0) \ominus_H \beta \cdot x(\sigma(T)) &= \int_0^{\sigma(T)} g(s)\cdot x(s)\delta_H s. \end{aligned} \tag{4.49}$$

Remark 4.7.2 *Suppose that x satisfies* (4.48). *Then*

$$\begin{aligned} x(t) &= \int_0^{\sigma(T)} G(t,s)\cdot h(s)\delta_H s + \sum_{j=1}^{m} G(t,t_j)\cdot v_j \\ &= \int_0^{t} G(t,s)\cdot h(s)\delta_H s + \int_t^{\sigma(T)} G(t,s)\cdot h(s)\delta_H s \\ &\quad + \sum_{\{j:t_j<t\}} G(t,t_j)\cdot v_j + \sum_{\{j:t\le t_j\}} G(t,t_j)\cdot v_j \\ &= \frac{1}{\Gamma}\cdot \int_0^{t} \left(e_{\ominus p}(t,s) \left(\alpha - \int_0^{\sigma(s)} e_{\ominus p}(\tau,0)g(\tau)\Delta\tau \right) \right) \cdot h(s)\delta_H s \\ &\quad + \frac{1}{\Gamma}\cdot \int_t^{\sigma(T)} \Bigg(e_{\ominus p}(t,s) \Bigg(\beta e_{\ominus p}(\sigma(T),0) \\ &\quad + \int_{\sigma(s)}^{\sigma(T)} g(r)e_{\ominus p}(r,0)\Delta r \Bigg) \Bigg) \cdot h(s)\delta_H s \\ &\quad + \sum_{\{j:t_j<t\}} \left(\frac{1}{\Gamma} e_{\ominus p}(t,t_j) \left(\alpha - \int_0^{\sigma(t_j)} g(\tau)e_{\ominus p}(\tau,0)\Delta\tau \right) \right) \cdot v_j \end{aligned}$$

$$+\sum_{\{j:t\leq t_j\}}\left(\frac{1}{\Gamma}e_{\ominus p}(t,t_j)\left(\beta e_{\ominus p}(\sigma(T),0)\right.\right.$$

$$\left.\left.+\int_{\sigma(t_j)}^{\sigma(T)} g(r)e_{\ominus p}(r,0)\Delta r\right)\right)\cdot v_j$$

$$= \frac{\alpha}{\Gamma}\cdot\int_0^t e_{\ominus p}(t,s)\cdot h(s)\delta_H s$$

$$\ominus_H\frac{1}{\Gamma}\int_0^t e_{\ominus p}(t,s)\cdot\int_0^{\sigma(s)}(g(\tau)e_{\ominus p}(\tau,0))\cdot h(s)\delta_H\tau\delta_H s$$

$$+\left(\frac{\beta}{\Gamma}e_{\ominus p}(\sigma(T),0)\right)\cdot\int_t^{\sigma(T)} e_{\ominus p}(t,s)\cdot h(s)\delta_H s$$

$$+\frac{1}{\Gamma}\cdot\int_t^{\sigma(T)}\int_{\sigma(s)}^{\sigma(T)}(e_{\ominus p}(t,s)g(r)e_{\ominus p}(r,0))\cdot h(s)\Delta r\delta_H s$$

$$+\frac{\alpha}{\Gamma}\cdot\sum_{\{j:t_j<t\}} e_{\ominus p}(t,t_j)\cdot v_j$$

$$\ominus_H\frac{1}{\Gamma}\cdot\sum_{\{j:t_j<t\}}\left(e_{\ominus p}(t,t_j)\left(\int_0^{\sigma(t_j)} g(\tau)e_{\ominus p}(\tau,0)\Delta\tau\right)\right)\cdot v_j$$

$$+\frac{\beta e_{\ominus p}(\sigma(T),0)}{\Gamma}\cdot\sum_{\{j:t\leq t_j\}} e_{\ominus p}(t,t_j)\cdot v_j$$

$$+\ \frac{1}{\Gamma}\cdot\sum_{\{j:t\leq t_j\}}\left(e_{\ominus p}(t,t_j)\left(\int_{\sigma(t_j)}^{\sigma(T)} g(r)e_{\ominus p}(r,0)\Delta r\right)\right)\cdot v_j$$

$$
\begin{aligned}
= \; & \frac{\alpha}{\Gamma} \cdot \int_{0}^{t} e_{\ominus p}(t,s) \cdot h(s) \delta_H s \\
& \ominus_H \frac{1}{\Gamma} \cdot \int_{0}^{t} e_{\ominus p}(t,s) \cdot \int_{0}^{\sigma(T)} (g(\tau) e_{\ominus p}(\tau,0)) \cdot h(s) \delta_H \tau \delta_H s \\
& + \frac{1}{\Gamma} \cdot \int_{0}^{t} e_{\ominus p}(t,s) \int_{\sigma(s)}^{\sigma(T)} (g(\tau) e_{\ominus p}(\tau,0)) \cdot h(s) \delta_H \tau \delta_H s \\
& + \frac{1}{\Gamma} \cdot \int_{t}^{\sigma(T)} e_{\ominus p}(t,s) \cdot \int_{\sigma(s)}^{\sigma(T)} (g(\tau) e_{\ominus p}(\tau,0)) \cdot h(s) \delta_H \tau \delta_H s \\
& + \left(\frac{\beta}{\Gamma} e_{\ominus p}(\sigma(T),0) \right) \cdot \int_{t}^{\sigma(T)} e_{\ominus p}(t,s) \cdot h(s) \delta_H s \\
& + \frac{\alpha}{\Gamma} \cdot \sum_{\{j:t_j<t\}} e_{\ominus p}(t,t_j) \cdot v_j \\
& \ominus_H \frac{1}{\Gamma} \cdot \sum_{\{j:t_j<t\}} \left(e_{\ominus p}(t,t_j) \left(\int_{0}^{\sigma(t_j)} g(\tau) e_{\ominus p}(\tau,0) \delta_H \tau \right) \right) \cdot v_j \\
& + \frac{\beta e_{\ominus p}(\sigma(T),0)}{\Gamma} \cdot \sum_{\{j:t\le t_j\}} e_{\ominus p}(t,t_j) \cdot v_j \\
+ \; & \frac{1}{\Gamma} \cdot \sum_{\{j:t\le t_j\}} \left(e_{\ominus p}(t,t_j) \left(\int_{\sigma(t_j)}^{\sigma(T)} g(r) e_{\ominus p}(r,0) \Delta r \right) \right) \cdot v_j \\
= \; & \frac{\alpha}{\Gamma} \cdot \int_{0}^{t} e_{\ominus p}(t,s) \cdot h(s) \delta_H s
\end{aligned}
$$

$$\ominus_H \frac{1}{\Gamma} \cdot \int_0^t e_{\ominus p}(t,s) \cdot \int_0^{\sigma(T)} (g(\tau) e_{\ominus p}(\tau,0)) \cdot h(s) \delta_H \tau \delta_H s$$

$$+\frac{1}{\Gamma} \cdot \int_0^{\sigma(T)} e_{\ominus p}(t,s) \cdot \int_{\sigma(s)}^{\sigma(T)} (g(\tau) e_{\ominus p}(\tau,0)) \cdot h(s) \delta_H \tau \delta_H s$$

$$+\left(\frac{\beta}{\Gamma} e_{\ominus p}(\sigma(T),0)\right) \cdot \int_t^{\sigma(T)} e_{\ominus p}(t,s) \cdot h(s) \delta_H s$$

$$+\frac{\alpha}{\Gamma} \cdot \sum_{\{j:t_j<t\}} e_{\ominus p}(t,t_j) \cdot v_j$$

$$\ominus_H \frac{1}{\Gamma} \cdot \sum_{\{j:t_j<t\}} \left(e_{\ominus p}(t,t_j) \left(\int_0^{\sigma(t_j)} g(\tau) e_{\ominus p}(\tau,0) \Delta\tau \right) \right) \cdot v_j$$

$$+\frac{\beta e_{\ominus p}(\sigma(T),0)}{\Gamma} \cdot \sum_{\{j:t \le t_j\}} e_{\ominus p}(t,t_j) \cdot v_j$$

$$+ \ \frac{1}{\Gamma} \cdot \sum_{\{j:t \le t_j\}} \left(e_{\ominus p}(t,t_j) \left(\int_{\sigma(t_j)}^{\sigma(T)} g(r) e_{\ominus p}(r,0) \Delta r \right) \right) \cdot v_j$$

$$= \ \frac{\alpha}{\Gamma} \cdot \int_0^t e_{\ominus p}(t,s) \cdot h(s) \delta_H s$$

$$\ominus_H \frac{1}{\Gamma} \cdot \int_0^t e_{\ominus p}(t,s) \cdot \int_0^{\sigma(T)} (g(\tau) e_{\ominus p}(\tau,0)) \cdot h(s) \delta_H \tau \delta_H s$$

$$+\frac{1}{\Gamma} \cdot \int_0^{\sigma(T)} e_{\ominus p}(t,s) \cdot \int_{\sigma(s)}^{\sigma(T)} (g(\tau) e_{\ominus p}(\tau,0)) \cdot h(s) \delta_H \tau \delta_H s$$

$$+\left(\frac{\beta}{\Gamma}e_{\ominus p}(\sigma(T),0)\right)\cdot\int_0^{\sigma(T)} e_{\ominus p}(t,s)\cdot h(s)\delta_H s$$

$$+\frac{\alpha}{\Gamma}\cdot\sum_{\{j:t_j<t\}} e_{\ominus p}(t,t_j)\cdot v_j$$

$$\ominus_H\frac{1}{\Gamma}\cdot\sum_{\{j:t_j<t\}} e_{\ominus p}(t,t_j)\cdot\left(\int_0^{\sigma(T)} g(\tau)e_{\ominus p}(\tau,0)\Delta\tau\right)\cdot v_j$$

$$+\frac{1}{\Gamma}\cdot\sum_{\{j:t_j<t\}} e_{\ominus p}(t,t_j)\cdot\left(\int_{\sigma(t_j)}^{\sigma(T)} g(\tau)e_{\ominus p}(\tau,0)\Delta\tau\right)\cdot v_j$$

$$+\frac{\beta e_{\ominus p}(\sigma(T),0)}{\Gamma}\cdot\sum_{\{j:t\le t_j\}} e_{\ominus p}(t,t_j)\cdot v_j$$

$$+\ \frac{1}{\Gamma}\cdot\sum_{\{j:t\le t_j\}} e_{\ominus p}(t,t_j)\cdot\left(\int_{\sigma(t_j)}^{\sigma(T)} g(r)e_{\ominus p}(r,0)\Delta r\right)\cdot v_j$$

$$=\ \frac{\alpha}{\Gamma}\cdot\int_0^{t} e_{\ominus p}(t,s)\cdot h(s)\delta_H s$$

$$\ominus_H\frac{1}{\Gamma}\cdot\int_0^{t} e_{\ominus p}(t,s)\cdot\int_0^{\sigma(T)}(g(\tau)e_{\ominus p}(\tau,0))\cdot h(s)\delta_H\tau\delta_H s$$

$$+\frac{1}{\Gamma}\cdot\int_0^{\sigma(T)} e_{\ominus p}(t,s)\cdot\int_{\sigma(s)}^{\sigma(T)}(g(\tau)e_{\ominus p}(\tau,0))\cdot h(s)\delta_H\tau\delta_H s$$

$$+\left(\frac{\beta}{\Gamma}e_{\ominus p}(\sigma(T),0)\right)\cdot\int_t^{\sigma(T)} e_{\ominus p}(t,s)\cdot h(s)\delta_H s$$

$$+\frac{\alpha}{\Gamma}\cdot\sum_{\{j:t_j<t\}} e_{\ominus p}(t,t_j)\cdot v_j$$

$$\ominus_H\frac{1}{\Gamma}\cdot\sum_{\{j:t_j<t\}} e_{\ominus p}(t,t_j)\cdot\left(\int_0^{\sigma(T)} g(\tau)e_{\ominus p}(\tau,0)\Delta\tau\right)\cdot v_j$$

$$+\frac{1}{\Gamma}\cdot\sum_{j=1}^{m} e_{\ominus p}(t,t_j)\cdot\left(\int_{\sigma(t_j)}^{\sigma(T)} g(\tau)e_{\ominus p}(\tau,0)\Delta\tau\right)\cdot v_j$$

$$+\frac{\beta e_{\ominus p}(\sigma(T),0)}{\Gamma}\cdot\sum_{\{j:t_j<t\}} e_{\ominus p}(t,t_j)\cdot v_j.$$

Proof 4.7.3 *By Theorem 3.2.1, it follows that $x\in PC([0,\sigma(T)]$ is a solution of the impulsive dynamic equation*

$$\begin{aligned}\delta_H x(t)+p(t)\cdot x(\sigma(t)) &= h(t),\quad t\in J,\\ x(t_j^+) &= x(t_j^-)+v_j,\quad j\in\{1,\ldots,m\},\end{aligned}\tag{4.50}$$

if and only if $x\in PC([0,\sigma(T)])$ and satisfies the integral equation

$$\begin{aligned}x(t) &= \int_0^t e_{\ominus p}(t,s)\cdot h(s)\delta_H s+e_{\ominus p}(t,0)\cdot x(0)\\ &\quad+\sum_{\{j:t_j<t\}} e_{\ominus p}(t,t_j)\cdot v_j,\quad t\in[0,\sigma(T)].\end{aligned}\tag{4.51}$$

1. *Let $x\in PC^{\Delta}([0,\sigma(T)])$ be a solution of* (4.49). *Then x can be represented in the form* (4.51). *Hence,*

$$\begin{aligned}x(\sigma(T)) &= \int_0^{\sigma(T)} e_{\ominus p}(\sigma(T),s)\cdot h(s)\delta_H s+e_{\ominus p}(\sigma(T),0)\cdot x(0)\\ &\quad+\sum_{\{j:t_j<\sigma(T)\}} e_{\ominus p}(\sigma(T),t_j)\cdot v_j,\end{aligned}$$

$$\begin{aligned}\beta x(\sigma(T)) &= \beta \cdot \int_0^{\sigma(T)} e_{\ominus p}(\sigma(T),s)\cdot h(s)\delta_H s \\ &\quad +\beta \cdot e_{\ominus p}(\sigma(T),0)\cdot x(0) \\ &\quad +\beta \cdot \sum_{\{j:t_j<\sigma(T)\}} e_{\ominus p}(\sigma(T),t_j)\cdot v_j,\end{aligned}$$

and

$$\begin{aligned}&\alpha\cdot x(0)\ominus_H \beta\cdot x(\sigma(T)) \\ &= \alpha\cdot x(0)\ominus_H \beta\cdot \int_0^{\sigma(T)} e_{\ominus p}(\sigma(T),s)\cdot h(s)\delta_H s \\ &\quad \ominus_H \beta e_{\ominus p}(\sigma(T),0)\cdot x(0) \\ &\quad \ominus_H \beta\cdot \sum_{\{j:t_j<\sigma(T)\}} e_{\ominus p}(\sigma(T),t_j)\cdot v_j \\ &= (\alpha-\beta e_{\ominus p}(\sigma(T),0))\cdot x(0) \\ &\quad \ominus_H \beta\cdot \int_0^{\sigma(T)} e_{\ominus p}(\sigma(T),s)\cdot h(s)\delta_H s \\ &\quad \ominus_H \beta\cdot \sum_{\{j:t_j<\sigma(T)\}} e_{\ominus p}(\sigma(T),t_j)\cdot v_j \\ &= \int_0^{\sigma(T)} g(s)\cdot x(s)\delta_H s \\ &= \int_0^{\sigma(T)} g(s)\cdot \Big(\int_0^{s} e_{\ominus p}(s,\tau)\cdot h(\tau)\delta_H \tau + e_{\ominus p}(s,0)\cdot x(0)\end{aligned} \tag{4.52}$$

$$+ \sum_{\{j:t_j<s\}} e_{\ominus p}(s,t_j) \cdot v_j \Bigg) \delta_H s$$

$$= \int_0^{\sigma(T)} \int_0^s (g(s)e_{\ominus p}(s,\tau)) \cdot h(\tau)\delta_H \tau \delta_H s$$

$$+ \int_0^{\sigma(T)} (g(s)e_{\ominus p}(s,0)) \cdot x(0)\delta_H s$$

$$+ \int_0^{\sigma(T)} g(s) \cdot \left(\sum_{\{j:t_j<s\}} e_{\ominus p}(s,t_j) \cdot v_j \right) \delta_H s,$$

whereupon

$$\left(\alpha - \beta e_{\ominus p}(\sigma(T),0) - \int_0^{\sigma(T)} g(s)e_{\ominus p}(s,0)\delta_H s \right) \cdot x(0)$$

$$= \int_0^{\sigma(T)} \int_0^s (g(s)e_{\ominus p}(s,\tau)) \cdot h(\tau)\delta_H \tau \delta_H s$$

$$+ \int_0^{\sigma(T)} g(s) \cdot \left(\sum_{\{j:t_j<s\}} e_{\ominus p}(s,t_j) \cdot v_j \right) \delta_H s$$

$$+\beta \cdot \int_0^{\sigma(T)} e_{\ominus p}(\sigma(T),s) \cdot h(s)\delta_H s$$

$$+\beta \cdot \sum_{\{j:t_j<\sigma(T)\}} e_{\ominus p}(\sigma(T),t_j) \cdot v_j,$$

and

$$\begin{aligned}
x(0) &= \frac{1}{\Gamma}\cdot\Bigg(\int_0^{\sigma(T)}\int_0^{s}(g(s)e_{\ominus p}(s,\tau))\cdot h(\tau)\delta_H\tau\delta_H s \\
&+ \int_0^{\sigma(T)} g(s)\cdot\left(\sum_{\{j:t_j<s\}} e_{\ominus p}(s,t_j)\cdot v_j\right)\delta_H s \\
&+ \beta\cdot\int_0^{\sigma(T)} e_{\ominus p}(\sigma(T),s)\cdot h(s)\delta_H s \\
&+ \beta\cdot\sum_{\{j:t_j<\sigma(T)\}} e_{\ominus p}(\sigma(T),t_j)\cdot v_j\Bigg) \\
&= \frac{1}{\Gamma}\cdot\Bigg(\int_0^{\sigma(T)}\int_{\sigma(s)}^{\sigma(T)}(g(s)e_{\ominus p}(s,\tau))\cdot h(\tau)\delta_H\tau\delta_H s \\
&+ \int_0^{\sigma(T)} g(s)\cdot\left(\sum_{\{j:t_j<s\}} e_{\ominus p}(s,t_j)\cdot v_j\right)\delta_H s \\
&+ \beta\cdot\int_0^{\sigma(T)} e_{\ominus p}(\sigma(T),s)\cdot h(s)\delta_H s \\
&+\beta\cdot\sum_{\{j:t_j<\sigma(T)\}} e_{\ominus p}(\sigma(T),t_j)\cdot v_j\Bigg).
\end{aligned}$$

Note that

$$\begin{aligned}
&\int_0^{\sigma(T)} g(s)\cdot\left(\sum_{\{j:t_j<s\}} e_{\ominus p}(s,t_j)\cdot v_j\right)\delta_H s \\
&= \int_0^{t_1} g(s)\cdot\left(\sum_{\{j:t_j<s\}} e_{\ominus p}(s,t_j)\cdot v_j\right)\delta_H s
\end{aligned}$$

$$
\begin{aligned}
& + \int_{t_1}^{t_2} g(s) \cdot \left(\sum_{\{j:t_j<s\}} e_{\ominus p}(s,t_j) \cdot v_j \right) \delta_H s \\
& + \int_{t_2}^{t_3} g(s) \cdot \left(\sum_{\{j:t_j<s\}} e_{\ominus p}(s,t_j) \cdot v_j \right) \delta_H s \\
& + \cdots \\
& + \int_{t_m}^{\sigma(T)} g(s) \cdot \left(\sum_{\{j:t_j<s\}} e_{\ominus p}(s,t_j) \cdot v_j \right) \delta_H s \\
= & \int_{t_1}^{t_2} (g(s) e_{\ominus p}(s,t_1)) \cdot v_1 \delta_H s \\
& + \int_{t_2}^{t_3} g(s) \cdot (e_{\ominus p}(s,t_1) \cdot v_1 + e_{\ominus p}(s,t_2) \cdot v_2) \delta_H s \\
& + \cdots \\
& + \int_{t_m}^{\sigma(T)} g(s) \cdot \Big(e_{\ominus p}(s,t_1) \cdot v_1 + e_{\ominus p}(s,t_2) \cdot v_2 \\
& + \cdots + e_{\ominus p}(s,t_m) \cdot v_m \Big) \delta_H s \\
= & \int_{t_1}^{\sigma(T)} (g(s) e_{\ominus p}(s,t_1)) \cdot v_1 \delta_H s \\
& + \int_{t_2}^{\sigma(T)} (g(s) e_{\ominus p}(s,t_2)) \cdot v_2 \delta_H s
\end{aligned}
$$

$$\begin{aligned}
&+\cdots \\
&+\int_{t_m}^{\sigma(T)} (g(s)e_{\ominus p}(s,t_m))\cdot v_m\delta_H s \\
&= \sum_{j=1}^{m}\int_{t_j}^{\sigma(T)} (g(s)e_{\ominus p}(s,t_j))\cdot v_j\delta_H s.
\end{aligned}$$

Therefore,

$$\begin{aligned}
x(0) &= \frac{1}{\Gamma}\cdot\Bigg(\int_0^{\sigma(T)}\int_{\sigma(s)}^{\sigma(T)} (g(s)e_{\ominus p}(s,\tau))\cdot h(\tau)\delta_H\tau\delta_H s \\
&+ \sum_{j=1}^{m}\int_{t_j}^{\sigma(T)} (g(s)e_{\ominus p}(s,t_j))\cdot v_j\delta_H s \\
&+\beta\cdot\int_0^{\sigma(T)} e_{\ominus p}(\sigma(T),s)\cdot h(s)\delta_H s \\
&+\beta\cdot\sum_{\{j:t_j<\sigma(T)\}} e_{\ominus p}(\sigma(T),t_j)\cdot v_j\Bigg) \\
&= \frac{1}{\Gamma}\cdot\Bigg(\beta\cdot\int_0^{\sigma(T)} e_{\ominus p}(\sigma(T),s)\cdot h(s)\delta_H s \\
&+\int_0^{\sigma(T)}\int_0^{\sigma(T)} (g(s)e_{\ominus p}(s,\tau))\cdot h(\tau)\delta_H\tau\delta_H s\ominus_H \\
&+\int_0^{\sigma(T)}\int_0^{\sigma(s)} (g(s)e_{\ominus p}(s,\tau))\cdot h(\tau)\delta_H\tau\delta_H s
\end{aligned}$$

$$+\sum_{j=1}^{m}\left(\left(\int_{t_j}^{\sigma(T)} g(s)e_{\ominus p}(s,t_j)\Delta s+\beta e_{\ominus p}(\sigma(T),t_j)\right)\cdot v_j\right).$$

Then

$$\begin{aligned}
x(t) &= \frac{e_{\ominus p}(t,0)}{\Gamma}\cdot\Bigg(\beta\cdot\int_0^{\sigma(T)} e_{\ominus p}(\sigma(T),s)\cdot h(s)\delta_H s \\
&\quad+\int_0^{\sigma(T)}\int_0^{\sigma(T)} (g(s)e_{\ominus p}(s,\tau))\cdot h(\tau)\delta_H\tau\delta_H s\ominus_H \\
&\quad+\int_0^{\sigma(T)}\int_0^{\sigma(s)} (g(s)e_{\ominus p}(s,\tau))\cdot h(\tau)\delta_H\tau\delta_H s \\
&\quad+\sum_{j=1}^{m}\left(\left(\int_{t_j}^{\sigma(T)} g(s)e_{\ominus p}(s,t_j)\Delta s+\beta e_{\ominus p}(\sigma(T),t_j)\right)\cdot v_j\right) \\
&\quad+\int_0^{t} e_{\ominus p}(t,s)\cdot h(s)\delta_H s+\sum_{\{j:t_j<t\}} e_{\ominus p}(t,t_j)\cdot v_j \\
&= \frac{e_{\ominus p}(t,0)}{\Gamma}\cdot\Bigg(\beta\cdot\int_0^{\sigma(T)} e_{\ominus p}(\sigma(T),s)\cdot h(s)\delta_H s \\
&\quad+\int_0^{\sigma(T)}\int_0^{\sigma(T)} (g(s)e_{\ominus p}(s,\tau))\cdot h(\tau)\delta_H\tau\delta_H s\ominus_H \\
&\quad+\int_0^{\sigma(T)}\int_0^{\sigma(s)} (g(s)e_{\ominus p}(s,\tau))\cdot h(\tau)\delta_H\tau\delta_H s
\end{aligned}$$

$$+\sum_{j=1}^{m}\left(\int_{t_j}^{\sigma(T)} g(s)e_{\ominus p}(s,t_j)\Delta s+\beta e_{\ominus p}(\sigma(T),t_j)\right)\cdot v_j\Bigg)$$

$$+\frac{1}{\Gamma}\cdot\left(\alpha\cdot\int_{\sigma(T)}^{t} e_{\ominus p}(t,s)\cdot h(s)\delta_H s\ominus_H(\beta e_p(0,\sigma(T)))\cdot\right.$$

$$+\int_{0}^{t} e_{\ominus p}(t,s)\cdot h(s)\delta_H s$$

$$\ominus_H\left(\int_{0}^{\sigma(T)} g(s)e_p(0,s)\Delta s\right)\cdot\int_{0}^{t} e_{\ominus p}(t,s)\cdot h(s)\delta_H s$$

$$+\alpha\cdot\sum_{\{j:t_j<t\}} e_{\ominus p}(t,t_j)\cdot v_j$$

$$\ominus_H(\beta e_p(0,\sigma(T)))\cdot\sum_{\{j:t_j<t\}} e_{\ominus p}(t,t_j)\cdot v_j$$

$$\left.\ominus_H\left(\int_{0}^{\sigma(T)} g(s)e_p(0,s)\Delta s\right)\cdot\sum_{\{j:t_j<t\}} e_{\ominus p}(t,t_j)\cdot v_j\right)$$

$$=\frac{\alpha}{\Gamma}\cdot\int_{0}^{t} e_{\ominus p}(t,s)\cdot h(s)\delta_H s$$

$$\ominus_H\frac{1}{\Gamma}\cdot\int_{0}^{\sigma(T)}\int_{0}^{t}(e_{\ominus p}(t,s)e_{\ominus p}(\tau,0)g(\tau))\cdot h(s)\delta_H s\delta_H\tau$$

$$+\frac{1}{\Gamma}\cdot\int_{0}^{\sigma(T)}\int_{\sigma(s)}^{\sigma(T)}(g(s)e_{\ominus p}(s,0)e_{\ominus p}(t,\tau))\cdot h(\tau)\delta_H\tau\delta_H s$$

$$+\left(\frac{\beta}{\Gamma}e_{\ominus p}(\sigma(T),0)\right)\cdot\int_{0}^{\sigma(T)}e_{\ominus p}(t,s)\cdot h(s)\delta_H s$$

$$+\frac{\alpha}{\Gamma}\cdot\sum_{\{j:t_j<t\}}e_{\ominus p}(t,t_j)\cdot v_j$$

$$+\left(\frac{1}{\Gamma}e_{\ominus p}(t,0)\right)\cdot\sum_{j=1}^{m}\left(\int_{t_j}^{\sigma(T)}g(s)e_{\ominus p}(s,t_j)\delta_H s\right)\cdot v_j$$

$$+\left(\frac{\beta}{\Gamma}e_{\ominus p}(\sigma(T),0)\right)\cdot\sum_{\{j:t_j<t\}}e_{\ominus p}(t,t_j)\cdot v_j$$

$$\ominus_H\left(\frac{1}{\Gamma}\int_{0}^{\sigma(T)}g(s)e_p(0,s)\Delta s\right)\cdot\sum_{\{j:t_j<t\}}e_{\ominus p}(t,t_j)\cdot v_j$$

$$=\int_{0}^{\sigma(T)}G(t,s)\cdot h(s)\delta_H s+\sum_{j=1}^{m}G(t,t_j)\cdot v_j,$$

i.e., x satisfies (4.48).

2. *Let* $x\in PC([0,\sigma(T)])$ *satisfy* (4.48). *Define*

$$H(s)=e_p(t,0)G(t,s),\quad t,s\in[0,\sigma(T)].$$

Then

$$H(s)=\begin{cases}\frac{1}{\Gamma}e_p(s,0)\left(\alpha-\int_{0}^{\sigma(s)}g(r)e_p(0,r)\Delta r\right)\\ \text{for}\quad 0\le s\le t\le\sigma(T),\quad\text{and}\\ \frac{1}{\Gamma}e_p(s,0)\left(\beta e_p(0,\sigma(T))+\int_{\sigma(s)}^{\sigma(T)}g(r)e_p(0,r)\Delta r\right)\\ \text{for}\quad 0\le t\le s\le\sigma(T).\end{cases}$$

Then

$$e_p(t,0)\cdot x(t) = \int_0^{\sigma(T)} H(s)\cdot h(s)\delta_H s + \sum_{j=1}^{m} H(t_j)\cdot v_j, \quad t\in[0,\sigma(T)].$$

Hence,

$$\begin{aligned}
& \delta_H\left(e_p(\cdot,0)x\right)(t) \\
=\; & \delta_H\left(\int_0^{\sigma(T)} H(s)\cdot h(s)\delta_H s\right) \\
=\; & \delta_H\left(\int_0^{t} H(s)\cdot h(s)\delta_H s + \int_t^{\sigma(T)} H(s)\cdot h(s)\delta_H s\right) \\
=\; & \delta_H\left(\frac{1}{\Gamma}\cdot\int_0^{t}\left(e_p(s,0)\left(\alpha-\int_0^{\sigma(s)} g(r)e_p(0,r)\Delta r\right)\right)\cdot h(s)\delta_H s\right) \\
& +\delta_H\left(\frac{1}{\Gamma}\cdot\int_t^{\sigma(T)}\left(e_p(s,0)\Big(\beta e_p(0,\sigma(T))\right.\right. \\
& \left.\left.+\int_{\sigma(s)}^{\sigma(T)} g(r)e_p(0,r)\delta_H r\Big)\right)\cdot h(s)\delta_H s\right) \\
=\; & \left(\frac{1}{\Gamma}e_p(t,0)\left(\alpha-\int_0^{\sigma(t)} g(r)e_p(0,r)\Delta r\right)\right)\cdot h(t) \\
& \ominus_H\left(\frac{1}{\Gamma}e_p(t,0)\left(\beta e_p(0,\sigma(T))+\int_{\sigma(t)}^{\sigma(T)} g(r)e_p(0,r)\Delta r\right)\right)\cdot h(t) \\
=\; & \left(\frac{1}{\Gamma}e_p(t,0)\left(\alpha-\beta e_p(0,\sigma(T))-\int_0^{\sigma(T)} g(r)e_p(0,r)\Delta r\right)\right)\cdot h(t) \\
=\; & e_p(t,0)\cdot h(t), \quad t\in[0,\sigma(T)].
\end{aligned}$$

Hence and

$$e_p(t,0)\cdot\delta_H x(t) + (p(t)e_p(t,0))\cdot x(\sigma(t)) = e_p(t,0)\cdot h(t), \quad t\in[0,\sigma(T)],$$

or

$$\delta_H x(t) + p(t)\cdot x(\sigma(t)) = h(t), \quad t \in [0, \sigma(T)].$$

Next,

$$\begin{aligned}
x(t_k^+) &= \int_0^{\sigma(T)} G(t_k^+, s)\cdot h(s)\delta_H s + \sum_{j=1}^{m} G(t_k^+, t_j)\cdot v_j \\
&= \int_0^{t_k} G(t_k^+, s)\cdot h(s)\delta_H s + \int_{t_k}^{\sigma(T)} G(t_k^+, s)\cdot h(s)\delta_H s \\
&\quad + \sum_{\{j:t_j<t_k^+\}} G(t_k^+, t_j)\cdot v_j + \sum_{\{j:t_j\geq t_k^+\}} G(t_k^+, t_j)\cdot v_j, \\
x(t_k^-) &= \int_0^{\sigma(T)} G(t_k^-, s)\cdot h(s)\delta_H s + \sum_{j=1}^{m} G(t_k^-, t_j)\cdot v_j \\
&= \int_0^{t_k} G(t_k^-, s)\cdot h(s)\delta_H s + \int_{t_k}^{\sigma(T)} G(t_k^-, s)\cdot h(s)\delta_H s \\
&\quad + \sum_{\{j:t_j<t_k^-\}} G(t_k^-, t_j)\cdot v_j + \sum_{\{j:t_j\geq t_k^-\}} G(t_k^-, t_j)\cdot v_j, \\
&\quad x(t_k^+) \ominus_H x(t_k^-) \\
&= \left(G(t_k^+, t_k) - G(t_k^-, t_k)\right)\cdot v_k \\
&= \left(\frac{1}{\Gamma}\left(\alpha - \int_0^{\sigma(t_k)} g(r)e_p(0,r)\Delta r \right.\right. \\
&\quad \left.\left. -\beta e_p(0,\sigma(T)) - \int_{\sigma(t_k)}^{\sigma(T)} g(r)e_p(0,r)\Delta r\right)\right)\cdot v_k \\
&= \left(\frac{1}{\Gamma}\left(\alpha - \int_0^{\sigma(T)} g(r)e_p(0,r)\Delta r - \beta e_p(0,\sigma(T))\right)\right)\cdot v_k
\end{aligned}$$

$$= \quad v_k.$$

Also, by (4.52), *we get*

$$\alpha \cdot x(0) \ominus_H \beta \cdot x(\sigma(T))$$

$$= \int_0^{\sigma(T)} g(s) \cdot \left(\int_0^s e_{\ominus p}(s,\tau) \cdot h(\tau) \delta_H \tau + e_{\ominus p}(s,0) \cdot x(0) \right.$$

$$\left. + \sum_{\{j: t_j < s\}} e_{\ominus p}(s,t_j) \cdot v_j \right) \delta_H s$$

$$= \int_0^{\sigma(T)} g(s) \cdot x(s) \delta_H s.$$

This completes the proof.

Lemma 4.7.4 *Suppose* $(H1)$, $(H3)$ *and* $H(4)$. *Then*

$$A \le G(t,s) \le B, \quad t,s \in [0,\sigma(T)],$$

where

$$A = \frac{1}{\Gamma} (e_p(0,\sigma(T)))^2,$$

$$B = \frac{1}{\Gamma} e_p(\sigma(T),0) \left(\alpha + \beta e_p(0,\sigma(T)) + \int_0^{\sigma(T)} g(s) e_p(0,s) \Delta s \right).$$

Proof 4.7.5 *By* $(H4)$, *we get*

$$\alpha - \int_0^{\sigma(T)} g(s) e_p(0,s) \Delta s \ge \beta e_p(0,\sigma(T)).$$

By $(H1)$, *we obtain*

$$e_p(s,0) \ \ge \ 1, \quad s \in [0,\sigma(T)],$$

$$e_p(0,t) \;\geq\; e_p(0,\sigma(T)), \quad t\in[0,\sigma(T)].$$

Then

$$
\begin{aligned}
G(t,s) \;&=\; \begin{cases} \dfrac{1}{\Gamma}e_{\ominus p}(t,s)\left(\alpha-\displaystyle\int_0^{\sigma(s)} g(\tau)e_{\ominus p}(\tau,0)\Delta\tau\right) \\ \text{for} \quad 0\leq s\leq t\leq\sigma(T) \quad \text{and} \\ \dfrac{1}{\Gamma}e_{\ominus p}(t,s)\left(\beta e_{\ominus p}(\sigma(T),0)+\displaystyle\int_{\sigma(s)}^{\sigma(T)} g(r)e_{\ominus p}(r,0)\Delta r\right) \\ \text{for} \quad 0\leq t\leq s\leq\sigma(T) \end{cases} \\
&\geq\; \begin{cases} \dfrac{1}{\Gamma}e_p(s,0)e_p(0,t)\left(\alpha-\displaystyle\int_0^{\sigma(T)} g(\tau)e_{\ominus p}(\tau,0)\Delta\tau\right) \\ \text{for} \quad 0\leq s<t\leq\sigma(T) \quad \text{and} \\ \dfrac{1}{\Gamma}e_p(s,0)e_p(0,t)\left(\beta e_{\ominus p}(\sigma(T),0)\right) \\ \text{for} \quad 0\leq t\leq s\leq\sigma(T) \end{cases} \\
&\geq\; \begin{cases} \dfrac{1}{\Gamma}e_p(0,\sigma(T))(\beta e_p(0,\sigma(T))) \\ \text{for} \quad 0\leq s<t\leq\sigma(T) \quad \text{and} \\ \dfrac{\beta}{\Gamma}\left(e_p(0,\sigma(T))\right)^2 \\ \text{for} \quad 0\leq t\leq s\leq\sigma(T) \end{cases} \\
&=\; A, \quad t,s\in[0,\sigma(T)].
\end{aligned}
$$

Next, by $(H1)$, we have the following.

$$
\begin{aligned}
e_p(s,0) \;&\leq\; e_p(\sigma(T),0), \quad s\in[0,\sigma(T)], \\
e_p(0,t) \;&\leq\; 1, \quad t\in[0,\sigma(T)],t
\end{aligned}
$$

and

$$G(t,s) = \begin{cases} \frac{1}{\Gamma} e_{\ominus p}(t,s) \left(\alpha - \int\limits_0^{\sigma(s)} g(\tau) e_{\ominus p}(\tau,0) \Delta\tau \right) \\ \text{for} \quad 0 \le s \le t \le \sigma(T) \quad \text{and} \\ \frac{1}{\Gamma} e_{\ominus p}(t,s) \left(\beta e_{\ominus p}(\sigma(T),0) + \int\limits_{\sigma(s)}^{\sigma(T)} g(r) e_{\ominus p}(r,0) \Delta r \right) \\ \text{for} \quad 0 \le t \le s \le \sigma(T) \end{cases}$$

$$= \begin{cases} \frac{1}{\Gamma} e_p(s,0) e_p(0,t) \left(\alpha - \int\limits_0^{\sigma(s)} g(r) e_p(0,r) \Delta r \right) \\ \text{for} \quad 0 \le s < t \le \sigma(T) \quad \text{and} \\ \frac{1}{\Gamma} e_p(s,0) e_p(0,t) \left(\beta e_p(0,\sigma(T)) + \int\limits_{\sigma(s)}^{\sigma(T)} g(r) e_p(0,r) \Delta r \right) \\ \text{for} \quad 0 \le t \le s \le \sigma(T) \end{cases}$$

$$\le \begin{cases} \frac{1}{\Gamma} e_p(\sigma(T),0) \left(\alpha + \int\limits_0^{\sigma(T)} g(r) e_p(0,r) \Delta r + \beta e_p(0,\sigma(T)) \right) \\ \text{for} \quad 0 \le s < t \le \sigma(T) \quad \text{and} \\ \frac{1}{\Gamma} e_p(\sigma(T),0) \left(\alpha + \int\limits_0^{\sigma(T)} g(r) e_p(0,r) \Delta r + \beta e_p(0,\sigma(T)) \right) \\ \text{for} \quad 0 \le t \le s \le \sigma(T) \end{cases}$$

$$= B, \quad t,s \in [0,\sigma(T)].$$

This completes the proof.

For $x \in PC([0,\sigma(T)])$, define the operator

$$\Phi x(t) = \int_0^{\sigma(T)} G(t,s)\cdot f(s,x(\sigma(s)))\delta_H s + \sum_{j=1}^m G(t,t_j)\cdot I_j(x(t_j)),$$

$t \in [0,\sigma(T)]$.

Lemma 4.7.6 *The operator $\Phi : PC([0,\sigma(T)]) \to PC([0,\sigma(T)])$ is completely continuous.*

Proof 4.7.7 *Let $\{x_n\}_{n\in\mathbb{N}}$ be a sequence of elements of $PC([0,\sigma(T)])$ such that $x_n \to x$, as $n \to \infty$, in $PC([0,\sigma(T)])$, for some $x \in PC([0,\sigma(T)])$. Then*

$$D(\Phi x_n(t),\Phi x(t))$$

$$\begin{aligned}
&= D\Bigg(\int_0^{\sigma(T)} G(t,s)\cdot f(s,x_n(\sigma(s)))\delta_H s + \sum_{j=1}^m G(t,t_j)\cdot I_j(x_n(t_j)),\\
&\quad \int_0^{\sigma(T)} G(t,s)\cdot f(s,x(\sigma(s)))\delta_H s + \sum_{j=1}^m G(t,t_j)\cdot I_j(x(t_j))\Bigg)\\
&= D\Bigg(\int_0^{\sigma(T)} G(t,s)\cdot (f(s,x_n(\sigma(s)))\ominus_H f(s,x(\sigma(s))))\delta_H s\\
&\quad + \sum_{j=1}^m G(t,t_j)\cdot (I_j(x_n(t_j))\ominus_H I_j(x(t_j))),\tilde{0}\Bigg)\\
&\le D\Bigg(\int_0^{\sigma(T)} G(t,s)\cdot (f(s,x_n(\sigma(s)))\ominus_H f(s,x(\sigma(s))))\delta_H s,\tilde{0}\Bigg)\\
&\quad + D\Bigg(\sum_{j=1}^m G(t,t_j)\cdot (I_j(x_n(t_j))\ominus_H I_j(x(t_j))),\tilde{0}\Bigg)
\end{aligned}$$

$$\leq \int_0^{\sigma(T)} G(t,s)D(f(s,x_n(\sigma(s))),f(s,x(\sigma(s))))\Delta s$$

$$+\sum_{j=1}^{m} G(t,t_j)D(I_j(x_n(t_j)),I_j(x(t_j)))$$

$$\leq B\Bigg(\int_0^{\sigma(T)} D(f(s,x_n(\sigma(s))),f(s,x(\sigma(s))))\Delta s$$

$$+\sum_{j=1}^{m} D(I_j(x_n(t_j)),I_j(x(t_j)))\Bigg)$$

$$\to 0, \quad as \quad n\to\infty.$$

Hence,

$$\|\Phi x_n - \Phi x\| \to 0, \quad as \quad n\to\infty.$$

Let

$$C = \{x \in PC([0,\sigma(T)]) : \|x\| \leq \gamma\},$$

for some positive constant γ. Then there exists a positive constant γ_1 such that, for $x \in C$, we have

$$D(f(t,x(\sigma(t))),\widetilde{0}) \;\leq\; \gamma_1, \quad t\in[0,\sigma(T)],$$

$$D(I_j(x(t_j)),\widetilde{0}) \;\leq\; \gamma_1, \quad j\in\{1,\ldots,m\}.$$

Hence, for $x \in C$, we have

$$D(\Phi x(t),\widetilde{0})$$

$$= D\Bigg(\int_0^{\sigma(T)} G(t,s)\cdot f(s,x(\sigma(s)))\delta_H s + \sum_{j=1}^{m} G(t,t_j)\cdot I_j(x(t_j)),\widetilde{0}\Bigg)$$

$$\leq D\Bigg(\int_0^{\sigma(T)} G(t,s)\cdot f(s,x(\sigma(s)))\delta_H s,\widetilde{0}\Bigg)$$

$$+D\left(\sum_{j=1}^{m} G(t,t_j)\cdot I_j(x(t_j)),\widetilde{0}\right)$$

$$\leq \int_0^{\sigma(T)} G(t,s)D(f(s,x(\sigma(s))),\widetilde{0})\Delta s$$

$$+\sum_{j=1}^{m} G(t,t_j)D(I_j(x(t_j)),\widetilde{0})$$

$$\leq B\gamma_1\sigma(T)+Bm\gamma_1, \quad t\in[0,\sigma(T)].$$

Then, for $x\in C$, we have

$$\|\Phi x\|\leq B\gamma_1(m+\sigma(T)).$$

Let $t^1,t^2\in[0,\sigma(T)]$ be arbitrarily chosen. Then, for $x\in PC([0,\sigma(T)])$, we have

$$D(\Phi x(t^1),\Phi x(t^2))$$

$$= D\left(\int_0^{\sigma(T)} G(t^1,s)\cdot f(s,x(\sigma(s)))\delta_H s+\sum_{j=1}^{m} G(t^1,t_j)\cdot I_j(x(t_j^1)),\right.$$

$$\left.\int_0^{\sigma(T)} G(t^2,s)\cdot f(s,x(\sigma(s)))\delta_H s+\sum_{j=1}^{m} G(t^2,t_j)\cdot I_j(x(t_j))\right)$$

$$= D\left(\int_0^{\sigma(T)} (G(t^1,s)-G(t^2,s))\cdot f(s,x(\sigma(s)))\delta_H s\right.$$

$$\left.+\sum_{j=1}^{m}(G(t^1,t_j)-G(t^2,t_j))\cdot I_j(x(t_j)),\widetilde{0}\right)$$

$$\leq D\left(\int_0^{\sigma(T)} (G(t^1,s)-G(t^2,s))\cdot f(s,x(\sigma(s)))\delta_H s,\widetilde{0}\right)$$

$$+D\left(\sum_{j=1}^{m}(G(t^1,t_j)-G(t^2,t_j))\cdot I_j(x(t_j)),\widetilde{0}\right)$$

$$\leq D\left(\int_0^{\sigma(T)}(G(t^1,s)-G(t^2,s))\cdot f(s,x(\sigma(s)))\delta_H s,\widetilde{0}\right)$$

$$+\sum_{j=1}^{m}D\left((G(t^1,t_j)-G(t^2,t_j))\cdot I_j(x(t_j)),\widetilde{0}\right)$$

$$\leq \int_0^{\sigma(T)}|G(t^1,s)-G(t^2,s)|D(f(s,x(\sigma(s))),\widetilde{0})\Delta s$$

$$+\sum_{j=1}^{m}|G(t^1,t_j)-G(t^2,t_j)|D(I_j(x(t_j)),\widetilde{0})$$

$$\to 0, \quad as \quad t^1\to t^2.$$

Now, using the Arzela-Ascoli theorem, we conclude that $\Phi : PC([0,\sigma(T)]) \to PC([0,\sigma(T)])$ *is completely continuous. This completes the proof.*

Let $\delta=\dfrac{A}{B}$ and

$$K=\{x\in PC([0,\sigma(T)]) : x(t)\geq \delta\|x\|, \quad t\in[0,\sigma(T)]\}.$$

Lemma 4.7.8 *We have* $\Phi : K\to K$.

Proof 4.7.9 *Let* $x\in K$ *be arbitrarily chosen. Then*

$$D(\Phi x(t),\widetilde{0})$$

$$= D\left(\int_0^{\sigma(T)}G(t,s)\cdot f(s,x(\sigma(s)))\delta_H s+\sum_{j=1}^{m}G(t,t_j)\cdot I_j(x(t_j)),\widetilde{0}\right)$$

$$\leq BD\left(\int_0^{\sigma(T)}f(s,x(\sigma(s)))\delta_H s+\sum_{j=1}^{m}I_j(x(t_j)),\widetilde{0}\right)$$

$$\leq B\left(D\left(\int_0^{\sigma(T)} f(s,x(\sigma(s)))\delta_H s,\widetilde{0}\right)+D\left(\sum_{j=1}^{m} I_j(x(t_j)),\widetilde{0}\right)\right)$$

$$\leq B\left(D\left(\int_0^{\sigma(T)} f(s,x(\sigma(s)))\delta_H s,\widetilde{0}\right)+\sum_{j=1}^{m} D\left(I_j(x(t_j)),\widetilde{0}\right)\right),$$

$t \in [0,\sigma(T)]$, *whereupon*

$$\|\Phi x\| \leq B\left(\int_0^{\sigma(T)} D(f(s,x(\sigma(s))),\widetilde{0})\delta_H s+\sum_{j=1}^{m} D(I_j(x(t_j)),\widetilde{0})\right).$$

Next,

$$D(\Phi x(t),\widetilde{0})$$

$$= D\left(\int_0^{\sigma(T)} G(t,s)f(s,x(\sigma(s)))\delta_H s+\sum_{j=1}^{m} G(t,t_j)I_j(x(t_j)),\widetilde{0}\right)$$

$$\geq A\left(\int_0^{\sigma(T)} D(f(s,x(\sigma(s))),\widetilde{0})\delta_H s+\sum_{j=1}^{m} D(I_j(x(t_j)),\widetilde{0})\right)$$

$$\geq \frac{A}{B}\|\Phi x\|$$

$$= \delta\|\Phi x\|, \quad t \in [0,\sigma(T)],$$

i.e., $\Phi x \in K$. *This completes the proof.*

Let

$$\max f_0 = \lim_{D(x,\widetilde{0})\to 0} \max_{t\in[0,T]} \frac{D(f(t,x),\widetilde{0})}{D(x,\widetilde{0})},$$

$$\min f_0 = \lim_{D(x,\widetilde{0})\to 0} \min_{t\in[0,T]} \frac{D(f(t,x),\widetilde{0})}{D(x,\widetilde{0})},$$

$$\begin{aligned}
\max f_\infty &= \lim_{D(x,\widetilde{0})\to\infty} \max_{t\in[0,T]} \frac{D(f(t,x),\widetilde{0})}{D(x,\widetilde{0})},\\
\min f_\infty &= \lim_{D(x,\widetilde{0})\to\infty} \min_{t\in[0,T]} \frac{D(f(t,x),\widetilde{0})}{D(x,\widetilde{0})},\\
I_{j0} &= \lim_{D(x,\widetilde{0})\to 0} \frac{I_j(x)}{x},\\
I_{j\infty} &= \lim_{D(x,\widetilde{0})\to\infty} \frac{D(I_j(x),\widetilde{0})}{D(x,\widetilde{0})}, \quad j\in\{1,\dots,m\}.
\end{aligned}$$

Theorem 4.7.10 *Assume* $(H1)$-$(H4)$ *and*

$$D\left(\int_0^{\sigma(T)} G(t,\tau)\cdot f(\tau,x(\sigma(\tau)))\delta_H\tau + \sum_{j=1}^m G(t,t_j)\cdot I_j(x(t_j)),\widetilde{0}\right)$$

$$\geq s\int_0^{\sigma(T)} G(t,\tau)D(f(\tau,x(\sigma(\tau))),\widetilde{0})\Delta\tau \tag{4.53}$$

for some $s\in(0,1)$ *and for any* $x\in K$*, and*

$$\max f_0 = \widetilde{0}, \quad \min f_\infty = \infty, \quad I_{j0} = \widetilde{0}, \quad j\in\{1,\dots,m\}.$$

Then problem (4.47) *has at least one positive solution.*

Proof 4.7.11 *By Lemma 4.7.6, it follows that* $\Phi : PC([0,\sigma(T)]) \to PC([0,\sigma(T)])$ *is completely continuous. From Lemma 4.7.8, we have* $\Phi : K\to K$*. Since*

$$\max f_0 = \widetilde{0} \quad \text{and} \quad I_{j0} = \widetilde{0}, \quad j\in\{1,\dots,m\},$$

for $\varepsilon\in\left(0,(B\sigma(T)+Bm)^{-1}\right]$*, there exists a positive constant* r_1 *such that*

$$D(f(t,x),\widetilde{0}) \leq \varepsilon D(x,\widetilde{0}), \quad D(I_j(x),\widetilde{0}) \leq \varepsilon D(x,\widetilde{0}), \quad j\in\{1,\dots,m\},$$

$x \in \Omega_{r_1}$, $t \in [0, \sigma(T)]$, *where*

$$\Omega_{r_1} = \{x \in K : \|x\| < r_1\}.$$

Because $\min f_\infty = \infty$, *for* $M \geq (sA\sigma(T)\delta)^{-1}$, *there exists a positive constant* $r_2 \geq \dfrac{r_1}{\delta}$ *such that*

$$D(f(t,x),\widetilde{0}) \geq MD(x,\widetilde{0}), \quad x \in \Omega_{r_2}, \quad t \in [0,\sigma(T)],$$

where

$$\Omega_{r_2} = \{x \in K : \|x\| < r_2\}.$$

For $x \in K \bigcap \partial\Omega_{r_1}$, *we have*

$$\begin{aligned}
& D(\Phi x(t),\widetilde{0}) \\
= \; & D\left(\int_0^{\sigma(T)} G(t,s)\cdot f(s,x(\sigma(s)))\delta_H s + \sum_{j=1}^m G(t,t_j)\cdot bI_j(x(t_j)),\widetilde{0}\right) \\
\leq \; & D\left(\int_0^{\sigma(T)} G(t,s)\cdot f(s,x(\sigma(s)))\delta_H s,\widetilde{0}\right) \\
& + D\left(\sum_{j=1}^m G(t,t_j)\cdot bI_j(x(t_j)),\widetilde{0}\right) \\
\leq \; & D\left(\int_0^{\sigma(T)} G(t,s)\cdot f(s,x(\sigma(s)))\delta_H s,\widetilde{0}\right) \\
& + \sum_{j=1}^m D\left(G(t,t_j)\cdot bI_j(x(t_j)),\widetilde{0}\right) \\
\leq \; & B\left(\int_0^{\sigma(T)} D(f(s,x(\sigma(s))),\widetilde{0})\delta_H s + \sum_{j=1}^m D(I_j(x(t_j)),\widetilde{0})\right)
\end{aligned}$$

$$\begin{aligned}
&\leq \varepsilon B \left(\int_0^{\sigma(T)} D(x(\sigma(s)),\widetilde{0})\delta_H s + \sum_{j=1}^{m} D(x(t_j),\widetilde{0}) \right) \\
&\leq \varepsilon B(\sigma(T)+m)\|x\| \\
&\leq \|x\|, \quad t \in [0,\sigma(T)].
\end{aligned}$$

Then

$$\|\Phi x\| \leq \|x\|, \quad x \in K \bigcap \partial\Omega_{r_1}.$$

For $x \in K \bigcap \partial\Omega_{r_2}$, *we have*

$$D(x(\sigma(s)),\widetilde{0}) \geq \delta\|x\|, \quad s \in [0,T],$$

and

$$\begin{aligned}
&D(\Phi x(t),\widetilde{0}) \\
&= D\left(\int_0^{\sigma(T)} G(t,s)\cdot f(s,x(\sigma(s)))\delta_H s + \sum_{j=1}^{m} G(t,t_j)\cdot I_j(x(t_j)),\widetilde{0} \right) \\
&\geq s \int_0^{\sigma(T)} G(t,s) D(f(s,x(\sigma(s))),\widetilde{0})\Delta s \\
&\geq sA \int_0^{\sigma(T)} D(f(s,x(\sigma(s))),\widetilde{0})\Delta s \\
&\geq sAM \int_0^{\sigma(T)} D(x(\sigma(s)),\widetilde{0})\Delta s \\
&\geq sAM\delta\|x\|\sigma(T) \\
&\geq \|x\|, \quad t \in [0,\sigma(T)].
\end{aligned}$$

Therefore

$$\|\Phi x\| \geq \|x\|, \quad x \in K \bigcap \partial\Omega_{r_2}.$$

From here and from Theorem 4.3.24, we conclude that Φ has at least one fixed point in $K\bigcap(\overline{\Omega_{r_2}}\backslash\Omega_{r_1})$. This completes the proof.

Theorem 4.7.12 *Assume* $(H1)$-$(H5)$, (4.53) *and*

$$\max f_\infty = \widetilde{0}, \quad \min f_0 = \infty, \quad I_{j\infty} = \widetilde{0}, \quad j \in \{1,\dots,m\}.$$

The problem (4.47) *has at least one positive solution.*

Proof 4.7.13 *By Lemma 4.7.6, we have $\Phi : PC([0,\sigma(T)]) \to PC([0,\sigma(T)])$ is completely continuous. From Lemma 4.7.8, it follows that $\Phi : K \to K$. Since*

$$\max f_\infty = \widetilde{0} \quad \textit{and} \quad I_{j\infty} = \widetilde{0}, \quad j \in \{1,\dots,m\},$$

for $\varepsilon \in \left(0, (B\sigma(T)+Bm)^{-1}\right]$, it follows that there exists a positive constant r_1 such that

$$\begin{aligned} D(f(t,x),\widetilde{0}) &\leq \varepsilon D(x,\widetilde{0}), \\ D(I_j(x),\widetilde{0}) &\leq \varepsilon D(x,\widetilde{0}), \quad j \in \{1,\dots,m\}, \\ & \qquad x \in [\delta r_1, \infty), \quad t \in [0,\sigma(T)]. \end{aligned}$$

Since $\min f_0 = \infty$, for $M > (sA\sigma(T)\delta)^{-1}$, there exists a positive constant r_2 such that $r_2 < \delta r_1$ and

$$D(f(t,x),\widetilde{0}) \geq MD(x,\widetilde{0}) \quad x \in (0,r_2], \quad t \in [0,\sigma(T)].$$

Let

$$\Omega_1 = \{x \in K : \|x\| \leq r_1\},$$

$$\Omega_2 = \{x \in K : \|x\| < r_2\}.$$

We have

$$\overline{\Omega_2} \subset \Omega_1.$$

For $x \in K \bigcap \partial\Omega_1$, *we get*

$$
\begin{aligned}
& D(\Phi x(t), \widetilde{0}) \\
= \; & D\left(\int_0^{\sigma(T)} G(t,s) \cdot f(s, x(\sigma(s)))\delta_H s + \sum_{j=1}^{m} G(t,t_j) \cdot bI_j(x(t_j)), \widetilde{0} \right) \\
\le \; & D\left(\int_0^{\sigma(T)} G(t,s) \cdot f(s, x(\sigma(s)))\delta_H s, \widetilde{0} \right) \\
& + D\left(\sum_{j=1}^{m} G(t,t_j) \cdot bI_j(x(t_j)), \widetilde{0} \right) \\
\le \; & D\left(\int_0^{\sigma(T)} G(t,s) \cdot f(s, x(\sigma(s)))\delta_H s, \widetilde{0} \right) \\
& + \sum_{j=1}^{m} D\left(G(t,t_j) \cdot bI_j(x(t_j)), \widetilde{0} \right) \\
\le \; & B\left(\int_0^{\sigma(T)} D(f(s, x(\sigma(s))), \widetilde{0})\delta_H s + \sum_{j=1}^{m} D(I_j(x(t_j)), \widetilde{0}) \right) \\
\le \; & B\left(\varepsilon \int_0^{\sigma(T)} D(x(\sigma(s)), \widetilde{0})\delta_H s + \varepsilon \sum_{j=1}^{m} D(x(t_j), \widetilde{0}) \right) \\
\le \; & B\varepsilon(\sigma(T) + m)\|x\| \\
\le \; & \|x\|, \quad t \in [0, \sigma(T)],
\end{aligned}
$$

whereupon

$$\|\Phi x\| \le \|x\|.$$

For $x \in K \bigcap \partial\Omega_2$, we obtain

$$D(x(\sigma(s)), \widetilde{0}) \geq \delta \|x\|, \quad s \in [0, T],$$

and

$$\begin{aligned}
&D(\Phi x(t), \widetilde{0}) \\
&= D\left(\int_0^{\sigma(T)} G(t,s) \cdot f(s, x(\sigma(s)))\delta_H s + \sum_{j=1}^{m} G(t, t_j) \cdot I_j(x(t_j)), \widetilde{0} \right) \\
&\geq s \int_0^{\sigma(T)} G(t,s) D(f(s, x(\sigma(s))), \widetilde{0}) \Delta s \\
&\geq sA \int_0^{\sigma(T)} D(f(s, x(\sigma(s))), \widetilde{0}) \Delta s \\
&\geq sAM \int_0^{\sigma(T)} D(x(\sigma(s)), \widetilde{0}) \Delta s \\
&\geq sMA\delta\sigma(T)\|x\| \\
&\geq \|x\|, \quad t \in [0, \sigma(T)],
\end{aligned}$$

whereupon

$$\|\Phi x\| \geq \|x\|.$$

Hence and with Theorem 4.3.24, it follows that the operator Φ has a fixed point in $K \bigcap (\overline{\Omega_2} \backslash \Omega_1)$. This completes the proof.

Theorem 4.7.14 *Assume $(H1)$-$(H5)$, (4.53) and*

$$\min f_0 = \infty, \quad \min f_\infty = \infty,$$

there exists a positive constant R such that

$$D(f(t,x),\widetilde{0}) < \frac{R}{2B\sigma(T)}, \quad t \in [0,\sigma(T)], \quad 0 < D(x,\widetilde{0}) \leq R,$$

$$D(I_j(x),\widetilde{0}) \leq \frac{D(x,\widetilde{0})}{2Bm}, \quad D(x,\widetilde{0}) \in (0,\infty), \quad j \in \{1,\ldots,m\}.$$

The problem (4.47) *has at least two positive solutions.*

Proof 4.7.15 *Let*

$$\Omega_R = \{x \in K : \|x\| < R\}.$$

Then, for $x \in K \bigcap \partial\Omega_R$, we get

$$D\left(\Phi x(t),\widetilde{0}\right)$$

$$\begin{aligned}
&= D\left(\int_0^{\sigma(T)} G(t,s)\cdot f(s,x(\sigma(s)))\delta_H s + \sum_{j=1}^m G(t,t_j)\cdot I_j(x(t_j)),\widetilde{0}\right) \\
&\leq D\left(\int_0^{\sigma(T)} G(t,s)\cdot f(s,x(\sigma(s)))\delta_H s,\widetilde{0}\right) \\
&\quad + D\left(\sum_{j=1}^m G(t,t_j)\cdot I_j(x(t_j)),\widetilde{0}\right) \\
&\leq D\left(\int_0^{\sigma(T)} G(t,s)\cdot f(s,x(\sigma(s)))\delta_H s,\widetilde{0}\right) \\
&\quad + \sum_{j=1}^m D\left(G(t,t_j)\cdot I_j(x(t_j)),\widetilde{0}\right) \\
&\leq BD\left(\int_0^{\sigma(T)} f(s,x(\sigma(s)))\delta_H s,\widetilde{0}\right) + \sum_{j=1}^m D\left(I_j(x(t_j)),\widetilde{0}\right)
\end{aligned}$$

$$\begin{aligned}
&\leq B\int_0^{\sigma(T)} D(f(s,x(\sigma(s))),\widetilde{0})\delta_H s + B\sum_{j=1}^{m} D(I_j(x(t_j)),\widetilde{0})\\
&\leq \frac{R}{2} + \frac{1}{2m}\sum_{j=1}^{m} D(x(t_j),\widetilde{0})\\
&\leq \frac{R}{2} + \frac{1}{2}\|x\|\\
&\leq \frac{R}{2} + \frac{R}{2}\\
&= R\\
&= \|x\|, \quad t\in[0,\sigma(T)],
\end{aligned}$$

whereupon

$$\|\Phi x\| \leq \|x\|.$$

Since $\min f_0 = \infty$, *for* $M > (sA\sigma(T)\delta)^{-1}$, *there exists a positive constant* $R_1 < \delta R$ *such that*

$$D(f(t,x),\widetilde{0}) \geq MD(x,\widetilde{0}), \quad D(x,\widetilde{0}) \in [0,R_1], \quad t\in[0,\sigma(T)].$$

Let

$$\Omega_{R_1} = \{x\in K : \|x\| < R_1\}.$$

Then, for $x \in K\bigcap \partial\Omega_{R_1}$, *we have*

$$D(x(t),\widetilde{0}) \geq \delta\|x\|, \quad t\in[0,\sigma(T)],$$

and

$$\begin{aligned}
&D(\Phi x(t),\widetilde{0})\\
&= D(\int_0^{\sigma(T)} G(t,s)\cdot f(s,x(\sigma(s)))\delta_H s + \sum_{j=1}^{m} G(t,t_j)\cdot I_j(x(t_j)),\widetilde{0})
\end{aligned}$$

$$
\begin{aligned}
&\geq s \int_0^{\sigma(T)} G(t,s) D(f(s,x(\sigma(s))),\widetilde{0}) \Delta s \\
&\geq sA \int_0^{\sigma(T)} D(f(s,x(\sigma(s))),\widetilde{0}) \Delta s \\
&\geq sAM \int_0^{\sigma(T)} D(x(\sigma(s)),\widetilde{0}) \Delta s \\
&\geq sAM\delta\sigma(T)\|x\| \\
&\geq \|x\|, \quad t \in [0,\sigma(T)],
\end{aligned}
$$

whereupon

$$\|\Phi x\| \geq \|x\|.$$

Because $\min f_\infty = \infty$, *for* $N \geq (sA\sigma(T)\delta)^{-1}$, *there exists a positive constant* $R_2 > \dfrac{R}{\delta}$ *such that*

$$D(f(t,x),\widetilde{0}) \geq ND(x,\widetilde{0}), \quad t \in [0,\sigma(T)], \quad x \in [\delta R_2, \infty).$$

Let

$$\Omega_{R_2} = \{x \in K : \|x\| < R_2\}.$$

Then, for $x \in K \bigcap \partial\Omega_{R_2}$, *we obtain*

$$D(x(t),\widetilde{0}) \geq \delta\|x\|, \quad t \in [0,\sigma(T)],$$

and

$$
\begin{aligned}
&D(\Phi x(t),\widetilde{0}) \\
&= D\left(\int_0^{\sigma(T)} G(t,s) \cdot f(s,x(\sigma(s)))\delta_H s + \sum_{j=1}^{m} G(t,t_j) \cdot I_j(x(t_j)),\widetilde{0} \right)
\end{aligned}
$$

$$\geq s\int_0^{\sigma(T)} G(t,s)D(f(s,x(\sigma(s))),\widetilde{0})\Delta s$$

$$\geq sA\int_0^{\sigma(T)} D(f(s,x(\sigma(s))),\widetilde{0})\Delta s$$

$$\geq sAN\int_0^{\sigma(T)} D(x(\sigma(s)),\widetilde{0})\Delta s$$

$$\geq sAN\delta\sigma(T)\|x\|$$

$$\geq \|x\|, \quad t\in[0,\sigma(T)],$$

whereupon

$$\|\Phi x\| \geq \|x\|.$$

Note that

$$\overline{\Omega_{R_1}} \subset \overline{\Omega_R} \subset \Omega_{R_2}.$$

Hence and with Theorem 4.3.24, it follows that Φ *has at least one fixed point in* $K\bigcap(\overline{\Omega_R}\backslash\Omega_{R_1})$ *and* Φ *has at least one fixed point in* $K\bigcap(\overline{\Omega_{R_2}}\backslash\Omega_R)$. *This completes the proof.*

Theorem 4.7.16 *Assume* $(H1)$-$(H5)$, (4.53) *and there exists a positive constant r such that*

$$D(f(t,x),\widetilde{0}) \geq \frac{r}{sA\sigma(T)}, \quad D(x,\widetilde{0})\in(0,r], \quad t\in[0,\sigma(T)],$$

$$\max f_0=\widetilde{0}, \quad \max f_\infty=\widetilde{0}, \quad I_{j0}=\widetilde{0}, \quad I_{j\infty}=\widetilde{0}, \quad j\in\{1,\dots,m\}.$$

The problem (4.47) *has at least two positive solutions.*

Proof 4.7.17 *Let*

$$\Omega_r=\{x\in K:\|x\|<r\}.$$

Then, for $x \in K\bigcap \partial\Omega_r$, *we get*

$$\begin{aligned} & D(\Phi x(t),\widetilde{0}) \\ = & D\left(\int\limits_0^{\sigma(T)} G(t,s)\cdot f(s,x(\sigma(s)))\delta_H s + \sum_{j=1}^{m} G(t,t_j)\cdot I_j(x(t_j)),\widetilde{0}\right) \\ \geq & s\int\limits_0^{\sigma(T)} G(t,s)D(f(s,x(\sigma(s))),\widetilde{0})\Delta s \\ \geq & As\int\limits_0^{\sigma(T)} D(f(s,x(\sigma(s))),\widetilde{0})\Delta s \\ \geq & As\sigma(T)\frac{r}{A\sigma(T)s} \\ = & r \\ = & \|x\|, \quad t\in[0,\sigma(T)], \end{aligned}$$

whereupon

$$\|\Phi x\| \geq \|x\|.$$

Since

$$\max f_0 = \widetilde{0}, \quad I_{j0} = \widetilde{0},$$

$j\in\{1,\ldots,m\}$, *for* $\varepsilon \leq (B\sigma(T)+Bm)^{-1}$, *there exists a positive constant* $r_1 < \delta r$ *such that*

$$D(f(t,x),\widetilde{0}) \leq \varepsilon D(x,\widetilde{0}), \quad D(I_j(x),\widetilde{0}) \leq \varepsilon D(x,\widetilde{0}),$$

$$D(x,\widetilde{0}) \in (0,r_1], \quad t\in[0,\sigma(T)], \quad j\in\{1,\ldots,m\}.$$

Let

$$\Omega_{r_1} = \{x\in K : \|x\| < r_1\}.$$

Then, for $x \in K \bigcap \partial\Omega_{r_1}$, *we have*

$$
\begin{aligned}
& D(\Phi x(t),\widetilde{0}) \\
= \; & D\left(\int_0^{\sigma(T)} G(t,s)\cdot f(s,x(\sigma(s)))\delta_H s + \sum_{j=1}^{m} G(t,t_j)\cdot I_j(x(t_j)),\widetilde{0}\right) \\
\leq \; & D\left(\int_0^{\sigma(T)} G(t,s)\cdot f(s,x(\sigma(s)))\delta_H s,\widetilde{0}\right) \\
& + D\left(\sum_{j=1}^{m} G(t,t_j)\cdot I_j(x(t_j)),\widetilde{0}\right) \\
\leq \; & D\left(\int_0^{\sigma(T)} G(t,s)\cdot f(s,x(\sigma(s)))\delta_H s,\widetilde{0}\right) \\
& + \sum_{j=1}^{m} D\left(G(t,t_j)\cdot I_j(x(t_j)),\widetilde{0}\right) \\
\leq \; & BD\left(\int_0^{\sigma(T)} f(s,x(\sigma(s)))\delta_H s,\widetilde{0}\right) + \sum_{j=1}^{m} D\left(I_j(x(t_j)),\widetilde{0}\right) \\
\leq \; & B\left(\int_0^{\sigma(T)} D(f(s,x(\sigma(s))),\widetilde{0})\Delta s + \sum_{j=1}^{m} D(I_j(x(t_j)),\widetilde{0})\right) \\
\leq \; & \varepsilon B\left(\int_0^{\sigma(T)} D(x(\sigma(s)),\widetilde{0})\Delta s + \sum_{j=1}^{m} D(x(t_j),\widetilde{0})\right) \\
\leq \; & \varepsilon B(\sigma(T)+m)\|x\| \\
\leq \; & \|x\|, \quad t \in [0,\sigma(T)],
\end{aligned}
$$

whereupon

$$\|\Phi x\| \leq \|x\|.$$

Because

$$\max f_\infty = \widetilde{0}, \quad I_{j\infty} = \widetilde{0}, \quad j \in \{1, \dots, m\},$$

for $\varepsilon_1 \leq (B\sigma(T) + Bm)^{-1}$, *there exists a positive constant* $r_2 > \dfrac{r}{\delta}$ *such that*

$$D(f(t,x),\widetilde{0}) \leq \varepsilon_1 D(x,\widetilde{0}),\ D(I_j(x),\widetilde{0}) \leq \varepsilon_1 D(x,\widetilde{0}),\ D(x,\widetilde{0}) \in [\delta r_2, \infty),$$

$j \in \{1, \dots, m\}, t \in [0, \sigma(T)]$. *Let*

$$\Omega_{r_2} = \{x \in K : \|x\| < r_2\}.$$

Then, for $x \in K \bigcap \partial\Omega_{r_2}$, *we get*

$$
\begin{aligned}
& D(\Phi x(t), \widetilde{0}) \\
= \ & D\left(\int_0^{\sigma(T)} G(t,s) \cdot f(s, x(\sigma(s)))\delta_H s + \sum_{j=1}^m G(t,t_j) \cdot I_j(x(t_j)), \widetilde{0} \right) \\
\leq \ & D\left(\int_0^{\sigma(T)} G(t,s) \cdot f(s, x(\sigma(s)))\delta_H s, \widetilde{0} \right) \\
& + D\left(\sum_{j=1}^m G(t,t_j) \cdot I_j(x(t_j)), \widetilde{0} \right) \\
\leq \ & D\left(\int_0^{\sigma(T)} G(t,s) \cdot f(s, x(\sigma(s)))\delta_H s, \widetilde{0} \right) \\
& + \sum_{j=1}^m D\left(G(t,t_j) \cdot I_j(x(t_j)), \widetilde{0} \right) \\
\leq \ & BD\left(\int_0^{\sigma(T)} f(s, x(\sigma(s)))\delta_H s, \widetilde{0} \right) + \sum_{j=1}^m D\left(I_j(x(t_j)), \widetilde{0} \right)
\end{aligned}
$$

$$\leq B\left(\int_0^{\sigma(T)} D(f(s,x(\sigma(s))),\widetilde{0})\Delta s + \sum_{j=1}^{m} D(I_j(x(t_j)),\widetilde{0})\right)$$

$$\leq \varepsilon_1 B\left(\int_0^{\sigma(T)} D(x(\sigma(s)),\widetilde{0})\Delta s + \sum_{j=1}^{m} D(x(t_j),\widetilde{0})\right)$$

$$\leq \varepsilon_1 B(\sigma(T)+m)\|x\|$$

$$\leq \|x\|, \quad t\in[0,\sigma(T)],$$

whereupon

$$\|\Phi x\| \leq \|x\|.$$

Note that

$$\overline{\Omega_{r_1}} \subset \overline{\Omega_r} \subset \Omega_{r_2}.$$

Hence and with Theorem 4.3.24, it follows that the operator Φ has at least one fixed point in $K\bigcap(\overline{\Omega_r}\backslash\Omega_{r_1})$ and Φ has at least one fixed point in $K\bigcap(\overline{\Omega_{r_2}}\backslash\Omega_r)$. This completes the proof.

Exercise 4.7.18 *Assume* $(H1)$-$(H5)$, (4.53) *and there exists a positive constant r such that*

$$D(f(t,x),\widetilde{0}) > \frac{r}{A\sigma(T)}$$

for $0 < D(x,\widetilde{0}) \leq r$, $t\in[0,\sigma(T)]$,

$$\max f_0 = \widetilde{0}, \quad \max f_\infty = \widetilde{0}, \quad I_{j0} = \widetilde{0}, \quad j\in\{1,\ldots,m\},$$

there exists a positive constant d such that

$$D(I_j(x),\widetilde{0}) \leq \frac{x}{2Bm},$$

$D(x,\widetilde{0}) \geq d$, $j\in\{1,\ldots,m\}$. *Prove that the problem* (4.47) *has at least two positive solutions.*

Definition 4.7.19 *Let E be a real Banach space with a cone K. A map $\alpha : K \to F([0,\infty))$ is said to be a continuous concave functional on K if α is continuous and*

$$\alpha(tx+(1-t)y) \geq t\alpha(x)+(1-t)\alpha(y)$$

for all $x,y \in K$ and $t \in [0,1]$. For $0 < a < b$ and α a non-negative continuous concave functional on K, define

$$\begin{aligned} K_a &= \{x \in K : \|x\| < a\}, \\ K(\alpha,a,b) &= \{x \in K : a \leq \alpha(x), \quad \|x\| < b\}. \end{aligned}$$

In our next result, we will use the following fixed-point theorem.

Theorem 4.7.20 (Legget-Williams fixed-point Theorem) *Let $\Phi : \overline{K_c} \to \overline{K_c}$ be completely continuous and α be a non-negative continuous concave functional on K such that $\alpha(x) \leq \|x\|$ for all $x \in K_c$. Suppose that there exist $0 < d < a < b \leq c$ such that*

1. $\{x \in K(\alpha,a,b) : \alpha(x) > a\} \neq \emptyset$, *$\alpha(\Phi x) > a$ for all $x \in K(\alpha,a,b)$,*
2. $\|\Phi x\| < d$ *for all* $\|x\| \leq d$,
3. $\alpha(\Phi x) > a$ *for all $x \in K(\alpha,a,c)$ with $\|\Phi x\| > b$.*

Then, Φ has at least three fixed points $x_1, x_2, x_3 \in \overline{K_c}$ satisfying $\|x_1\| < d$, $0 < \alpha(x_2)$, $\|x_3\| > d$, $\alpha(x_3) < a$.

Theorem 4.7.21 *Assume* $(H1)$-$(H5)$, (4.53) *and there exist numbers d, a and c with $0 < d < a < \frac{a}{\delta} < c$ such that*

$$\max_{t\in[0,\sigma(T)]} D(f(t,x),\widetilde{0}) \leq \frac{d}{2B\sigma(T)},$$

$$D(I_j(x),\widetilde{0}) \leq \frac{d}{2Bm}, \quad j \in \{1,\ldots,m\}, \quad D(x,\widetilde{0}) \in (0,d],$$

$$\max_{t\in[0,\sigma(T)]} D(f(t,x),\widetilde{0}) \leq \frac{c}{2B\sigma(T)},$$

$$D(I_j(x),\widetilde{0}) \leq \frac{c}{2Bm}, \quad j\in\{1,\ldots,m\}, \quad D(x,\widetilde{0})\in(0,c],$$

$$\min_{t\in[0,\sigma(T)]} D(f(t,x),\widetilde{0}) > \frac{a}{sA\sigma(T)},$$

$$D(I_j(x),\widetilde{0}) > \frac{a}{2Am}, \quad j\in\{1,\ldots,m\}, \quad D(x,\widetilde{0})\in\left[a,\frac{a}{\delta}\right].$$

The problem (4.47) *has at least three positive solutions.*

Proof 4.7.22 *For $x \in K$, define*

$$\alpha(x) = \min_{t\in[0,\sigma(T)]} D(x(t),\widetilde{0}).$$

We have

$$\alpha(x) \geq 0, \quad x\in K.$$

For $\lambda \in [0,1]$ and for any $x,y \in K$, we obtain

$$\begin{aligned}
&\alpha(\lambda D(x,\widetilde{0})+(1-\lambda)D(y,\widetilde{0}))\\
&= \min_{t\in[0,\sigma(T)]}(\lambda D(x(t),\widetilde{0})+(1-\lambda)D(y(t),\widetilde{0}))\\
&\geq \lambda \min_{t\in[0,\sigma(T)]} D(x(t),\widetilde{0})+(1-\lambda)\min_{t\in[0,\sigma(T)]} D(y(t),\widetilde{0})\\
&= \lambda\alpha(x)+(1-\lambda)\alpha(y).
\end{aligned}$$

Also, if $\{x_n\}_{n\in\mathbb{N}} \subset K$ and $x_n \to x$, as $n\to\infty$, in K, then

$$\begin{aligned}
\alpha(x_n) &= \min_{t\in[0,\sigma(T)]} D(x_n(t),\widetilde{0})\\
&\to \min_{t\in[0,\sigma(T)]} D(x(t),\widetilde{0})\\
&= \alpha(x), \quad as \quad n\to\infty.
\end{aligned}$$

Therefore, α is a non-negative continuous concave functional on K and $\alpha(x) \leq \|x\|$, $x \in K$. Assume that there exists a positive constant r such that

$$\max_{t\in[0,\sigma(T)]} D(f(t,x),\widetilde{0}) \leq \frac{r}{2B\sigma(T)},$$

$$D(I_j(x),\widetilde{0}) \leq \frac{r}{2Bm}, \quad j \in \{1,\ldots,m\}, \quad D(x,\widetilde{0}) \in (0,r].$$

Then, for $x \in K_r$, we get

$$
\begin{aligned}
& D(\Phi x(t),\widetilde{0}) \\
= \ & D\left(\int_0^{\sigma(T)} G(t,s)\cdot f(s,x(\sigma(s)))\delta_H s + \sum_{j=1}^{m} G(t,t_j)\cdot I_j(x(t_j)),\widetilde{0}\right) \\
\leq \ & D\left(\int_0^{\sigma(T)} G(t,s)\cdot f(s,x(\sigma(s)))\delta_H s,\widetilde{0}\right) \\
& + D\left(\sum_{j=1}^{m} G(t,t_j)\cdot I_j(x(t_j)),\widetilde{0}\right) \\
\leq \ & D\left(\int_0^{\sigma(T)} G(t,s)\cdot f(s,x(\sigma(s)))\delta_H s,\widetilde{0}\right) \\
& + \sum_{j=1}^{m} D\left(G(t,t_j)\cdot I_j(x(t_j)),\widetilde{0}\right) \\
\leq \ & BD\left(\int_0^{\sigma(T)} f(s,x(\sigma(s)))\delta_H s,\widetilde{0}\right) + \sum_{j=1}^{m} D\left(I_j(x(t_j)),\widetilde{0}\right) \\
\leq \ & B\frac{r}{2B\sigma(T)}\sigma(T) + Bm\frac{r}{2Bm} \\
= \ & \frac{r}{2} + \frac{r}{2}
\end{aligned}
$$

$$= r, \quad t \in [0, \sigma(T)],$$

whereupon

$$\|\Phi x\| \leq r \quad or \quad \Phi x \in K_r.$$

Therefore $\Phi : \overline{K_c} \to K_c$ *and* $\Phi : \overline{K_d} \to K_d$. *Let* $b = \frac{a}{\delta}$. *Note that*

$$a < \frac{(1+\delta)a}{2\delta} < \frac{a}{\delta}$$

and

$$\begin{aligned}\alpha\left(\frac{(1+\delta)a}{2\delta}\right) &= \frac{(1+\delta)a}{2\delta}\\ &> a.\end{aligned}$$

Consequently

$$\frac{(1+\delta)a}{2\delta} \in \{x \in K(\alpha, a, b) : \alpha(x) > a\},$$

i.e.,

$$\{x \in K(\alpha, a, b) : \alpha(x) > a\} \neq \emptyset.$$

Next, for $x \in K(\alpha, a, b)$, *we get*

$$\begin{aligned}&D(\Phi x(t), \widetilde{0})\\ &= D\left(\int_0^{\sigma(T)} G(t,s) \cdot f(s, x(\sigma(s)))\delta_H s + \sum_{j=1}^{m} G(t, t_j) \cdot I_j(x(t_j)), \widetilde{0}\right)\\ &\geq s \int_0^{\sigma(T)} G(t,s) D(f(s, x(\sigma(s))), \widetilde{0})\Delta s\\ &\geq sA \int_0^{\sigma(T)} D(f(s, x(\sigma(s))), \widetilde{0})\delta_H s\end{aligned}$$

$$
\begin{aligned}
&> As\frac{a}{sA\sigma(T)}\sigma(T)\\
&= a, \quad t\in[0,\sigma(T)],
\end{aligned}
$$

whereupon

$$\alpha(\Phi x) > a.$$

Now, suppose that $x\in K(\alpha,a,c)$ *and* $\|\Phi x\| > b = \frac{a}{\delta}$. *Then*

$$
\begin{aligned}
\alpha(\Phi(x)) &\geq \frac{A}{B}\max_{t\in[0,\sigma(T)]} D(\Phi x(t),\widetilde{0})\\
&= \frac{A}{B}\|\Phi x\|\\
&> \delta b\\
&= a.
\end{aligned}
$$

From here and from Theorem 4.7.20, it follows that the problem (4.47) *has at least three positive solutions* x_1, x_2, x_3 *such that*

$$\|x_1\| < d, \quad \min_{t\in[0,\sigma(T)]} D(x_2(t),\widetilde{0}) > a, \quad \|x_3\| > d, \quad \min_{t\in[0,\sigma(T)]} D(x_3(t),\widetilde{0}) < a.$$

This completes the proof.

Exercise 4.7.23 *Let* $\mathbb{T}=[0,1]\bigcup[2,3]$, *where* $[0,1]$ *and* $[2,3]$ *are the real-valued intervals. Prove that the PBVP*

$$
\begin{aligned}
\delta_H x(t)+x(\sigma(t)) &= f(t,x(\sigma(t))), \quad t\in[0,3]\setminus\left\{\frac{1}{3},\frac{1}{2},\frac{5}{2}\right\},\\
x\left(\frac{1}{3}^+\right)\ominus_H x\left(\frac{1}{3}^-\right) &= I\left(x\left(\frac{1}{3}\right)\right)\\
x\left(\frac{1}{2}^+\right)\ominus_H x\left(\frac{1}{2}^-\right) &= I\left(x\left(\frac{1}{2}\right)\right)\\
x\left(\frac{5}{2}^+\right)\ominus_H x\left(\frac{5}{2}^-\right) &= I\left(x\left(\frac{5}{2}\right)\right)\\
x(0)\ominus_H x(3) &= 0,
\end{aligned}
$$

where

$$f(t,x(\sigma(t))) =$$

$$= \begin{cases} \left((x(\sigma(t)))^2 + \frac{e^2-1}{5184e^4(e^2+2)^2}, x(\sigma(t)), (x(\sigma(t)))^2 \right), \\ D(x,\widetilde{0}) \in \left[0, \frac{e^2(e^4-1)}{48}\right] \\ \left(\frac{e^2-2}{18e^6(e^2+2)} x(\sigma(t)) + \frac{e^4(e^4-4)^2}{48}, x(\sigma(t)), x(\sigma(t)) \right), \\ D(x,\widetilde{0}) \in \left[\frac{e^2(e^4-1)}{48}, \infty\right), \end{cases}$$

$$I(x) = \begin{cases} \left(x^2 + \frac{e^2-1}{5184e^4(e^2+2)^2}, x^2, x \right), \\ D(x,\widetilde{0}) \in \left[0, \frac{e^2(e^4-1)}{48}\right] \\ \left(\frac{e^2-2}{18e^6(e^2+2)} x + \frac{e^4(e^4-4)^2}{48}, 1, 0 \right), \\ D(x,\widetilde{0}) \in \left[\frac{e^2(e^4-1)}{48}, \infty\right), \end{cases}$$

has at least three positive solutions.

Hint. Choose

$$d = \frac{e^2-2}{24e^2(e^2+2)}, \quad a = \frac{e^2-2}{24}$$

and c sufficiently large.

4.8 ADVANCED PRACTICAL PROBLEMS

Problem 4.8.1 *Let* $\mathbb{T} = [0,1]\bigcup[3,4]$, *where* $[0,1]$ *and* $[3,4]$ *are the real value intervals. Investigate for existence of solutions the following impulsive dynamic equation*

$$\delta_H y(t) = (t^2+t+1+4(y(t)), y(t), -y(t)), \quad t \in [0,1]\bigcup(3,4],$$

$$y(3^+) = 2 \cdot y(3) + (1,1,1),$$

$$B(y(0), y(4)) = 0.$$

Problem 4.8.2 *Let* $\mathbb{T} = [0,1] \bigcup [2,4] \bigcup [5,6] \bigcup [7,8]$, *where* $[0,1]$, $[2,4]$, $[5,6]$, *and* $[7,8]$ *are the real-value intervals. Consider the periodic boundary value problem*

$$\begin{aligned}
\delta_H x(t) &= \ominus_H(2t) \cdot x^{\sigma}(t) + \left(\frac{1}{t^2+1}, \frac{1}{t}, 0\right), \quad t \in \mathbb{T} \backslash \{2,4\}, \\
x(2+) &= \left(\frac{1}{1+(x(2))^2}, x(2), x(2) + (-1,2,1)\right) \\
x(5+) &= \left(\frac{2}{4+2(x(5))^2}, (x(5))^2, -x(5)\right) \\
x(7+) &= \left(\frac{1}{1+3(x(7))^2}, -1, 2\right), \\
x(0) &= x(8).
\end{aligned}$$

Find constants l_1 *and* l_2 *as in Theorem 4.2.6 so that the PBVP considered has a unique solution.*

Problem 4.8.3 *Prove that the problem*

$$\delta_H x(t) + p(t) \cdot x(\sigma(t)) = \left(12(x(\sigma(t)))^{\frac{1}{21}} + (x(\sigma(t)))^{\frac{1}{7}}, 0, 0\right), \; t \in J_0,$$

$$\begin{aligned}
x(t_k^+) &= x(t_k^-) + \left(4(x(t_k^-))^{\frac{1}{17}} + 7(x(t_k^-))^{\frac{1}{11}}, (x(t_k^-))^3, 1\right), \\
x(0) &= x(\sigma(T)),
\end{aligned}$$

$k \in \{1, \ldots, m\}$, *where* p *satisfies (H2), has at least one positive solution.*

Problem 4.8.4 *Find conditions for f, I_k, $k \in \{1,\ldots,m\}$, so that the problem*

$$\begin{aligned} \delta_H x(t) &= f(t,x(\sigma(t))), \quad t \in J_0, \\ x(t_k^+) &= x(t_k) + I_k(x(t_k)), \quad k \in \{1,\ldots,m\}, \\ & x(0) = x(\sigma(T)), \end{aligned}$$

has at least one positive solution.

Hint. Take a function $a \in \mathscr{R}$ arbitrarily and let

$$g(t,x(\sigma(t))) = a(t)\cdot x(\sigma(t)) + f(t,x(\sigma(t))), \quad t \in J_0.$$

Thus we get the problem

$$\begin{aligned} \delta_H x(t) + a(t)\cdot x(\sigma(t)) &= g(t,x(\sigma(t))), \quad t \in J_0, \\ x(t_k^+) &= x(t_k) + I_k(x(t_k)), \quad k \in \{1,\ldots,m\}, \\ & x(0) = x(\sigma(T)), \end{aligned}$$

i.e., the considered problem can be reduced to the problem (4.14), (4.15).

Problem 4.8.5 *Let* $\mathbb{T} = \bigcup\limits_{n=0}^{\infty} [2n, 2n+1]$. *Prove that the following PBVP*

$$\begin{aligned} \delta_H x(t) + p(t)\cdot x(\sigma(t)) &= f(t,x(\sigma(t))), \quad t \in [0,3] \backslash \{\tfrac{1}{2}\}, \\ x\left(\frac{1}{2}^+\right) \ominus_H x\left(\frac{1}{2}^-\right) &= I\left(x\left(\frac{1}{2}\right)\right), \\ \alpha\cdot x(0) \ominus_H \beta x(4) &= \int\limits_0^4 g(s)\cdot x(s)\delta_H s, \end{aligned}$$

has at least one positive solution, where

1.

$$T = 3, \quad p(t) = t,$$

$$f(t, x(\sigma(t))) = \left((t+1)(x(\sigma(t)))^4, (x(\sigma(t)))^2, 1\right),$$

$$I(x) = (x^3, x, x^2), \quad \alpha = 1, \quad \beta = \frac{1}{2},$$

$$g(t) = \begin{cases} (t, -t, 1) & t \in [0,1] \\ \widetilde{0} & t \notin [0,1], \end{cases}$$

2.

$$T = 3, \quad p(t) = t,$$

$$f(t, x(\sigma(t))) = \left((t+1)(x(\sigma(t)))^{\frac{1}{3}}, (x(\sigma(t)))^{\frac{1}{7}}, x(\sigma(t))\right),$$

$$I(x) = \left(x^{\frac{1}{2}}, x, x^{\frac{1}{9}}\right), \quad \alpha = 1, \quad \beta = \frac{1}{2},$$

$$g(t) = \begin{cases} (1, -1, 1) & t \in \{1, 3\} \\ \widetilde{0} & \textit{otherwise}. \end{cases}$$

CHAPTER 5

Existence of Solutions of Second-Order Fuzzy Impulsive Dynamic Equations

Let $\mathbb{T}$ be a time scale with a forward jumx p operator and delta differentiation operator σ and Δ, respectively. Let also, $0, T \in \mathbb{T}$, $\{t_j\}_{j=1}^m \subset \mathbb{T}$,

$$0 < t_1 < t_2 \ldots < t_m < T,$$

t_j, $j \in \{1, \ldots, m\}$, are right-dense. Set $t_0 = 0$ and $t_{m+1} = T$.

5.1 LINEAR SECOND-ORDER FUZZY IMPULSIVE DYNAMIC EQUATIONS

Consider the following linear impulsive dynamic equation

$$\begin{aligned} \delta_H^2 y(t) + \alpha \cdot \delta_H y(t) + \beta \cdot y(t) &= g(t), \quad t \in [0,T] \backslash \{t_j\}_{j=1}^m, \\ y(t_j^+) \ominus_H y(t_j^-) &= I_j(t_j, y(t_j)), \\ \delta_H y(t_j^+) \ominus_H \delta_H y(t_j^-) &= L_j(t_j, y(t_j)), \quad j \in \{1, \ldots, m\}, \\ y(0) &= y_0, \\ \delta_H y(0) &= y_1, \end{aligned} \tag{5.1}$$

DOI: 10.1201/9781003655367-5

where $\alpha,\beta \in \mathbb{R}$, $y_0, y_1 \in F(\mathbb{R})$, $I_j, L_j \in \mathscr{C}(\mathbb{T}\times F(\mathbb{R}))$, $j \in \{1,\ldots,m\}$, $g \in \mathscr{C}_{frd}(\mathbb{T})$,

$$\begin{aligned} y(t_j^+) &= \lim_{h\to 0+} y(t_j+h), \\ y(t_j^-) &= y(t_j), \\ \delta_H y(t_j^+) &= \lim_{h\to 0+} \delta_H y(t_j+h), \\ \delta_H y(t_j^-) &= \delta_H y(t_j), \quad j\in\{1,\ldots,m\}. \end{aligned}$$

1. Let

$$\begin{aligned} \alpha^2 - 4\beta &\neq 0, \\ \mu\beta - \alpha &\in \mathscr{R}. \end{aligned}$$

Set

$$\begin{aligned} \lambda_1 &= \frac{-\alpha - \sqrt{\alpha^2-4\beta}}{2}, \\ \lambda_2 &= \frac{-\alpha + \sqrt{\alpha^2-4\beta}}{2}. \end{aligned}$$

Then the system of functions

$$e_{\lambda_1}(t,t_j) \quad \text{and} \quad e_{\lambda_2}(t,t_j), \quad j\in\{0,\ldots,m+1\},$$

forms a fundamental system for the equation

$$\delta_H^2 y(t) + \alpha\cdot\delta_H y(t) + \beta\cdot y(t) = \widetilde{0}, \quad t\in[0,T].$$

For the Wronskian of $y_1(\cdot)$ and $y_2(\cdot)$, we get

$$W(y_1,y_2)(t) = \begin{vmatrix} y_1(t) & y_2(t) \\ y_1^{\Delta}(t) & y_2^{\Delta}(t) \end{vmatrix}$$

$$= \begin{vmatrix} e_{\lambda_1}(t,t_j) & e_{\lambda_2}(t,t_j) \\ \lambda_1 e_{\lambda_1}(t,t_j) & \lambda_2 e_{\lambda_2}(t,t_j) \end{vmatrix}$$

$$= \lambda_2 e_{\lambda_1}(t,t_j)e_{\lambda_2}(t,t_j) - \lambda_1 e_{\lambda_1}(t,t_j)e_{\lambda_2}(t,t_j)$$

$$= (\lambda_2 - \lambda_1)e_{\lambda_1}(t,t_j)e_{\lambda_2}(t,t_j),$$

$$j \in \{0,\dots,m+1\}. \quad t \in [0,T].$$

(a) Let $t \in [0,t_1]$. Then we obtain the following IVP

$$\begin{aligned} \delta_H^2 y(t) + \alpha \cdot \delta_H y(t) + \beta \cdot y(t) &= g(t), \quad t \in (0,t_1], \\ y(0) &= y_0, \\ \delta_H y(0) &= y_1. \end{aligned}$$

By the variation of parameters, for the general solution we have the representation

$$\begin{aligned} y(t) &= e_{\lambda_1}(t,0)\cdot c_1 + e_{\lambda_2}(t,0)\cdot c_2 \\ &\quad + \int_0^t \frac{e_{\lambda_1}(\sigma(\tau),0)e_{\lambda_2}(t,0) - e_{\lambda_2}(\sigma(\tau),0)e_{\lambda_1}(t,0)}{(\lambda_2-\lambda_1)e_{\lambda_1}(\sigma(\tau),0)e_{\lambda_2}(\sigma(\tau),0)} \cdot g(\tau)\delta_H\tau \\ &= e_{\lambda_1}(t,0)\cdot c_1 + e_{\lambda_2}(t,0)\cdot c_2 \\ &\quad + \frac{1}{\lambda_2-\lambda_1} \\ &\quad \cdot \int_0^t \left(e_{\lambda_2}(t,0)e_{\lambda_2}(0,\sigma(\tau)) - e_{\lambda_1}(t,0)e_{\lambda_1}(0,\sigma(\tau))\right)\cdot g(\tau)\delta_H\tau \\ &= e_{\lambda_1}(t,0)\cdot c_1 + e_{\lambda_2}(t,0)\cdot c_2 \\ &\quad + \frac{1}{\lambda_2-\lambda_1}\cdot \int_0^t \left(e_{\lambda_2}(t,\sigma(\tau)) - e_{\lambda_1}(t,\sigma(\tau))\right)\cdot g(\tau)\delta_H\tau, \\ &\quad t \in [0,t_1], \end{aligned}$$

where c_1 and c_2 are fuzzy real constants which will be determined in the following. We have

$$\begin{aligned} y(0) &= c_1 + c_2 \\ &= y_0, \end{aligned}$$

$$\begin{aligned} \delta_H y(t) &= (\lambda_1 e_{\lambda_1}(t,0)) \cdot c_1 + (\lambda_2 e_{\lambda_2}(t,0)) \cdot c_2 \\ &+ \left(\frac{1}{\lambda_2 - \lambda_1}\left(e_{\lambda_2}(\sigma(t),\sigma(t)) - e_{\lambda_1}(\sigma(t),\sigma(t))\right)\right) \cdot g(t) \\ &+ \frac{1}{\lambda_2 - \lambda_1} \cdot \int_0^t (\lambda_2 e_{\lambda_2}(t,\sigma(\tau)) - \lambda_1 e_{\lambda_1}(t,\sigma(\tau))) \cdot g(\tau)\delta_H \tau \\ &= (\lambda_1 e_{\lambda_1}(t,0)) \cdot c_1 + (\lambda_2 e_{\lambda_2}(t,0)) \cdot c_2, \\ &+ \frac{1}{\lambda_2 - \lambda_1} \cdot \int_0^t (\lambda_2 e_{\lambda_2}(t,\sigma(\tau)) - \lambda_1 e_{\lambda_1}(t,\sigma(\tau))) \cdot g(\tau)\delta_H \tau \end{aligned}$$

$$\begin{aligned} \delta_H y(0) &= \lambda_1 \cdot c_1 + \lambda_2 \cdot c_2 \\ &= y_1. \end{aligned}$$

$t \in [0, t_1]$. Thus, we get the system.

$$\begin{aligned} c_1 + c_2 &= y_0 \\ \lambda_1 \cdot c_1 + \lambda_2 \cdot c_2 &= y_1. \end{aligned}$$

Hence,

$$c_1 = \frac{1}{\lambda_2 - \lambda_1} \cdot (\lambda_2 \cdot y_0 \ominus_H y_1),$$

$$c_2 = \frac{1}{\lambda_1 - \lambda_2} \cdot (\lambda_1 \cdot y_0 \ominus_H y_1).$$

Then

$$y(t) = \left(e_{\lambda_1}(t,0)\frac{1}{\lambda_2 - \lambda_1}\right) \cdot (\lambda_2 \cdot y_0 \ominus_H y_1)$$

$$+\frac{1}{\lambda_1 - \lambda_2} \cdot (\lambda_1 \cdot y_0 \ominus_H y_1)$$

$$+\frac{1}{\lambda_2 - \lambda_1} \cdot \int_0^t \left(e_{\lambda_2}(t,\sigma(\tau)) - e_{\lambda_1}(t,\sigma(\tau))\right) \cdot g(\tau)\delta_H\tau,$$

$$t \in [0,t_1].$$

From here,

$$y(t_1) = \left(\frac{1}{\lambda_2 - \lambda_1}e_{\lambda_1}(t_1,0)\right) \cdot (\lambda_2 \cdot y_0 \ominus_H y_1)$$

$$+\left(\frac{1}{\lambda_1 - \lambda_2}e_{\lambda_2}(t_1,0)\right) \cdot (\lambda_1 \cdot y_0 \ominus_H y_1)$$

$$+\frac{1}{\lambda_2 - \lambda_1} \cdot \int_0^{t_1} \left(e_{\lambda_2}(t_1,\sigma(\tau)) - e_{\lambda_1}(t_1,\sigma(\tau))\right) \cdot g(\tau)\delta_H\tau$$

$$= a_0,$$

$$\delta_H y(t) = \left(\lambda_1\frac{1}{\lambda_2 - \lambda_1}e_{\lambda_1}(t,0)\right) \cdot (\lambda_2 \cdot y_0 \ominus_H y_1)$$

$$+\left(\lambda_2\frac{1}{\lambda_1 - \lambda_2}e_{\lambda_2}(t,0)\right) \cdot (\lambda_1 \cdot y_0 - y_1)$$

$$+\frac{1}{\lambda_2-\lambda_1}\cdot\int_0^t (\lambda_2 e_{\lambda_2}(t,\sigma(\tau))-\lambda_1 e_{\lambda_1}(t,\sigma(\tau)))\cdot g(\tau)\delta_H\tau,$$

$$\begin{aligned}\delta_H y(t_1) &= \left(\lambda_1\frac{1}{\lambda_2-\lambda_1}e_{\lambda_1}(t,0)\right)\cdot(\lambda_2\cdot y_0\ominus_H y_1)\\ &\quad+\left(\lambda_2\frac{1}{\lambda_1-\lambda_2}e_{\lambda_2}(t,0)\right)\cdot(\lambda_1\cdot y_0-y_1)\\ &\quad+\frac{1}{\lambda_2-\lambda_1}\cdot\int_0^{t_1}(\lambda_2 e_{\lambda_2}(t_1,\sigma(\tau))-\lambda_1 e_{\lambda_1}(t_1,\sigma(\tau)))\cdot g(\tau)\delta_H\tau\\ &= a_1.\end{aligned}$$

Set

$$\begin{aligned}b_0 &= a_0+I_1(t_1,y(t_1)),\\ b_1 &= a_1+L_1(t_1,y(t_1)).\end{aligned}$$

(b) Let $t\in[t_1,t_2]$. We get the following IVP

$$\begin{aligned}\delta_H^2 y(t)+\alpha\cdot\delta_H y(t)+\beta\cdot y(t) &= g(t),\quad t\in(t_1,t_2],\\ y(t_1) &= b_0,\\ \delta_H y(t_1) &= b_1.\end{aligned}$$

Note that

$$\left(\frac{1}{\lambda_2-\lambda_1}e_{\lambda_1}(t,t_1)\right)\cdot(\lambda_2\cdot a_0\ominus_H a_1)$$
$$+\left(\frac{1}{\lambda_1-\lambda_2}e_{\lambda_2}(t,t_1)\right)\cdot(\lambda_1\cdot a_0\ominus_H a_1)$$

$$
\begin{aligned}
& + \frac{1}{\lambda_2-\lambda_1} \cdot \int_{t_1}^{t} \left(e_{\lambda_2}(t,\sigma(\tau)) - e_{\lambda_1}(t,\sigma(\tau))\right) \cdot g(\tau)\delta_H \tau \\
= & \left(\frac{\lambda_2}{\lambda_2-\lambda_1} e_{\lambda_1}(t,t_1)\right) \cdot a_0 \ominus_H \left(\frac{1}{\lambda_2-\lambda_1} e_{\lambda_1}(t,t_1)\right) \cdot a_1 \\
& + \left(\frac{\lambda_1}{\lambda_1-\lambda_2} e_{\lambda_2}(t,t_1)\right) \cdot a_0 \ominus_H \left(\frac{1}{\lambda_1-\lambda_2} e_{\lambda_2}(t,t_1)\right) \cdot a_1 \\
& + \frac{1}{\lambda_2-\lambda_1} \cdot \int_{t_1}^{t} \left(e_{\lambda_2}(t,\sigma(\tau)) - e_{\lambda_1}(t,\sigma(\tau))\right) \cdot g(\tau)\delta_H \tau \\
= & \frac{\lambda_2}{\lambda_2-\lambda_1} \cdot \left(\left(\frac{1}{\lambda_2-\lambda_1} e_{\lambda_1}(t,0)\right) \cdot (\lambda_2 \cdot y_0 \ominus_H y_1) \right. \\
& + \left(\frac{1}{\lambda_2-\lambda_1} e_{\lambda_1}(t,t_1) e_{\lambda_2}(t_1,0)\right) \cdot (\lambda_1 \cdot y_0 \ominus_H y_1) \\
& + \frac{1}{\lambda_2-\lambda_1} \\
& \left. \cdot \int_{0}^{t_1} \left(e_{\lambda_1}(t,t_1) e_{\lambda_2}(t_1,\sigma(\tau)) - e_{\lambda_1}(t,\sigma(\tau))\right) \cdot g(\tau)\delta_H \tau \right) \\
& \ominus_H \frac{1}{\lambda_2-\lambda_1} \cdot \left(\left(\lambda_1 \frac{1}{\lambda_2-\lambda_1} e_{\lambda_1}(t,0)\right) \cdot (\lambda_2 \cdot y_0 \ominus_H y_1) \right. \\
& + \left(\lambda_2 \frac{1}{\lambda_1-\lambda_2} e_{\lambda_1}(t,t_1) e_{\lambda_2}(t_1,0)\right) \cdot (\lambda_1 \cdot y_0 \ominus_H y_1) \\
& + \frac{1}{\lambda_2-\lambda_1} \\
& \left. \cdot \int_{0}^{t_1} \left(\lambda_2 e_{\lambda_1}(t,t_1) e_{\lambda_2}(t_1,\sigma(\tau)) - \lambda_1 e_{\lambda_1}(t,\sigma(\tau))\right) \cdot g(\tau)\delta_H \tau \right) \\
& + \frac{\lambda_1}{\lambda_1-\lambda_2} \cdot \left(\left(\frac{1}{\lambda_2-\lambda_1} e_{\lambda_1}(t_1,0) e_{\lambda_2}(t,t_1)\right) \cdot (\lambda_2 \cdot y_0 \ominus_H y_1) \right. \\
& + \left(\frac{1}{\lambda_2-\lambda_1} e_{\lambda_2}(t,0)\right) \cdot (\lambda_1 \cdot y_0 \ominus_H y_1) \\
& + \frac{1}{\lambda_2-\lambda_1} \\
& \left. \cdot \int_{0}^{t_1} \left(e_{\lambda_2}(t,\sigma(\tau)) - e_{\lambda_2}(t,t_1) e_{\lambda_1}(t_1,\sigma(\tau))\right) \cdot g(\tau)\delta_H \tau \right)
\end{aligned}
$$

$$\ominus_H \frac{1}{\lambda_1-\lambda_2}$$

$$\cdot\left(\left(\lambda_1 \frac{1}{\lambda_2-\lambda_1} e_{\lambda_2}(t,t_1)e_{\lambda_1}(t_1,0)\right)\cdot(\lambda_2\cdot y_0 \ominus_H y_1)\right.$$

$$+\left(\lambda_2 \frac{1}{\lambda_1-\lambda_2} e_{\lambda_2}(t,0)\right)\cdot(\lambda_1\cdot y_0 \ominus_H y_1)$$

$$+\frac{1}{\lambda_2-\lambda_1}$$

$$\left.\cdot\int_0^{t_1}\left(\lambda_2 e_{\lambda_2}(t,\sigma(\tau))-\lambda_1 e_{\lambda_2}(t,t_1)e_{\lambda_1}(t_1,\sigma(\tau))\right)\cdot g(\tau)\delta_H\tau\right)$$

$$+\frac{1}{\lambda_2-\lambda_1}\cdot\int_{t_1}^{t}\left(e_{\lambda_2}(t_1,\sigma(\tau))-e_{\lambda_1}(t_1,\sigma(\tau))\right)\cdot g(\tau)\delta_H\tau$$

$$= \left(\frac{1}{\lambda_2-\lambda_1} e_{\lambda_1}(t,0)\right)\cdot(\lambda_2\cdot y_0 \ominus_H y_1)$$

$$+\left(\frac{1}{\lambda_1-\lambda_2} e_{\lambda_2}(t,0)\right)\cdot(\lambda_1\cdot y_0 \ominus_H y_1)$$

$$+\frac{\lambda_2}{(\lambda_2-\lambda_1)^2}$$

$$\cdot\int_0^{t_1}\left(e_{\lambda_1}(t,t_1)e_{\lambda_2}(t_1,\sigma(\tau))-e_{\lambda_1}(t,\sigma(\tau))\right)\cdot g(\tau)\delta_H\tau$$

$$\ominus_H \frac{1}{(\lambda_2-\lambda_1)^2}$$

$$\cdot\int_0^{t_1}\left(\lambda_2 e_{\lambda_1}(t,t_1)e_{\lambda_2}(t_1,\sigma(\tau))-\lambda_1 e_{\lambda_1}(t,\sigma(\tau))\right)\cdot g(\tau)\delta_H\tau$$

$$\ominus_H \frac{\lambda_1}{(\lambda_2-\lambda_1)^2}$$

$$\cdot\int_0^{t_1}\left(e_{\lambda_2}(t,\sigma(\tau))-e_{\lambda_2}(t,t_1)e_{\lambda_1}(t_1,\sigma(\tau))\right)\cdot g(\tau)\delta_H\tau$$

$$+\frac{1}{(\lambda_2-\lambda_1)^2}$$

$$
\begin{aligned}
&\cdot \int_0^{t_1} (\lambda_2 e_{\lambda_2}(t,\sigma(\tau)) - \lambda_1 e_{\lambda_2}(t,t_1) e_{\lambda_1}(t_1,\sigma(\tau))) \cdot g(\tau)\delta_H\tau \\
&+\frac{1}{\lambda_2-\lambda_1} \cdot \int_{t_1}^{t} (e_{\lambda_2}(t,\sigma(\tau)) - e_{\lambda_1}(t,\sigma(\tau))) \cdot g(\tau)\delta_H\tau \\
= \;& \left(\frac{1}{\lambda_2-\lambda_1} e_{\lambda_1}(t,0)\right) \cdot (\lambda_2 \cdot y_0 \ominus_H y_1) \\
&+\left(\frac{1}{\lambda_1-\lambda_2} e_{\lambda_2}(t,0)\right) \cdot (\lambda_1 \cdot y_0 \ominus_H y_1) \\
&\ominus_H \frac{1}{\lambda_2-\lambda_1} \cdot \int_0^{t_1} e_{\lambda_1}(t,\sigma(\tau)) \cdot g(\tau)\delta_H\tau \\
&+\frac{1}{\lambda_2-\lambda_1} \cdot \int_0^{t_1} e_{\lambda_2}(t,\sigma(\tau)) \cdot g(\tau)\delta_H\tau \\
&+\frac{1}{\lambda_2-\lambda_1} \cdot \int_{t_1}^{t} (e_{\lambda_2}(t,\sigma(\tau)) - e_{\lambda_1}(t,\sigma(\tau))) \cdot g(\tau)\delta_H\tau \\
= \;& \left(\frac{1}{\lambda_2-\lambda_1} e_{\lambda_1}(t,0)\right) \cdot (\lambda_2 \cdot y_0 \ominus_H y_1) \\
&+\left(\frac{1}{\lambda_1-\lambda_2} e_{\lambda_2}(t,0)\right) \cdot (\lambda_1 \cdot y_0 \ominus_H y_1) \\
&+\frac{1}{\lambda_2-\lambda_1} \cdot \int_0^{t} (e_{\lambda_2}(t,\sigma(\tau)) - e_{\lambda_1}(t,\sigma(\tau))) \cdot g(\tau)\delta_H\tau, \\
& t \in (t_1,t_2].
\end{aligned}
$$

Hence,

$$
\begin{aligned}
y(t) \;=\;& \left(\frac{1}{\lambda_2-\lambda_1} e_{\lambda_1}(t,t_1)\right) \cdot (\lambda_2 \cdot b_0 \ominus_H b_1) \\
&+\left(\frac{1}{\lambda_1-\lambda_2} e_{\lambda_2}(t,t_1)\right) \cdot (\lambda_1 \cdot b_0 \ominus_H b_1) \\
&+\frac{1}{\lambda_2-\lambda_1} \cdot \int_{t_1}^{t} (e_{\lambda_2}(t,\sigma(\tau)) - e_{\lambda_1}(t,\sigma(\tau))) \cdot g(\tau)\delta_H\tau
\end{aligned}
$$

$$
\begin{aligned}
&= \left(\frac{1}{\lambda_2-\lambda_1}e_{\lambda_1}(t,t_1)\right)\cdot(\lambda_2\cdot a_0\ominus_H a_1) \\
&\quad +\left(\frac{1}{\lambda_1-\lambda_2}e_{\lambda_2}(t,t_1)\right)\cdot(\lambda_1\cdot a_0\ominus_H a_1) \\
&\quad +\frac{1}{\lambda_2-\lambda_1}\cdot\int_{t_1}^{t}\left(e_{\lambda_2}(t,\sigma(\tau))-e_{\lambda_1}(t,\sigma(\tau))\right)\cdot g(\tau)\delta_H\tau \\
&\quad +\left(\frac{1}{\lambda_2-\lambda_1}e_{\lambda_1}(t,t_1)\right)\cdot(\lambda_2\cdot I_1(t_1,y(t_1))\ominus_H L_1(t_1,y(t_1))) \\
&\quad +\left(\frac{1}{\lambda_1-\lambda_2}e_{\lambda_2}(t,t_1)\right)\cdot(\lambda_1\cdot I_1(t_1,y(t_1))\ominus_H L_1(t_1,y(t_1))) \\
&= \left(\frac{1}{\lambda_2-\lambda_1}e_{\lambda_1}(t,t_1)\right)\cdot(\lambda_2\cdot y_0\ominus_H y_1) \\
&\quad +\left(\frac{1}{\lambda_1-\lambda_2}e_{\lambda_2}(t,t_1)\right)\cdot(\lambda_1\cdot y_0\ominus_H y_1) \\
&\quad +\frac{1}{\lambda_2-\lambda_1}\cdot\int_{0}^{t}\left(e_{\lambda_2}(t,\sigma(\tau))-e_{\lambda_1}(t,\sigma(\tau))\right)\cdot g(\tau)\delta_H\tau \\
&\quad +\left(\frac{1}{\lambda_2-\lambda_1}e_{\lambda_1}(t,t_1)\right)\cdot(\lambda_2\cdot I_1(t_1,y(t_1))\ominus_H L_1(t_1,y(t_1))) \\
&\quad +\left(\frac{1}{\lambda_1-\lambda_2}e_{\lambda_2}(t,t_1)\right)\cdot(\lambda_1\cdot I_1(t_1,y(t_1))\ominus_H L_1(t_1,y(t_1))), \\
&\quad t\in[0,t_2].
\end{aligned}
$$

Continuing as above, we arrive at

$$
\begin{aligned}
y(t) &= \left(\frac{1}{\lambda_2-\lambda_1}e_{\lambda_1}(t,t_1)\right)\cdot(\lambda_2\cdot y_0\ominus_H y_1) \\
&+\left(\frac{1}{\lambda_1-\lambda_2}e_{\lambda_2}(t,t_1)\right)\cdot(\lambda_1\cdot y_0\ominus_H y_1) \\
&+\frac{1}{\lambda_2-\lambda_1}\cdot\int_{0}^{t}\left(e_{\lambda_2}(t,\sigma(\tau))-e_{\lambda_1}(t,\sigma(\tau))\right)\cdot g(\tau)\delta_H\tau \\
&+\sum_{\{j:0<t_j<t\}}\left(\frac{1}{\lambda_2-\lambda_1}e_{\lambda_1}(t,t_j)\right)\cdot(\lambda_2\cdot I_j(t_j,y(t_j))\ominus_H L_j(t_j,y(t_j)))
\end{aligned}
$$

$$+ \sum_{\{j:0<t_j<t\}} \left(\frac{1}{\lambda_1 - \lambda_2} e_{\lambda_2}(t,t_j) \right) \cdot (\lambda_1 \cdot I_j(t_j y(t_j)) \ominus_H L_j(t_j, y(t_j)))),$$

for $\quad t \in [0,T]$.

Exercise 5.1.1 *Let* $\mathbb{T} = [0,1] \bigcup [2,3]$. *Find a solution of the fuzzy impulsive dynamic equation*

$$\begin{aligned}
\delta_H^2 y(t) + 3 \cdot \delta_H y(t) + 2 \cdot y(t) &= \widetilde{0}, \quad t \in \mathbb{T} \backslash \{2\}, \\
y(0) &= (-1,0,1), \\
\delta_H y(0) &= (0,2,-3), \\
y(2^+) \ominus_H y(2) &= (4,-1,2), \\
\delta_H y(2^+) \ominus_H \delta_H y(2) &= (3,1,4).
\end{aligned}$$

Exercise 5.1.2 *Let* $\mathbb{T} = [0,2] \bigcup [3,4] \bigcup [5,6]$. *Find a solution of the fuzzy impulsive dynamic equation*

$$\begin{aligned}
\delta_H^2 y(t) \ominus_H 5 \cdot \delta_H y(t) + 6 \cdot y(t) &= (t, t+1, 2t), \quad t \in \mathbb{T} \backslash \{3,5\}, \\
y(0) &= (1,-1,2), \\
\delta_H y(0) &= (-1,0,3), \\
y(3^+) \ominus_H y(3) &= (1,-1,2), \\
\delta_H y(3^+) \ominus_H \delta_H y(3) &= (0,0,1), \\
y(5^+) \ominus_H y(5) &= (-1,2,1), \\
\delta_H y(5^+) \ominus_H \delta_H y(5) &= (2,-1,3).
\end{aligned}$$

2. Now we consider the fuzzy impulsive dynamic equation

$$\begin{aligned} \delta_H^2 y(t) \ominus_H (2\alpha)\cdot\delta_H y(t) + \alpha^2\cdot y(t) &= g(t), \ \ t\in[0,T]\backslash\{t_j\}_{j=1}^m,\\ y(t_j^+)\ominus_H y(t_j^-) &= I_j(t_j, y(t_j)),\\ \delta_H y(t_j^+)\ominus_H \delta_H y(t_j^-) &= L_j(t_j, y(t_j)),\\ & \quad\ j\in\{1,\ldots,m\},\\ y(0) &= y_0,\\ \delta_H y(0) &= y_1, \end{aligned} \tag{5.2}$$

where $\alpha \in \mathbb{R}$, $y_0, y_1 \in F(\mathbb{R})$, $I_j, L_j \in \mathscr{C}(\mathbb{T}\times F(\mathbb{R}))$, $j \in \{1,\ldots,m\}$, $g \in \mathscr{C}_{frd}(\mathbb{T})$. Suppose that $\alpha \in \mathscr{R}$. Note that

$$e_\alpha(t,t_j) \quad \text{and} \quad e_\alpha(t,t_j)\int_{t_j}^{t}\frac{\Delta s}{1+\alpha\mu(s)}, \qquad t\in[0,T],$$

$j\in\{1,\ldots,m\}$, forms a fundamental system for the equation

$$\delta_H^2 y(t)\ominus_H (2\alpha)\cdot\delta_H y(t) + \alpha^2\cdot y(t) = 0, \quad t\in[0,T].$$

Let

$$\begin{aligned} y_1(t,t_j) &= e_\alpha(t,t_j),\\ y_2(t,t_j) &= e_\alpha(t,t_j)\int_{t_j}^{t}\frac{\Delta s}{1+\alpha\mu(s)}, \quad t\in[0,T],\ j\in\{1,\ldots,m\}. \end{aligned}$$

We have

$$y_2^\Delta(t,t_j) = e_\alpha^\Delta(t,t_j)\int_{t_j}^{t}\frac{\Delta s}{1+\alpha\mu(s)} + e_\alpha(\sigma(t),t_j)\frac{1}{1+\alpha\mu(t)}$$

$$= \alpha e_\alpha(t,t_j)\int_{t_j}^{t}\frac{\Delta s}{1+\alpha\mu(s)}+(1+\alpha\mu(t))e_\alpha(t,t_j)\frac{1}{1+\alpha\mu(t)}$$

$$= \alpha y_2(t,t_j)+y_1(t,t_j), \quad t\in[0,T], \quad j\in\{1,\ldots,m\}.$$

Next,

$$\begin{aligned}
W(y_1(\cdot,t_j),y_2(\cdot,t_j))(t) &= \begin{vmatrix} y_1(t,t_j) & y_2(t,t_j) \\ y_1^\Delta(t,t_j) & y_2^\Delta(t,t_j)\end{vmatrix} \\
&= \begin{vmatrix} y_1(t,t_j) & y_2(t,t_j) \\ \alpha y_1(t,t_j) & \alpha y_2(t,t_j)+y_1(t,t_j)\end{vmatrix} \\
&= \alpha y_1(t,t_j)y_2(t,t_j)+(y_1(t,t_j))^2-\alpha y_1(t,t_j)y_2(t,t_j) \\
&= (y_1(t,t_j))^2.
\end{aligned}$$

(a) Let $t\in[0,t_1]$. Then we get the IVP

$$\begin{aligned}
\delta_H^2 y(t)\ominus_H(2\alpha)\cdot\delta_H y(t)+\alpha^2\cdot y(t) &= g(t), \quad t\in(0,t_1], \\
y(0) &= y_0, \\
\delta_H y(0) &= y_1.
\end{aligned}$$

Using the variation of parameters, we obtain

$$\begin{aligned}
y(t) &= y_1(t,0)\cdot c_1+y_2(t,0)\cdot c_2 \\
&+\int_0^t\frac{y_1(\sigma(\tau),0)y_2(t,0)-y_2(\sigma(\tau),0)y_1(t,0)}{(y_1(\sigma(\tau),0))^2}\cdot g(\tau)\delta_H\tau,
\end{aligned}$$

where c_1 and c_2 are fuzzy real constants which will be determined below. We have

$$y(0) = y_1(0,0)\cdot c_1+y_2(0,0)\cdot c_2$$

$$
\begin{aligned}
&= c_1 \\
&= y_0, \\
\delta_H y(t) &= y_1^{\Delta}(t,0)\cdot c_1 + y_2^{\Delta}(t,0)\cdot c_2
\end{aligned}
$$

$$
+\frac{y_1(\sigma(t),0)y_2(\sigma(t),0)-y_2(\sigma(t),0)y_1(\sigma(t),0)}{(y_1(\sigma(t),0))^2}\cdot g(t)
$$

$$
+\int_0^t \frac{y_1(\sigma(\tau),0)y_2^{\Delta}(t,0)-y_2(\sigma(\tau),0)y_1^{\Delta}(t,0)}{(y_1(\sigma(\tau),0))^2}\cdot g(\tau)\delta_H\tau
$$

$$
=(\alpha y_1(t,0))\cdot c_1+(\alpha y_2(t,0)+y_1(t,0))\cdot c_2
$$

$$
[6pt]+\int_0^t \frac{y_1(\sigma(\tau),0)y_2^{\Delta}(t,0)-y_2(\sigma(\tau),0)y_1^{\Delta}(t,0)}{(y_1(\sigma(\tau),0))^2}\cdot g(\tau)\delta_H\tau,
$$

$t\in[0,t_1]$, and

$$
\begin{aligned}
\delta_H y(0) &= y_1^{\Delta}(0,0)\cdot c_1 + y_1(0,0)\cdot c_2 \\
&= \alpha\cdot c_1 + c_2 \\
&= \alpha\cdot y_0 + c_2 \\
&= y_1, \\
c_2 &= y_1 \ominus_H \alpha\cdot y_0.
\end{aligned}
$$

Thus,

$$
\begin{aligned}
y(t) &= y_1(t,0)\cdot y_0 + y_2(t,0)\cdot(y_1\ominus_H \alpha\cdot y_0) \\
&\quad +\int_0^t \frac{y_1(\sigma(\tau),0)y_2(t,0)-y_2(\sigma(\tau),0)y_1(t,0)}{(y_1(\sigma(\tau),0))^2}\cdot g(\tau)\delta_H\tau,
\end{aligned}
$$

$t\in[0,t_1]$. Hence,

$$
y(t_1) = y_1(t_1,0)\cdot y_0 + y_2(t_1,0)\cdot(y_1\ominus_H \alpha\cdot y_0)
$$

$$
+\int_0^{t_1} \frac{y_1(\sigma(\tau),0)y_2(t_1,0)-y_2(\sigma(\tau),0)y_1(t_1,0)}{(y_1(\sigma(\tau),0))^2}\cdot g(\tau)\delta_H\tau,
$$

$$\begin{aligned}
\delta_H y(t_1) &= (\alpha y_1(t_1,0))\cdot y_0+(\alpha y_2(t_1,0)\\
&\quad +y_1(t_1,0))\cdot(y_1\ominus_H \alpha\cdot y_0)\\
&\quad +\int_0^{t_1}\frac{y_1(\sigma(\tau),0)y_2^{\Delta}(t_1,0)-y_2(\sigma(\tau),0)y_1^{\Delta}(t_1,0)}{(y_1(\sigma(\tau),0))^2}\cdot g(\tau)\delta_H\tau.
\end{aligned}$$

(b) Let $t\in[t_1,t_2]$. Then we get the following IVP

$$\begin{aligned}
\delta_H^2 y(t)\ominus_H(2\alpha)\cdot\delta_H y(t)+\alpha^2\cdot y(t) &= g(t),\quad t\in(t_1,t_2],\\
y(t_1^+)\ominus_H y(t_1^-) &= I_1(t_1,y(t_1)),\\
\delta_H y(t_1^+)\ominus_H \delta_H y(t_1^-) &= L_1(t_1,y(t_1)).
\end{aligned}$$

For its solution, we have the representation

$$\begin{aligned}
y(t) &= y_1(t,0)\cdot y_0+y_2(t,0)\cdot(y_1\ominus_H\alpha\cdot y_0)\\
&\quad +\int_0^{t}\frac{y_1(\sigma(\tau),0)y_2(t,0)-y_2(\sigma(\tau),0)y_1(t,0)}{(y_1(\sigma(\tau),0))^2}\cdot g(\tau)\delta_H\tau\\
&\quad +y_1(t,t_1)\cdot I_1(t_1,y(t_1))\\
&\quad +y_2(t,t_1)\cdot(L_1(t_1,y(t_1))\ominus_H\alpha\cdot I_1(t_1,y(t_1))),\quad t\in[t_1,t_2].
\end{aligned}$$

Continuing in this manner, we obtain that

$$\begin{aligned}
y(t) &= y_1(t,0)\cdot y_0+y_2(t,0)\cdot(y_1\ominus_H\alpha\cdot y_0)\\
&\quad +\int_0^{t}\frac{y_1(\sigma(\tau),0)y_2(t,0)-y_2(\sigma(\tau),0)y_1(t,0)}{(y_1(\sigma(\tau),0))^2}\cdot g(\tau)\delta_H\tau\\
&\quad +\sum_{\{j:t_j<t\}} y_1(t,t_j)\cdot I_j(t_j,y(t_j))\\
&\quad +\sum_{\{j:t_j<t\}} y_2(t,t_j)\cdot(L_j(t_j,y(t_j))\ominus_H\alpha\cdot I_j(t_j,y(t_j))),\ t\in[0,T].
\end{aligned}$$

Exercise 5.1.3 *Let* $\mathbb{T} = [0,1]\bigcup[2,3]\bigcup[4,5]$. *Find a solution of the impulsive dynamic equation*

$$\begin{aligned} \delta_H^2 y(t) \ominus_H 4\cdot\delta_H y(t) + 4\cdot y(t) &= (t+1,t,t-1), \;\; t\in\mathbb{T}\backslash\{2,4\}, \\ y(0) &= (0,1,0), \\ \delta_H y(0) &= 1, \\ y(2^+) &= y(2), \\ \delta_H y(2^+)\ominus_H \delta_H y(2) &= (5,2,0), \\ y(4^+)\ominus_H y(4) &= (3,0,1), \\ \delta_H y(4^+)\ominus_H \delta_H y(4) &= (-7,1,0). \end{aligned}$$

5.2 NON-LINEAR SECOND-ORDER FUZZY IMPULSIVE DYNAMIC EQUATIONS

In this section, we will investigate the existence of solutions to the following IVP for a class of non-linear impulsive dynamic equations

$$\begin{aligned} \delta_H^2 y(t) &= f(t,y(t)), \quad t\in[0,T]\backslash\{t_j\}_{j=1}^m, \\ y(0) &= y_0, \\ \delta_H y(0) &= y_1, \\ y(t_j^+) &= y(t_j^-) + I_j(t_j,y(t_j)), \\ \delta_H y(t_j^+) &= \delta_H y(t_j^-) + L_j(t_j,y(t_j)), \quad j\in\{1,\ldots,m\}, \end{aligned} \tag{5.3}$$

where $y_0, y_1 \in F(\mathbb{R})$,

(H1) $f\in\mathscr{C}(\mathbb{T}\times F(\mathbb{R}))$,

$$D(f(t,x),f(t,y)) \le FD(x,y), \quad t\in\mathbb{T}, \quad x,y\in F(\mathbb{R}),$$

for some positive constant F,

(H2) $I_j\in\mathscr{C}(\mathbb{T}\times F(\mathbb{R}))$,

$$D(I_j(t,x),I_j(t,y)) \le AD(x,y), \quad t\in\mathbb{T}, \quad x,y\in F(\mathbb{R}),$$

$j\in\{1,\ldots,m\}$, for some positive constant A,

(H3) $L_j \in \mathscr{C}(\mathbb{T} \times F(\mathbb{R}))$,

$$D(L_j(t,x), L_j(t,y)) \le BD(x,y), \quad t \in \mathbb{T}, \quad x,y \in F(\mathbb{R}),$$

$j \in \{1, \ldots, m\}$, for some positive constant B.

Define

$$\begin{aligned} PC([0,T]) &= \Big\{ x : [0,T] \to \mathbb{R}, \quad x \quad \text{is} \quad \text{continuous} \\ &\qquad \text{on} \quad [0,T] \backslash \{t_k\}_{k=1}^m, \quad \text{ld-continuous} \quad \text{at} \quad t_k, \\ &\qquad \exists \lim_{t \to t_k+} x(t), \quad k \in \{1, \ldots, m\} \Big\}. \end{aligned}$$

$PC([0,T])$ is a Banach space with the norm

$$\|x\| = \max_{j \in \{1,\ldots,m\}} \sup_{t \in [t_j, t_{j+1}]} D(x(t), \widetilde{0}).$$

Definition 5.2.1 *A function* $y \in PC([0,T]) \bigcap \bigcup_{j=1}^{m} \mathscr{C}^2((t_j, t_{j+1})$ *is said to be a solution of the equation* (5.3) *if it satisfies the first equation of* (5.3) *on* $[0,T] \backslash \{t_j\}_{j=1}^m$ *and it satisfies the second, third, fourth, and fifth equations of* (5.3).

For $y \in PC([0,T])$, define the operator

$$\begin{aligned} \Phi y(t) &= y_0 + t \cdot y_1 + \int_0^t (t - \sigma(\tau)) \cdot f(\tau, y(\tau)) \delta_H \tau \\ &\quad + \sum_{\{j : t_j < t\}} I_j(t_j, y(t_j)) \\ &\quad + \sum_{\{j : t_j < t\}} (t - t_j) \cdot L_j(t_j, y(t_j)), \quad t \in [0,T]. \end{aligned}$$

Exercise 5.2.2 *Prove that any fixed point* $y \in PC([0,T])$ *of the operator* Φ *is a solution of the equation* (5.3).

Theorem 5.2.3 *Assume* $(H1)$-$(H3)$ *and*

$$FT^2 + Am + BmT < 1.$$

Then the equation (5.3) *has a unique solution.*

Proof 5.2.4 *Note that* $\Phi : PC([0,T]) \to PC([0,T])$. *For* $x, y \in PC([0,T])$, *we obtain*

$$\begin{aligned} D(\Phi y(t), \Phi x(t)) &= D\Bigg(y_0 + t \cdot y_1 + \int_0^t (t - \sigma(\tau)) \cdot f(\tau, y(\tau)) \delta_H \tau \\ &\quad + \sum_{\{j:t_j<t\}} I_j(t_j, y(t_j)) \\ &\quad + \sum_{\{j:t_j<t\}} (t - t_j) \cdot L_j(t_j, y(t_j)), \\ &\quad y_0 + t \cdot y_1 + \int_0^t (t - \sigma(\tau)) \cdot f(\tau, x(\tau)) \delta_H \tau \\ &\quad + \sum_{\{j:t_j<t\}} I_j(t_j, x(t_j)) \\ &\quad + \sum_{\{j:t_j<t\}} (t - t_j) \cdot L_j(t_j, x(t_j)) \Bigg) \\ &= D\Bigg(\int_0^t (t - \sigma(\tau)) \cdot (f(\tau, y(\tau)) \ominus_H f(\tau, x(\tau))) \delta_H \tau \\ &\quad + \sum_{\{j:t_j<t\}} (I_j(t_j, y(t_j)) \ominus_H I_j(t_j, x(t_j))) \\ &\quad + \sum_{\{j:t_j<t\}} (t - t_j) \cdot (L_j(t_j, y(t_j)) \ominus_H L_j(t_j, x(t_j))), \widetilde{0} \Bigg) \end{aligned}$$

$$
\begin{aligned}
&\le D\Big(\int_0^t (t-\sigma(\tau))\cdot(f(\tau,y(\tau))\ominus_H f(\tau,x(\tau)))\delta_H\tau,\widetilde{0}\Big)\\
&\quad +D\Big(\sum_{\{j:t_j<t\}}(I_j(t_j,y(t_j))\ominus_H I_j(t_j,x(t_j))),\widetilde{0}\Big)\\
&\quad +D\Big(\sum_{\{j:t_j<t\}}(t-t_j)\cdot(L_j(t_j,y(t_j))\ominus_H L_j(t_j,x(t_j))),\widetilde{0}\Big)\\
&\le D\Big(\int_0^t (t-\sigma(\tau))\cdot(f(\tau,y(\tau))\ominus_H f(\tau,x(\tau)))\delta_H\tau,\widetilde{0}\Big)\\
&\quad +\sum_{\{j:t_j<t\}} D\Big((I_j(t_j,y(t_j))\ominus_H I_j(t_j,x(t_j))),\widetilde{0}\Big)\\
&\quad +\sum_{\{j:t_j<t\}} D\Big((t-t_j)\cdot(L_j(t_j,y(t_j))\ominus_H L_j(t_j,x(t_j))),\widetilde{0}\Big)\\
&\le \int_0^t (t-\sigma(\tau))D(f(\tau,y(\tau)),f(\tau,x(\tau)))\Delta\tau\\
&\quad +\sum_{\{j:t_j<t\}} D(I_j(t_j,y(t_j)),I_j(t_j,x(t_j)))\\
&\quad +\sum_{\{j:t_j<t\}} (t-t_j)D(L_j(t_j,y(t_j)),L_j(t_j,x(t_j)))\\
&\le F\int_0^T (T-\sigma(\tau))D(y(\tau),x(\tau))\Delta\tau\\
&\quad +A\sum_{j=1}^m D(y(t_j),x(t_j))\\
&\quad +B\sum_{j=1}^m (T-t_j)D(y(t_j),x(t_j))
\end{aligned}
$$

$$\leq \left(FT^2 + Am + BmT\right) \|y - x\|, \quad t \in [0,T],$$

whereupon

$$\begin{aligned} \|\Phi y - \Phi x\| &\leq \left(FT^2 + Am + BmT\right) \|y - x\| \\ &< \|y - x\|. \end{aligned}$$

Consequently $\Phi : PC([0,T]) \to PC([0,T])$ is a contraction. Hence, the operator $\Phi : PC([0,T]) \to PC([0,T])$ has a unique fixed point in $PC([0,T])$. This completes the proof.

Now we suppose

(H4) there exist non-negative functions $g, h \in \mathscr{C}_{rd}([0,T])$ such that

$$\begin{aligned} \gamma_1 &= \max_{t \in [0,T]} \int_0^t (t - \sigma(\tau)) g(\tau) \Delta\tau < \infty, \\ \gamma_2 &= \max_{t \in [0,T]} \int_0^t (t - \sigma(\tau)) h(\tau) \Delta\tau < \infty, \\ D(f(t,x), f(t,y)) &\leq D(x,y) h(t), \quad t \in \mathbb{T}, \quad x, y \in \mathbb{R}, \\ D(f(t,y), \widetilde{0}) &\leq g(t) + D(y, \widetilde{0}) h(t), \quad t \in \mathbb{T}, \quad x, y \in F(\mathbb{R}). \end{aligned}$$

For $y \in PC([0,T])$, define the operators

$$\begin{aligned} \Phi_1 y(t) &= \int_0^t (t - \sigma(\tau)) \cdot f(\tau, y(\tau)) \delta_H \tau, \\ \Phi_2 y(t) &= y_0 + t \cdot y_1 + \sum_{\{j : t_j < t\}} I_j(t_j, y(t_j)) \\ &\quad + \sum_{\{j : t_j < t\}} (t - t_j) \cdot L_j(t_j, y(t_j)), \quad t \in [0,T]. \end{aligned}$$

Lemma 5.2.5 *Assume (H4). Then the operator* $\Phi_1 : PC([0,T]) \to PC([0,T])$ *is completely continuous.*

Proof 5.2.6 *Let* $y \in PC([0,T])$. *Then*

$$\begin{aligned} D(\Phi_1 y(t), \widetilde{0}) &= D\left(\int_0^t (t-\sigma(\tau))\cdot f(\tau, y(\tau))\delta_H \tau, \widetilde{0}\right) \\ &\leq \int_0^t (t-\sigma(\tau))D(f(\tau,y(\tau)),\widetilde{0})\Delta\tau \\ &\leq \int_0^t (t-\sigma(\tau))(g(\tau)+D(y(\tau),\widetilde{0})h(\tau)\Delta\tau \\ &= \int_0^t (t-\sigma(\tau))g(\tau)\Delta\tau \\ &\quad + \int_0^t (t-\sigma(\tau))h(\tau)D(y(\tau),\widetilde{0})\Delta\tau \\ &\leq \gamma_1 + \gamma_2\|y\|, \quad t \in [0,T], \end{aligned}$$

whereupon

$$\|\Phi_1 y\| \leq \gamma_1 + \gamma_2\|y\|.$$

Thus, $\Phi_1 : PC([0,T]) \to PC([0,T])$ *is bounded. Let* $\{y_n\}_{n\in\mathbb{N}} \subset PC([0,T])$ *and* $y_n \to y$, *as* $n \to \infty$, *in* $PC([0,T])$, *for some* $y \in PC([0,T])$. *Then*

$$D(\Phi_1 y_n(t), \Phi_1 y(t)) = D\Bigg(\int_0^t (t-\sigma(\tau))\cdot f(\tau, y_n(\tau))\delta_H \tau,$$

$$\int_0^t (t-\sigma(\tau))\cdot f(\tau,y(\tau))\delta_H\tau\Bigg)$$

$$
\begin{aligned}
&= D\left(\int_0^t (t-\sigma(\tau))\cdot(f(\tau,y_n(\tau))\ominus_H f(\tau,y(\tau)))\delta_H\tau,\widetilde{0}\right)\\
&\le \int_0^t (t-\sigma(\tau))D(f(\tau,y_n(\tau)),f(\tau,y(\tau)))\Delta\tau\\
&\le \int_0^t (t-\sigma(\tau))h(\tau)D(y_n(\tau),y(\tau))\Delta\tau\\
&\le \gamma_2\|y_n-y\|,\quad t\in[0,T],
\end{aligned}
$$

from where

$$\|\Phi_1 y_n-\Phi_1 y\|\le\gamma_2\|y_n-y\|.$$

Therefore $\Phi_1 : PC([0,T])\to PC([0,T])$ *is continuous. Let* $y\in PC([0,T])$ *and* $t^1,t^2\in[0,T]$ *be arbitrarily chosen. Without loss of generality, suppose that* $t^1<t^2$. *Then*

$$
\begin{aligned}
D(\Phi_1 y(t^2),\Phi_1 y(t^1)) &= D\left(\int_0^{t^2}(t^2-\sigma(\tau))\cdot f(\tau,y(\tau))\delta_H\tau,\right.\\
&\qquad\left.\int_0^{t^1}(t^1-\sigma(\tau))\cdot f(\tau,y(\tau))\delta_H\tau\right)
\end{aligned}
$$

$$
\begin{aligned}
&= D\left(\int_0^{t^1}(t^2-\sigma(\tau))\cdot f(\tau,y(\tau))\delta_H\tau+\int_{t^1}^{t^2}(t^2-\sigma(\tau))\cdot f(\tau,y(\tau))\delta_H\tau,\right.\\
&\qquad\left.\int_0^{t^1}(t^1-\sigma(\tau))\cdot f(\tau,y(\tau))\delta_H\tau\right)\\
&= D\left(\int_0^{t^1}(t^2-t^1)\cdot f(\tau,y(\tau))\delta_H\tau,\int_{t^1}^{t^2}(t^2-\sigma(\tau))\cdot f(\tau,y(\tau))\delta_H\tau\right)
\end{aligned}
$$

$$
\begin{aligned}
&\leq D\left(\int_0^{t^1}(t^2-t^1)\cdot f(\tau,y(\tau))\delta_H\tau,\widetilde{0}\right) \\
&\quad +D\left(\int_{t^1}^{t^2}(t^2-\sigma(\tau))\cdot f(\tau,y(\tau))\delta_H\tau,\widetilde{0}\right) \\
&\leq \int_0^{t^1}(t^2-t^1)D(f(\tau,y(\tau)),\widetilde{0})\Delta\tau \\
&\quad +\int_{t^1}^{t^2}(t^2-\sigma(\tau))D(f(\tau,y(\tau)),\widetilde{0})\Delta\tau \\
&\to 0, \quad as \quad t^1\to t^2.
\end{aligned}
$$

Hence, by the Arzela-Ascoli theorem, we conclude that the operator $\Phi_1 : PC([0,T]) \to PC([0,T])$ is completely continuous. This completes the proof.

Lemma 5.2.7 *Assume $(H2)$ - $(H4)$ and*

$$Am+BmT<1. \tag{5.4}$$

Then $\Phi_2 : PC([0,T]) \to PC([0,T])$ is a contraction.

Proof 5.2.8 *Let $x,y\in PC([0,T])$ be arbitrarily chosen. Then*

$$
\begin{aligned}
D(\Phi_2y(t),\Phi_2x(t)) \;=\; D\Bigg(& y_0+t\cdot y_1+\sum_{\{j:t_j<t\}} I_j(t_j,y(t_j)) \\
& +\sum_{\{j:t_j<t\}}(t-t_j)\cdot L_j(t_j,y(t_j)), \\
& y_0+t\cdot y_1+\sum_{\{j:t_j<t\}} I_j(t_j,x(t_j))
\end{aligned}
$$

$$
\begin{aligned}
&+ \sum_{\{j:t_j<t\}} (t-t_j)\cdot L_j(t_j,x(t_j))\Big) \\
&= D\Big(\sum_{\{j:t_j<t\}} (I_j(t_j,y(t_j))\ominus_H I_j(t_j,x(t_j))) \\
&+ \sum_{\{j:t_j<t\}} (t-t_j)\cdot (L_j(t_j,y(t_j))\ominus_H L_j(t_j,x(t_j))),\widetilde{0}\Big) \\
&\leq D\Big(\sum_{\{j:t_j<t\}} (I_j(t_j,y(t_j))\ominus_H I_j(t_j,x(t_j))),\widetilde{0}\Big) \\
&+D\Big(\sum_{\{j:t_j<t\}} (t-t_j)\cdot (L_j(t_j,y(t_j))\ominus_H L_j(t_j,x(t_j))),\widetilde{0}\Big) \\
&\leq \sum_{\{j:t_j<t\}} D\Big((I_j(t_j,y(t_j))\ominus_H I_j(t_j,x(t_j))),\widetilde{0}\Big) \\
&+ \sum_{\{j:t_j<t\}} D\Big((t-t_j)\cdot (L_j(t_j,y(t_j))\ominus_H L_j(t_j,x(t_j))),\widetilde{0}\Big) \\
&= \sum_{\{j:t_j<t\}} D(I_j(t_j,y(t_j)),I_j(t_j,x(t_j))) \\
&+ \sum_{\{j:t_j<t\}} (t-t_j)D(L_j(t_j,y(t_j)),L_j(t_j,x(t_j))) \\
&\leq \sum_{\{j:t_j<t\}} AD(y(t_j),x(t_j)) + \sum_{\{j:t_j<t\}} B(t-t_j)D(y(t_j),x(t_j)) \\
&\leq (Am+BmT)\|y-x\|, \quad t\in[0,T],
\end{aligned}
$$

from where

$$
\begin{aligned}
\|\Phi_2 y - \Phi_2 x\| &\leq (Am+BmT)\|y-x\| \\
&< \|y-x\|.
\end{aligned}
$$

This completes the proof.

Theorem 5.2.9 *Assume* (*H*2)–(*H*4) *and C is a positive constant such that*

$$\gamma_1 + |y_0| + |y_1|T + \sum_{j=1}^{m} \left(D(I_j(t_j,0),\widetilde{0}) + TD(L_j(t_j,0),\widetilde{0}) \right) \tag{5.5}$$
$$+(Am + BmT + \gamma_2)C \leq C.$$

Then the equation (5.3) *has a solution* $y \in PC([0,T])$ *such that*

$$\|y\| \leq C.$$

Proof 5.2.10 *Define*

$$M = \{y \in PC([0,T]) : \|y\| \leq C\}.$$

By Lemma 5.2.5, we have $\Phi_1 : PC([0,T]) \to PC([0,T])$ *is completely continuous. By Lemma 5.2.7, we get* $\Phi_2 : PC([0,T]) \to PC([0,T])$ *is a contraction. Let* $x, y \in M$. *Then*

$$\|x\| \leq C,$$

$$\|y\| \leq C,$$

and

$$\begin{aligned} D(\Phi_1 y(t) + \Phi_2 x(t),\widetilde{0}) &= D\Bigg(\int_0^t (t-\sigma(\tau)) \cdot f(\tau, y(\tau)) \delta_H \tau \\ &\quad + y_0 + t \cdot y_1 + \sum_{\{j:t_j<t\}} I_j(t_j, x(t_j)) + \sum_{\{j:t_j<t\}} (t-t_j) \cdot L_j(t_j, x(t_j)), \widetilde{0} \Bigg) \\ &\leq D\Bigg(\int_0^t (t-\sigma(\tau)) \cdot f(\tau, y(\tau)) \delta_H \tau, \widetilde{0} \Bigg) \\ &\quad + D(y_0,\widetilde{0}) + D(y_1,\widetilde{0})t + \sum_{\{j:t_j<t\}} D(I_j(t_j, x(t_j)),\widetilde{0}) \end{aligned}$$

$$
\begin{aligned}
& + \sum_{\{j:t_j<t\}} (t-t_j)D(L_j(t_j,x(t_j)),\widetilde{0}) \\
\leq\ & \int_0^t (t-\sigma(\tau))D(f(\tau,y(\tau)),\widetilde{0})\Delta\tau \\
& + D(y_0,\widetilde{0}) + D(y_1,\widetilde{0})t + \sum_{\{j:t_j<t\}} D(I_j(t_j,x(t_j)) \ominus_H I_j(t_j,0) + I_j(t_j,0),\widetilde{0}) \\
& + \sum_{\{j:t_j<t\}} (t-t_j)D(L_j(t_j,x(t_j)) \ominus_H L_j(t_j,0) + L_j(t_j,0),\widetilde{0}) \\
\leq\ & \int_0^t (t-\sigma(\tau))D(f(\tau,y(\tau)),\widetilde{0})\Delta\tau \\
& + D(y_0,\widetilde{0}) + D(y_1,\widetilde{0})t + \sum_{\{j:t_j<t\}} D(I_j(t_j,x(t_j)),I_j(t_j,0)) \\
& + \sum_{\{j:t_j<t\}} D(I_j(t_j,0),\widetilde{0}) \\
& + \sum_{\{j:t_j<t\}} (t-t_j)D(L_j(t_j,x(t_j)),L_j(t_j,0)) \\
& + \sum_{\{j:t_j<t\}} (t-t_j)D(L_j(t_j,0),\widetilde{0}) \\
\leq\ & \int_0^t (t-\sigma(\tau))(g(\tau)+h(\tau)D(y(\tau),\widetilde{0}))\Delta\tau \\
& + D(y_0,\widetilde{0}) + D(y_1,\widetilde{0})t + \sum_{\{j:t_j<t\}} D\Big(I_j(t_j,x(t_j)),I_j(t_j,0)\Big) \\
& + \sum_{\{j:t_j<t\}} D\Big(I_j(t_j,0)),\widetilde{0}\Big) \\
& + \sum_{\{j:t_j<t\}} (t-t_j)D\Big(L_j(t_j,x(t_j)),L_j(t_j,0)\Big) \\
& + \sum_{\{j:t_j<t\}} (t-t_j)D\Big(L_j(t_j,0),\widetilde{0}\Big) \\
\leq\ & \gamma_1 + \gamma_2\|y\| + D(y_0,\widetilde{0}) + D(y_1,\widetilde{0})T \\
& + A\sum_{\{j:t_j<t\}} D\Big(x(t_j),\widetilde{0}\Big) + \sum_{\{j:t_j<t\}} D\Big(I_j(t_j,0),\widetilde{0}\Big)
\end{aligned}
$$

$$
\begin{aligned}
&+B\sum_{\{j:t_j<t\}}(t-t_j)D\Big(x(t_j),\widetilde{0}\Big)+\sum_{\{j:t_j<t\}}(t-t_j)D\Big(L_j(t_j,0),\widetilde{0}\Big)\\
\leq\ & \gamma_1+D(y_0,\widetilde{0})+D(y_1,\widetilde{0})T\\
&+\sum_{j=1}^{m}\Big(D(I_j(t_j,0),\widetilde{0})+TD(L_j(t_j,0)|,\widetilde{0})\Big)\\
&+Am\|x\|+BmT\|x\|+\gamma_2\|y\|\\
\leq\ & \gamma_1+D(y_0,\widetilde{0})+D(y_1,\widetilde{0})T\\
&+\sum_{j=1}^{m}\Big(D(I_j(t_j,0),\widetilde{0})+TD(L_j(t_j,0),\widetilde{0})\Big)\\
&+(Am+BmT+\gamma_2)C\\
\leq\ & C,\quad t\in[0,T].
\end{aligned}
$$

Hence,

$$\|\Phi_1 y+\Phi_2 x\|\leq C,$$

i.e., $\Phi_1 y+\Phi_2 x\in M$*. From here and from the Kransole'skii fixed-point theorem, it follows that the operator* $\Phi_1+\Phi_2$ *has a fixed in* M*. This completes the proof.*

Theorem 5.2.11 *Assume* $(H1)$–$(H4)$, (5.4), (5.5) *and*

$$FT^2+Am+BmT<1.$$

Then the equation (5.3) *has a unique solution* $y\in PC([0,T])$ *such that*

$$\|y\|\leq C.$$

Proof 5.2.12 *By Theorem 5.2.9, it follows that the problem* (5.3) *has a solution* $y\in M$*, where the set* M *is defined in the proof of Theorem 5.2.9. Suppose that the operator* $\Phi=\Phi_1+\Phi_2$ *has two fixed points* $x,y\in M$*. Then*

$$D(\Phi y(t),\Phi x(t)) = D\Big(y_0+y_1 t+\int_0^t(t-\sigma(\tau))\cdot f(\tau,y(\tau))\delta_H\tau$$

$$
\begin{aligned}
& + \sum_{\{j:t_j<t\}} I_j(t_j, y(t_j)) \\
& + \sum_{\{j:t_j<t\}} (t-t_j) \cdot L_j(t_j, y(t_j)), \\
& y_0 + t \cdot y_1 + \int_0^t (t-\sigma(\tau)) \cdot f(\tau, x(\tau)) \delta_H \tau \\
& + \sum_{\{j:t_j<t\}} I_j(t_j, x(t_j)) \\
& + \sum_{\{j:t_j<t\}} (t-t_j) \cdot L_j(t_j, x(t_j)) \Bigg) \\
= {} & D\Bigg(\int_0^t (t-\sigma(\tau)) \cdot f(\tau, y(\tau)) \delta_H \tau \\
& + \sum_{\{j:t_j<t\}} I_j(t_j, y(t_j)) \\
& + \sum_{\{j:t_j<t\}} (t-t_j) \cdot L_j(t_j, y(t_j)), \\
& + \int_0^t (t-\sigma(\tau)) \cdot f(\tau, x(\tau)) \delta_H \tau \\
& + \sum_{\{j:t_j<t\}} I_j(t_j, x(t_j)) \\
& + \sum_{\{j:t_j<t\}} (t-t_j) \cdot L_j(t_j, x(t_j)) \Bigg) \\
\leq {} & D\Bigg(\int_0^t (t-\sigma(\tau)) \cdot f(\tau, y(\tau)) \delta_H \tau, \int_0^t (t-\sigma(\tau)) \cdot f(\tau, x(\tau)) \delta_H \tau \Bigg)
\end{aligned}
$$

$$
\begin{aligned}
&+D\Bigg(\sum_{\{j:t_j<t\}} I_j(t_j,y(t_j)), \sum_{\{j:t_j<t\}} I_j(t_j,x(t_j))\Bigg) \\
&+D\Bigg(\sum_{\{j:t_j<t\}} (t-t_j)\cdot L_j(t_j,y(t_j)), \sum_{\{j:t_j<t\}} (t-t_j)\cdot L_j(t_j,x(t_j))\Bigg) \\
\leq\ & \int_0^t (t-\sigma(\tau))D(f(\tau,y(\tau)),f(\tau,x(\tau)))\Delta\tau \\
&+ \sum_{\{j:t_j<t\}} D(I_j(t_j,y(t_j)),I_j(t_j,x(t_j))) \\
&+ \sum_{\{j:t_j<t\}} (t-t_j)D(L_j(t_j,y(t_j)),L_j(t_j,x(t_j))) \\
\leq\ & F\int_0^T (T-\sigma(\tau))D(y(\tau),x(\tau))\Delta\tau \\
&+A\sum_{j=1}^{m} D(y(t_j),x(t_j)) \\
&+B\sum_{j=1}^{m} (T-t_j)D(y(t_j),x(t_j)) \\
\leq\ & \left(FT^2+Am+BmT\right)\|y-x\|, \quad t\in[0,T],
\end{aligned}
$$

whereupon

$$
\begin{aligned}
\|y-x\| &= \|\Phi y-\Phi x\| \\
&\leq \left(FT^2+Am+BmT\right)\|y-x\| \\
&< \|y-x\|.
\end{aligned}
$$

This is a contradiction. This completes the proof.

Exercise 5.2.13 *Find conditions for* f_1, f_2, I_j, L_j, $j \in \{1,\dots,m\}$, *so that the impulsive dynamic equation*

$$\begin{aligned}
\delta_H^2 y(t) &= f_1(t) + f_2(y(t)), \quad t \in [0,T]\backslash\{t_j\}_{j=1}^m, \\
y(0) &= y_0, \\
\delta_H y(0) &= y_1, \\
y(t_j^+) &= y(t_j^-) + I_j(t_j, y(t_j)), \\
\delta_H y(t_j^+) &= \delta_H y(t_j^-) + L_j(t_j, y(t_j)), \quad j \in \{1,\dots,m\},
\end{aligned}$$

has a unique solution in $PC([0,T])$.

5.3 EXTREMAL SOLUTIONS OF NON-LINEAR SECOND-ORDER FUZZY IMPULSIVE DYNAMIC EQUATIONS

In this section we investigate the following fuzzy impulsive dynamic equation

$$\begin{aligned}
\ominus_H \delta_H^2 y(t) &= f(t, y(t)), \quad t \in [0,T]\backslash\{t_j\}_{j=1}^m, \\
y(t_j^+) \ominus_H y(t_j^-) &= I_j(y(t_j)), \\
\delta_H y(t_j^+) \ominus_H \delta_H y(t_j^-) &= L_j(y(t_j)), \quad j \in \{1,\dots,m\}, \qquad (5.6) \\
y(0) &= y_0, \\
\delta_H y(0) &= y_1,
\end{aligned}$$

where $y_0, y_1 \in F(\mathbb{R})$, $f : [0,T] \times F(\mathbb{R}) \to F(\mathbb{R})$, $I_j, L_j : F(\mathbb{R}) \to F(\mathbb{R})$, $j \in \{1,\dots,m\}$, are given functions. With $AC^j([0,T])$ we denote the space of j-time differentiable functions $y : [0,T] \to F(\mathbb{R})$ whose jth delta derivative, $\delta_H^j y$, is absolutely continuous.

Definition 5.3.1 *A function $y \in PC([0,T]) \bigcap AC^1([0,T] \backslash \{t_j\}_{j=1}^m)$ is said to be a solution of* (5.6) *if it satisfies the fuzzy dynamic equation*

$$\ominus_H \delta_H^2 y(t) = f(t, y(t)) \quad on \quad [0,T] \backslash \{t_j\}_{j=1}^m,$$

and for each $j \in \{1, \dots, m\}$ the function y satisfies the second, third, fourth, and fifth conditions of (5.6).

Definition 5.3.2 *A function $\alpha \in PC([0,T])$ is said to be a lower solution of* (5.6) *if*

$$\begin{aligned} \ominus_H \delta_H^2 \alpha(t) &\leq f(t, \alpha(t)) \quad on \quad J \backslash \{t_j\}_{j=1}^m, \\ \alpha(t_j^+) \ominus_H \alpha(t_j^-) &= I_j(\alpha(t_j)), \\ \delta_H \alpha(t_j^+) \ominus_H \delta_H \alpha(t_j^-) &\geq L_j(\alpha(t_j^-)), \quad j \in \{1, \dots, m\}, \\ \alpha(0) &\geq y_0, \\ \delta_H \alpha(0) &\geq y_1. \end{aligned}$$

Definition 5.3.3 *A function $\beta \in PC([0,T])$ is said to be an upper solution of* (5.6) *if*

$$\begin{aligned} \ominus_H \delta_H^2 \beta(t) &\geq f(t, \beta(t)) \quad on \quad J \backslash \{t_j\}_{j=1}^m, \\ \beta(t_j^+) \ominus_H \beta(t_j^-) &= I_j(\beta(t_j)), \\ \delta_H \beta(t_j^+) \ominus_H \delta_H \beta(t_j^-) &\leq L_j(\beta(t_j^-)), \quad j \in \{1, \dots, m\}, \\ \beta(0) &\leq y_0, \\ \delta_H \beta(0) &\leq y_1. \end{aligned}$$

By Section 5.2, it follows that y is a solution of the equation (5.6) if and only if $y \in PC([0,T])$ satisfies the integral equation

$$\begin{aligned} y(t) &= y_0 + t \cdot y_1 \ominus_H \int_0^t (t-\sigma(s)) \cdot f(s, y(s)) \delta_H s \\ &+ \sum_{\{j: t_j < t\}} \left(I_j(y(t_j)) + (t-t_j) \cdot L_j(y(t_j))\right), \quad t \in [0,T]. \end{aligned}$$

In our next result, we will use the following fixed-point theorem of Heikkilä and Lakshmikantham.

Theorem 5.3.4 (fixed-point Theorem of Heikkilä and Lakshmikantham) Let $[a,b]$ be an ordered interval in a subset Y of an ordered Banach space X and let $Q : [a,b] \to [a,b]$ be a non-decreasing mapping. If each sequence $\{Qx_n\}_{n\in\mathbb{N}} \subset Q([a,b])$ converges, whenever $\{x_n\}_{n\in\mathbb{N}}$ is a monotone sequence in $[a,b]$, then the sequence of Q-iteration of a converges to the least fixed point x_* of Q and the sequence Q-iteration of b converges to the greatest fixed point x^* of Q. Moreover,

$$\begin{aligned} x_* &= \min\{y \in [a,b] : y \geq Qy\}, \\ x^* &= \max\{y \in [a,b] : y \leq Qy\}. \end{aligned}$$

Definition 5.3.5 *A function $f : [0,T] \times F(\mathbb{R})$ is said to be $\mathbb{L}^1$-Chandrabhan if*

1. *$t \to f(t,x)$ is measurable for each $x \in F(\mathbb{R})$,*
2. *$x \to f(t,x)$ is non-increasing in x for almost every $t \in [0,T]$,*
3. *for each $q > 0$, there exists a non-negative function $h_q \in \mathbb{L}^1([0,T])$ such that*

$$D(f(t,x), \widetilde{0}) \leq h_q(t),$$

for all $D(x, \widetilde{0}) \leq q$ and for almost each $t \in [0,T]$.

Theorem 5.3.6 *Let*

1. *the function f is $\mathbb{L}^1$-Chandrabhan,*
2. *the functions I_j, L_j, $j \in \{1, \ldots, m\}$, are continuous and non-decreasing,*
3. *there exist α, $\beta \in PC([0,T])$, lower and upper solutions, respectively, for the equation* (5.6) *such that $\beta \leq \alpha$.*

Then the equation (5.6) *has a minimal and a maximal solution.*

Proof 5.3.7 *For $y \in PC([0,T])$, define the operator*

$$\Phi y(t) = y_0 + t \cdot y_1 \ominus_H \int_0^t (t - \sigma(s)) \cdot f(s, y(s)) \delta_H s$$

$$+ \sum_{\{j: t_j < t\}} \left(I_j(y(t_j)) \ominus_H (t - t_j) \cdot L_j(y(t_j)) \right), \quad t \in [0,T].$$

Let $y \in [\beta, \alpha]$ and $t \in [0,T]$, and

$$t_i = \max\{t_j : t_j < t, \quad j \in \{1, \ldots, m\}\}.$$

Note that

$$\delta_H(s \cdot \delta_H \beta(s)) = \delta_H \beta(s) + \sigma(s) \cdot \delta_H^2 \beta(s),$$

$$\delta_H(s \cdot \delta_H \alpha(s)) = \delta_H \alpha(s) + \sigma(s) \cdot \delta_H^2 \alpha(s),$$

whereupon

$$\sigma(s) \cdot \delta_H^2 \beta(s) = \delta_H(s \cdot \delta_H \beta(s)) \ominus_H \delta_H \beta(s),$$

$$\sigma(s) \cdot \delta_H^2 \alpha(s) = \delta_H(s \cdot \delta_H \alpha(s)) \ominus_H \delta_H \alpha(s), \quad s \in [0,T].$$

By the definition of the upper solution, we obtain

$$\Phi y(t) \geq \Phi \beta(t)$$

$$
\begin{aligned}
&\geq \beta(0)+t\cdot\delta_H\beta(t)\ominus_H\int_0^t(t-\sigma(s))\cdot f(s,\beta(s))\delta_H s\\
&\quad+\sum_{\{j:0<t_j<t\}}(I_j(\beta(t_j))+(t-t_j)\cdot L_j(\beta(t_j)))\\
&\geq \beta(0)+t\cdot\delta_H\beta(0)\ominus_H\int_0^{t_1}(t-\sigma(s))\cdot f(s,\beta(s))\delta_H s\\
&\quad\ominus_H\int_{t_1}^{t_2}(t-\sigma(s))\cdot f(s,\beta(s))\delta_H s\\
&\quad\ominus_H\cdots\\
&\quad\ominus_H\int_{t_i}^{t}(t-\sigma(s))\cdot f(s,\beta(s))\delta_H s\\
&\quad+\sum_{k=1}^{i}(I_k(\beta(t_k))+(t-t_k)\cdot L_k(\beta(t_k)))\\
&\geq \beta(0)+t\cdot\delta_H\beta(0)+\int_0^{t_1}t\cdot\delta_H^2\beta(s)\delta_H s\\
&\quad+\int_0^{t_1}(\delta_H\beta(s)\ominus_H\delta_H(s\cdot\delta_H\beta(s)))\delta_H s\\
&\quad+\int_{t_1}^{t_2}t\cdot\delta_H^2\beta(s)\delta_H s+\int_{t_1}^{t_2}(\delta_H\beta(s)\ominus_H\delta_H(s\cdot\delta_H\beta(s)))\delta_H s\\
&\quad+\cdots\\
&\quad+\int_{t_i}^{t}t\cdot\delta_H^2\beta(s)\delta_H s+\int_{t_1}^{t}(\delta_H\beta(s)\ominus_H\delta_H(s\cdot\delta_H\beta(s)))\delta_H s\\
&= \beta(0)+t\cdot\delta_H\beta(0)+t\cdot\delta_H\beta(t_1^-)\ominus_H t\cdot\delta_H\beta(0)\\
&\quad+\beta(t_1^-)\ominus_H\beta(0)\ominus_H t_1\cdot\delta_H\beta(t_1^-)
\end{aligned}
$$

$$
\begin{aligned}
& +t\cdot\delta_H\beta(t_2^-)\ominus_H t\cdot\delta_H\beta(t_1^+)+\beta(t_2^-)\ominus_H\beta(t_1^+)\\
& \ominus_H t_2\cdot\delta_H\beta(t_2^-)+t_1\cdot\delta_H\beta(t_1^+)\\
& +\cdots\\
& +t\cdot\delta_H\beta(t)\ominus_H t\delta_H\beta(t_i^+)\\
& +\beta(t)\ominus_H\beta(t_i^+)\ominus_H t\cdot\delta_H\beta(t)+t_i\cdot\delta_H\beta(t_i^+)\\
& +\sum_{k=1}^{i}(I_k(\beta(t_k))+(t-t_k)\cdot L_k(\beta(t_k)))\\
= & (t-t_1)\cdot\delta_H\beta(t_1^-)+\beta(t_1^-)\ominus_H\beta(t_1^+)\\
& +(t-t_2)\cdot\delta_H\beta(t_2^-)\ominus_H(t-t_1)\cdot\delta_H\beta(t_1^+)+\beta(t_2^-)\\
& +\cdots\\
& +\beta(t)\ominus_H(t-t_i)\cdot(\delta_H\beta(t_i^+)\ominus_H\delta_H\beta(t_i^-))+\beta(t_i^-)\ominus_H\beta(t_i^+)\\
& +\sum_{k=1}^{i}(I_k(\beta(t_k))+(t-t_k)\cdot L_k(\beta(t_k)))\\
= & \ominus_H(t-t_1)\cdot(\delta_H\beta(t_1^+)\ominus_H\delta_H\beta(t_1^-))\ominus_H(\beta(t_1^+)\ominus_H\beta(t_1^-))\\
& \ominus_H(t-t_2)\cdot(\delta_H\beta(t_2^+)\ominus_H\delta_H\beta(t_2^-))\ominus_H(\beta(t_2^+)\ominus_H\beta(t_2^-))\\
& \ominus_H\cdots\\
& \ominus_H(t-t_i)\cdot(\delta_H\beta(t_i^+)\ominus_H\delta_H\beta(t_i^-))\ominus_H(\beta(t_i^+)\ominus_H\beta(t_i^-))\\
& +\sum_{k=1}^{i}(I_k(\beta(t_k))+(t-t_k)\cdot L_k(\beta(t_k)))\\
\geq & \ominus_H(t-t_1)\cdot L_1(\beta(t_1^-))\ominus_H I_1(\beta(t_1^-))\\
& \ominus_H(t-t_2)L_2(\beta(t_2^-))\ominus_H I_2(\beta(t_2^-))
\end{aligned}
$$

$$
\begin{aligned}
&\ominus_H \cdots \\
&+\beta(t) \ominus_H (t-t_i)\cdot L_i(\beta(t_i^-)) \ominus_H I_i(\beta(t_i^-)) \\
&+\sum_{k=1}^{i}(I_k(\beta(t_k)) + (t-t_k)\cdot L_k(\beta(t_k))) \\
= \;& \beta(t).
\end{aligned}
$$

By the definition of the lower solution, we get

$$
\begin{aligned}
\Phi y(t) \;\leq\;& \Phi\alpha(t) \\
\leq\;& \alpha(0) + t\cdot\delta_H\alpha(t) \ominus_H \int_0^t (t-\sigma(s))\cdot f(s,\alpha(s))\delta_H s \\
&+ \sum_{\{j:0<t_j<t\}} (I_j(\alpha(t_j)) + (t-t_j)\cdot L_j(\alpha(t_j))) \\
\leq\;& \alpha(0) + t\cdot\delta_H\alpha(0) \ominus_H \int_0^{t_1} (t-\sigma(s))\cdot f(s,\alpha(s))\delta_H s \\
&\ominus_H \int_{t_1}^{t_2} (t-\sigma(s))\cdot f(s,\alpha(s))\delta_H s \\
&\ominus_H \cdots \\
&\ominus_H \int_{t_i}^{t} (t-\sigma(s))\cdot f(s,\alpha(s))\delta_H s \\
&+\sum_{k=1}^{i}(I_k(\alpha(t_k)) + (t-t_k)\cdot L_k(\alpha(t_k))) \\
\leq\;& \alpha(0) + t\cdot\delta_H\alpha(0) + \int_0^{t_1} t\cdot\delta_H^2\alpha(s)\delta_H s
\end{aligned}
$$

$$
\begin{aligned}
& +\int_{0}^{t_1}(\delta_H\alpha(s)\ominus_H \delta_H(s\cdot\delta_H\alpha(s)))\delta_H s \\
& +\int_{t_1}^{t_2} t\cdot\delta_H^2\alpha(s)\delta_H s+\int_{t_1}^{t_2}(\delta_H\alpha(s)\ominus_H \delta_H(s\cdot\delta_H\alpha(s)))\delta_H s \\
& +\cdots \\
& +\int_{t_i}^{t} t\cdot\delta_H^2\alpha(s)\delta_H s+\int_{t_1}^{t}(\delta_H\alpha(s)\ominus_H \delta_H(s\cdot\delta_H\alpha(s)))\delta_H s \\
= \; & \alpha(0)+t\cdot\delta_H\alpha(0)+t\cdot\delta_H\alpha(t_1^-)\ominus_H t\cdot\delta_H\alpha(0) \\
& +\alpha(t_1^-)\ominus_H \alpha(0)\ominus_H t_1\cdot\delta_H\alpha(t_1^-) \\
& +t\cdot\delta_H\alpha(t_2^-)\ominus_H t\cdot\delta_H\alpha(t_1^+) \\
& +\alpha(t_2^-)\ominus_H \alpha(t_1^+)\ominus_H t_2\cdot\delta_H\alpha(t_2^-)+t_1\cdot\delta_H\alpha(t_1^+) \\
& +\cdots \\
& +t\cdot\delta_H\alpha(t)\ominus_H t\cdot\delta_H\alpha(t_i^+) \\
& +\alpha(t)\ominus_H \alpha(t_i^+)\ominus_H t\cdot\delta_H\alpha(t)+t_i\cdot\delta_H\alpha(t_i^+) \\
& +\sum_{k=1}^{i}(I_k(\alpha(t_k))+(t-t_k)\cdot L_k(\alpha(t_k))) \\
= \; & (t-t_1)\cdot\delta_H\alpha(t_1^-)+\alpha(t_1^-)\ominus_H \alpha(t_1^+) \\
& +(t-t_2)\cdot\delta_H\alpha(t_2^-)\ominus_H (t-t_1)\cdot\delta_H\alpha(t_1^+)+\alpha(t_2^-) \\
& +\cdots \\
& +\alpha(t)\ominus_H (t-t_i)\cdot(\delta_H\alpha(t_i^+)\ominus_H \delta_H\alpha(t_i^-)) \\
& +\alpha(t_i^-)\ominus_H \alpha(t_i^+) \\
& +\sum_{k=1}^{i}(I_k(\alpha(t_k))+(t-t_k)\cdot L_k(\alpha(t_k)))
\end{aligned}
$$

$$\begin{aligned}
&= \ominus_H(t-t_1)\cdot(\delta_H\alpha(t_1^+)\ominus_H\delta_H\alpha(t_1^-))\ominus_H(\alpha(t_1^+)\ominus_H\alpha(t_1^-)) \\
&\quad \ominus_H(t-t_2)\cdot(\delta_H\alpha(t_2^+)\ominus_H\delta_H\alpha(t_2^-))\ominus_H(\alpha(t_2^+)\ominus_H\alpha(t_2^-)) \\
&\quad \ominus_H\cdots \\
&\quad \ominus_H(t-t_i)\cdot(\delta_H\alpha(t_i^+)\ominus_H\delta_H\alpha(t_i^-))\ominus_H(\alpha(t_i^+)\ominus_H\alpha(t_i^-)) \\
&\quad +\sum_{k=1}^{i}(I_k(\alpha(t_k))+(t-t_k)\cdot L_k(\alpha(t_k))) \\
&\leq \ominus_H(t-t_1)\cdot L_1(\alpha(t_1^-))\ominus_H I_1(\alpha(t_1^-)) \\
&\quad \ominus_H(t-t_2)\cdot L_2(\alpha(t_2^-))\ominus_H I_2(\alpha(t_2^-)) \\
&\quad \ominus_H\cdots \\
&\quad +\alpha(t)\ominus_H(t-t_i)\cdot L_i(\alpha(t_i^-))\ominus_H I_i(\alpha(t_i^-)) \\
&\quad +\sum_{k=1}^{i}(I_k(\alpha(t_k))+(t-t_k)\cdot L_k(\alpha(t_k))) \\
&= \alpha(t).
\end{aligned}$$

Thus,

$$\beta(t)\leq\Phi y(t)\leq\alpha(t),\quad t\in[0,T].$$

Hence,

$$\beta\leq\Phi y\leq\alpha,$$

and

$$\|\Phi y\|\leq\max\{\|\alpha\|,\|\beta\|\}.$$

Let now, $\{y_n\}_{n\in\mathbb{N}}$ *be a monotone sequence in* $[\beta,\alpha]$*. We have that*

$\Phi y_n \in [\beta, \alpha]$ *and*

$$\begin{aligned}
\|\Phi y_n\| &\leq \max\{\sup_{t\in[0,T]} D(\alpha(t),\widetilde{0}), \sup_{t\in[0,T]} D(\beta(t),\widetilde{0})\} \\
&= q < \infty.
\end{aligned}$$

Let $y \in PC([0,T])$ *and* $t^1, t^2 \in [0,T]$. *Without loss of generality, we assume that* $t^1 < t^2$. *Then*

$$\begin{aligned}
\Phi y(t^2) \ominus_H \Phi y(t^1) &= y_0 + t^2 \cdot y_1 \ominus_H \int_0^{t^2} (t^2 - \sigma(s)) \cdot f(s, y(s))\delta_H s \\
&+ \sum_{\{j:0<t_j<t^2\}} (I_j(y(t_j^-)) + (t^2 - t_j)L_j(y(t_j^-))) \\
&\ominus_H y_0 \ominus_H t^1 \cdot y_1 + \int_0^{t^1} (t^1 - \sigma(s)) \cdot f(s, Dy(s))\delta_H s \\
&\ominus_H \sum_{\{j:0<t_j<t^1\}} (I_j(y(t_j^-)) + (t^1 - t_j)L_j(y(t_j^-))) \\
&= (t^2 - t^1) \cdot y_1 \ominus_H \int_0^{t^1} (t^2 - t^1) \cdot f(s, y(s))\delta_H s \\
&\ominus_H \int_{t^1}^{t^2} (t^2 - \sigma(s)) \cdot f(s, y(s))\delta_H s \\
&+ \sum_{\{j:0<t_j<t^1\}} (t^2 - t^1) \cdot L_j(y(t_j)) \\
&+ \sum_{\{j:t^1\leq t_j<t^2\}} (I_j(y(t_j)) + (t^2 - t_j) \cdot L_j(y(t_j^-))),
\end{aligned}$$

and

$$D(\Phi y(t^2), \Phi y(t^1)) = D\Big((t^2 - t^1) \cdot y_1 \ominus_H \int_0^{t^1} (t^2 - t^1) \cdot f(s, y(s))\delta_H s$$

$$
\begin{aligned}
&\ominus_H \int_{t^1}^{t^2} (t^2 - \sigma(s)) \cdot f(s, y(s)) \delta_H s \\
&+ \sum_{\{j:0<t_j<t^1\}} (t^2 - t^1) \cdot L_j(y(t_j)) \\
&+ \sum_{\{j:t^1 \le t_j < t^2\}} (I_j(y(t_j)) + (t^2 - t_j) \cdot L_j(y(t_j^-))), \widetilde{0} \Bigg) \\
\le\ & D\Big((t^2 - t^1) \cdot y_1, \widetilde{0} \Big) \\
&+ D\Big(\int_0^{t^1} (t^2 - t^1) \cdot f(s, y(s)) \delta_H s, \widetilde{0} \Big) \\
&+ D\Big(\int_{t^1}^{t^2} (t^2 - \sigma(s)) \cdot f(s, y(s)) \delta_H s, \widetilde{0} \Big) \\
&+ D\Big(\sum_{\{j:0<t_j<t^1\}} (t^2 - t^1) \cdot L_j(y(t_j)), \widetilde{0} \Big) \\
&+ D\Big(\sum_{\{j:t^1 \le t_j < t^2\}} (I_j(y(t_j)) + (t^2 - t_j) \cdot L_j(y(t_j^-))), \widetilde{0} \Big) \\
\le\ & D\Big((t^2 - t^1) \cdot y_1, \widetilde{0} \Big) \\
&+ D\Big(\int_0^{t^1} (t^2 - t^1) \cdot f(s, y(s)) \delta_H s, \widetilde{0} \Big) \\
&+ D\Big(\int_{t^1}^{t^2} (t^2 - \sigma(s)) \cdot f(s, y(s)) \delta_H s, \widetilde{0} \Big)
\end{aligned}
$$

$$+ \sum_{\{j:0<t_j<t^1\}} D\Big((t^2-t^1)\cdot L_j(y(t_j)),\widetilde{0}\Big)$$

$$+ \sum_{\{j:t^1\leq t_j<t^2\}} D\Big((I_j(y(t_j))+(t^2-t_j)\cdot L_j(y(t_j^-))),\widetilde{0}\Big)$$

$$\leq (t^2-t^1)D(y_1,\widetilde{0})+\int_0^{t^1}(t^2-t^1)\cdot D(f(s,y(s)),\widetilde{0})\Delta s$$

$$+\int_{t^1}^{t^2}(t^2-\sigma(s))D(f(s,y(s)),\widetilde{0})\Delta s$$

$$+ \sum_{\{j:0<t_j<t^1\}} (t^2-t^1)D(L_j(y(t_j)),\widetilde{0})$$

$$+ \sum_{\{j:t^1\leq t_j<t^2\}} D(I_j(y(t_j))+(t^2-t_j)\cdot L_j(y(t_j^-)),\widetilde{0})$$

$$\to 0, \quad as \quad t^1\to t^2,$$

whenever $D(y,\widetilde{0})\leq q$. *Next, if* $x\leq y$, *then* $\Phi x\leq \Phi y$. *Note that* $\{\Phi y_n\}_{n\in\mathbb{N}}$ *is an equicontinuous family. Hence, using the Arzela-Ascoli theorem, it follows that* $\{\Phi y_n\}_{n\in\mathbb{N}}$ *converges in* $\Phi([\beta,\alpha])$. *From here and from Theorem 5.3.4, we deduce that* Φ *has a least and a greatest fixed point in* $[\beta,\alpha]$. *This completes the proof.*

Exercise 5.3.8 *Find conditions for the function g so that the impulsive dynamic equation*

$$\begin{aligned}
\ominus_H \delta_H^2 y(t) &= \frac{t^2}{1+t^4+t^8}\cdot g(y(t)), \quad t\in[0,T]\backslash\{t_j\}_{j=1}^m,\\
y(t_j^+)\ominus_H y(t_j^-) &= 3\cdot y(t_j^-),\\
\delta_H y(t_j^+)\ominus_H \delta_H y(t_j^-) &= \ominus_H y(t_j^-), \quad j\in\{1,\dots,m\},\\
y(0) &= (1,0,1),\\
\delta_H y(0) &= (-1,0,1),
\end{aligned}$$

has a maximal and a minimal solution.

5.4 ADVANCED PRACTICAL PROBLEMS

Problem 5.4.1 *Let* $\mathbb{T} = [0,1]\bigcup[2,3]\bigcup[4,5]$. *Find a solution of the impulsive dynamic equation*

$$\begin{aligned}
\delta_H^2 y(t) \ominus_H 7 \cdot \delta_H y(t) + 10 \cdot y(t) &= (t+1,t,1) \quad t \in \mathbb{T}\backslash\{2,4\}, \\
y(0) &= (2,0,1), \\
\delta_H y(0) &= (-1,1,2), \\
y(2^+) &= 3 \cdot y(2), \\
\delta_H y(2^+) \ominus_H \delta_H y(2) &= (1,0,1), \\
y(4^+) \ominus_H y(4) &= (1,0,-1), \\
\delta_H y(4^+) \ominus_H \delta_H y(4) &= (-2,0,2).
\end{aligned}$$

Problem 5.4.2 *Let* $\mathbb{T} = [0,1]\bigcup[2,3]\bigcup[4,5]$. *Find a solution of the impulsive dynamic equation*

$$\begin{aligned}
\delta_H^2 y(t) \ominus_H 6 \cdot \delta_H y(t) + 9 \cdot y(t) &= (t^2+t+1,1,1+t), \; t \in \mathbb{T}\backslash\{2,4\}, \\
y(0) &= (1,0,-1), \\
\delta_H y(0) &= (1,2,-1), \\
y(2^+) &= \ominus_H y(2), \\
\delta_H y(2^+) + \delta_H y(2) &= (-1,0,1), \\
y(4^+) + y(4) &= (-2,1,2), \\
\delta_H y(4^+) + \delta_H y(4) &= (2,0,1).
\end{aligned}$$

Problem 5.4.3 *Find conditions for* f_1, f_2, I_j, L_j, $j \in \{1,\dots,m\}$, *so that the impulsive dynamic equation*

$$\begin{aligned}
\delta_H^2 y(t) &= f_1(t) \cdot f_2(y(t)), \quad t \in [0,T]\backslash\{t_j\}_{j=1}^m, \\
y(0) &= y_0, \\
\delta_H y(0) &= y_1, \\
y(t_j^+) &= y(t_j^-) + I_j(t_j, y(t_j)), \\
\delta_H y(t_j^+) &= \delta_H y(t_j^-) + L_j(t_j, y(t_j)), \quad j \in \{1,\dots,m\},
\end{aligned}$$

has a unique solution in $PC([0,T])$.

Problem 5.4.4 *Find conditions for the functions f and g so that the impulsive dynamic equation*

$$\begin{aligned}
-\delta_H^2 y(t) &= \left(\frac{t^4}{1+4t^8+9t^{10}} f(t)\right) \cdot g(y(t)), \\
& \quad t \in [0,T] \backslash \{t_j\}_{j=1}^m, \\
y(t_j^+) \ominus_H y(t_j^-) &= (-2) \cdot y(t_j^-), \\
\delta_H y(t_j^+) \ominus_H \delta_H y(t_j^-) &= 3 \cdot y(t_j^-), \quad j \in \{1,\ldots,m\}, \\
y(0) &= (-1,1,-2), \\
\delta_H y(0) &= (0,1,0),
\end{aligned}$$

has a maximal and a minimal solution.

Problem 5.4.5 *Find conditions for the functions f, g, and h so that the impulsive dynamic equation*

$$\begin{aligned}
\ominus_H \delta_H^2 y(t) &= \left(\frac{1}{1+t^9+2t^{12}} f(t)\right) \cdot g(y(t)) + h(y(t)), \\
& \quad t \in [0,T] \backslash \{t_j\}_{j=1}^m, \\
y(t_j^+) \ominus_H y(t_j^-) &= 4 \cdot y(t_j^-), \\
\delta_H y(t_j^+) \ominus_H \delta_H y(t_j^-) &= 7 \cdot y(t_j^-), \quad j \in \{1,\ldots,m\}, \\
y(0) &= (1,1,-1), \\
\delta_H y(0) &= (1,1,2),
\end{aligned}$$

has a maximal and a minimal solution.

CHAPTER 6

Boundary Value Problems for Second-Order Fuzzy Impulsive Dynamic Equations

Let $\mathbb{T}$ be a time scale with a forward jump operator and delta differentiation operator σ and Δ, respectively. Let also, $0, T \in \mathbb{T}$, $\{t_j\}_{j=1}^{m} \subset \mathbb{T}$, t_j are right-dense in $\mathbb{T}$, $j \in \{1, \ldots, m\}$,

$$0 = t_0 < t_1 < \ldots < t_m < t_{m+1} = T.$$

Set

$$J = [0, T], \quad J_k = [t_k, t_{k+1}], \quad k \in \{0, \ldots, m\}.$$

6.1 EXISTENCE OF SOLUTIONS FOR BOUNDARY VALUE PROBLEMS FOR SECOND-ORDER FUZZY IMPULSIVE DYNAMIC EQUATIONS

In this section we will investigate the following BVP

$$\begin{aligned}
\ominus_H \delta_H^2 y(t) &= f(t, y(t)), \quad t \in J \backslash \{t_j\}_{j=1}^{m}, \\
y(t_j^+) \ominus_H y(t_j^-) &= I_j(y(t_j)), \\
\delta_H y(t_j^+) \ominus_H \delta_H y(t_j^-) &= L_j(y(t_j)), \quad j \in \{1, \ldots, m\}, \\
y(0) &= y(T) = \widetilde{0},
\end{aligned} \tag{6.1}$$

DOI: 10.1201/9781003655367-6

where

(H1) $f \in \mathscr{C}([0,T] \times F(\mathbb{R}))$, $\sigma \in \mathscr{C}(\mathbb{T})$,

$$D(f(t,y),\widetilde{0}) \leq p(t), \quad (t,y) \in [0,T] \times F(\mathbb{R}),$$

for some non-negative function $p \in \mathscr{C}([0,T])$,

(H2) $I_j, L_j \in \mathscr{C}(F(\mathbb{R}))$,

$$D(I_j(y),\widetilde{0}) \leq c_j,$$

$$D(L_j(y),\widetilde{0}) \leq l_j, \quad j \in \{1,\ldots,m\}, \quad y \in F(\mathbb{R}),$$

c_j, l_j are non-negative constants, $j \in \{1,\ldots,m\}$.

Theorem 6.1.1 *A function $y \in PC([0,T]) \bigcap \mathscr{C}^2([0,T] \backslash \{t_j\}_{j=1}^m)$ is a solution of the BVP* (6.1) *if and only if $y \in PC([0,T])$ and it satisfies the integral equation*

$$y(t) = \int_0^T G(t,s) \cdot f(s,y(s))\delta_H s + \sum_{j=1}^m W_j(t,y(t_j)), \quad t \in [0,T], \quad (6.2)$$

where

$$G(t,s) = \begin{cases} \dfrac{T-t}{T}\sigma(s) & 0 \leq s \leq t \\ \\ \dfrac{T-\sigma(s)}{T}t & t \leq s \leq T, \end{cases}$$

$$W_j(t,y(t_j)) = \begin{cases} \dfrac{T-t}{T} \cdot I_j(y(t_j)) \ominus_H \left(\dfrac{T-t}{T}t_j\right) \cdot L_j(y(t_j)) & 0 \leq t \leq t_j \\ \\ \left(-\dfrac{t}{T}\right) \cdot I_j(y(t_j)) \ominus_H \left(\dfrac{T-t_j}{T}t\right) \cdot L_j(y(t_j)) & t_j < t \leq T. \end{cases}$$

Proof 6.1.2 *By Section 5.2, it follows that*

$$y \in PC([0,T]) \bigcap \mathscr{C}^2([0,T] \backslash \{t_j\}_{j=1}^m)$$

of the fuzzy impulsive dynamic equation

$$\begin{aligned} \ominus_H \delta_H^2 y(t) &= f(t,y(t)), \quad t \in J \backslash \{t_j\}_{j=1}^m, \\ y(t_j^+) \ominus_H y(t_j^-) &= I_j(y(t_j)), \\ \delta_H y(t_j^+) \ominus_H \delta_H y(t_j^-) &= L_j(y(t_j)), \quad j \in \{1,\dots,m\}, \end{aligned} \tag{6.3}$$

if and only if $y \in PC([0,T])$ *and it satisfies the integral equation*

$$\begin{aligned} y(t) &= y_0 + t \cdot y_1 \ominus_H \int_0^t (t-\sigma(s)) \cdot f(s,y(s)) \delta_H s \\ &\quad + \sum_{\{j:0<t_j<t\}} (I_j(y(t_j)) + (t-t_j) \cdot L_j(y(t_j))), \end{aligned} \tag{6.4}$$

where y_0 *and* y_1 *are fuzzy real constants.*

1. *Let* $y \in PC([0,T]) \bigcap \mathscr{C}^2([0,T] \backslash \{t_j\}_{j=1}^m)$ *satisfy the BVP* (6.1). *Then* $y \in PC([0,T])$ *satisfies* (6.4) *and*

$$\begin{aligned} \widetilde{0} &= y(\widetilde{0}) \\ &= y_0. \end{aligned}$$

Hence,

$$\begin{aligned} \widetilde{0} &= y(T) \\ &= T \cdot y_1 \ominus_H \int_0^T (T-\sigma(s)) \cdot f(s,y(s)) \delta_H s \\ &\quad + \sum_{j=1}^m (I_j(y(t_j)) + (T-t_j) \cdot L_j(y(t_j))), \end{aligned}$$

whereupon

$$\begin{aligned} T \cdot y_1 &= \int_0^T (T-\sigma(s)) \cdot f(s,y(s)) \delta_H s \\ &\quad \ominus_H \sum_{j=1}^m (I_j(y(t_j)) + (T-t_j) \cdot L_j(y(t_j))), \end{aligned}$$

or

$$\begin{aligned} y_1 &= \frac{1}{T} \cdot \int_0^T (T - \sigma(s)) \cdot f(s, y(s)) \delta_H s \\ &\ominus_H \frac{1}{T} \cdot \sum_{j=1}^m (I_j(y(t_j)) + (T - t_j) \cdot L_j(y(t_j))). \end{aligned}$$

Then

$$\begin{aligned} y(t) &= \frac{t}{T} \cdot \int_0^T (T - \sigma(s)) \cdot f(s, y(s)) \delta_H s \\ &\ominus_H \frac{t}{T} \cdot \sum_{\{j: t \le t_j < T\}} (I_j(y(t_j)) + (T - t_j) \cdot L_j(y(t_j))) \\ &\ominus_H \int_0^t (t - \sigma(s)) \cdot f(s, y(s)) \delta_H s \\ &+ \sum_{\{j: 0 < t_j < t\}} (I_j(y(t_j)) + (t - t_j) \cdot L_j(y(t_j))), \end{aligned}$$

i.e.,

$$\begin{aligned} y(t) &= \int_0^t \left(\frac{T - t}{T} \sigma(s) \right) \cdot f(s, y(s)) \delta_H s \\ &+ \int_t^T \frac{t(T - \sigma(s))}{T} \cdot f(s, y(s)) \delta_H s \\ &+ \sum_{\{j: 0 < t_j < t\}} \left(\frac{T - t}{T} \cdot I_j(y(t_j)) \ominus_H \left(\frac{T - t}{T} t_j \right) \cdot L_j(y(t_j)) \right) \\ &+ \sum_{\{j: t \le t_j < T\}} \left(-\frac{t}{T} \cdot I_j(y(t_j)) \ominus_H \frac{T - t_j}{T} \cdot L_j(y(t_j)) \right). \end{aligned} \tag{6.5}$$

2. *Let* $y \in PC([0, T])$ *satisfy* (6.2). *Then*

$$y \in PC([0, T]) \bigcap \mathscr{C}^2([0, T] \backslash \{t_j\}_{j=1}^m)$$

satisfies (6.3) *and* (6.5). *Hence, using* (6.5), *we arrive at*

$$y(0) = \widetilde{0},$$

$$y(T) = \widetilde{0}.$$

This completes the proof.

Note that, if $(H2)$ holds, then

$$D(W_j(t,y(t_j)),\widetilde{0}) \leq c_j + Tl_j, \quad j \in \{1,\dots,m\}.$$

Let

$$\sup_{(t,s)\in[0,T]\times[0,T]} G(t,s) = A.$$

For $y \in PC([0,T])$, define the operator

$$\Phi y(t) = \int_0^T G(t,s)\cdot f(s,y(s))\delta_H s + \sum_{j=1}^m W_j(t,y(t_j)), \quad t \in [0,T].$$

Note that $\Phi : PC([0,T]) \to PC([0,T])$.

Theorem 6.1.3 *Assume* $(H1)$–$(H2)$. *Then BVP* (6.1) *has at least one solution.*

Proof 6.1.4 *1.* $\Phi : PC([0,T]) \to PC([0,T])$ *is continuous. Let* $\{y_n\}_{n\in\mathbb{N}}$ *be a sequence of elements of* $PC([0,T])$ *such that* $y_n \to y$, *as* $n \to \infty$, *for some* $y \in PC([0,T])$. *Then*

$$D(\Phi y_n(t), \Phi y(t))$$

$$= D\Big(\int_0^T G(t,s)\cdot f(s,y_n(s))\delta_H s + \sum_{j=1}^m W_j(t,y_n(t_j)),$$

$$\int_0^T G(t,s)\cdot f(s,y(s))\delta_H s + \sum_{j=1}^m W_j(t,y(t_j))\Big)$$

$$
\begin{aligned}
&= D\Big(\int_0^T G(t,s)\cdot(f(s,y_n(s))\ominus_H f(s,y(s)))\delta_H s \\
&\quad + \sum_{j=1}^m (W_j(t,y_n(t_j))\ominus_H W_j(t,y(t_j))),\widetilde{0}\Big) \\
&\le D\Big(\int_0^T G(t,s)\cdot(f(s,y_n(s))\ominus_H f(s,y(s)))\delta_H s,\widetilde{0}\Big) \\
&\quad +D\Big(\sum_{j=1}^m (W_j(t,y_n(t_j))\ominus_H W_j(t,y(t_j))),\widetilde{0}\Big) \\
&\le D\Big(\int_0^T G(t,s)\cdot(f(s,y_n(s))\ominus_H f(s,y(s)))\delta_H s,\widetilde{0}\Big) \\
&\quad + \sum_{j=1}^m D\Big((W_j(t,y_n(t_j))\ominus_H W_j(t,y(t_j))),\widetilde{0}\Big) \\
&\le \int_0^T G(t,s)D(f(s,y_n(s)),f(s,y(s)))\Delta s \\
&\quad + \sum_{j=1}^m D(W_j(t,y_n(t_j)),W_j(t,y(t_j))) \\
&\le \int_0^T G(t,s)D(f(s,y_n(s)),f(s,y(s)))\Delta s \\
&\quad + \sum_{j=1}^m D(W_j(t,y_n(t_j)),W_j(t,y(t_j))) \\
&\to 0, \quad as \quad n\to\infty, \quad t\in[0,T],
\end{aligned}
$$

whereupon

$$
\|\Phi y_n \ominus_H \Phi y\| \to 0, \quad as \quad n\to\infty.
$$

2. *$\Phi : PC([0,T]) \to PC([0,T])$ maps a bounded set into a bounded set. Let*

$$B = \{y \in PC([0,t]) : \|y\| \leq C\},$$

for some positive constant C. Then, for $y \in B$, we get

$$D(\Phi y(t),\widetilde{0}) = D\Big(\int_0^T G(t,s)\cdot f(s,y(s))\delta_H s + \sum_{j=1}^m W_j(t,y(t_j)),\widetilde{0}\Big) \tag{6.6}$$

$$\leq D\Big(\int_0^T G(t,s)\cdot f(s,y(s))\delta_H s,\widetilde{0}\Big) + D\Big(\sum_{j=1}^m W_j(t,y(t_j)),\widetilde{0}\Big)$$

$$\leq D\Big(\int_0^T G(t,s)\cdot f(s,y(s))\delta_H s,\widetilde{0}\Big) + \sum_{j=1}^m D\Big(W_j(t,y(t_j)),\widetilde{0}\Big)$$

$$\leq \int_0^T G(t,s)D(f(s,y(s)),\widetilde{0})\Delta s + \sum_{j=1}^m D(W_j(t,y(t_j)),\widetilde{0})$$

$$\leq \int_0^T G(t,s)D(f(s,y(s)),\widetilde{0})\delta_H s + \sum_{j=1}^m D(W_j(t,y(t_j)),\widetilde{0})$$

$$\leq A\int_0^T p(s)\delta_H s + \sum_{j=1}^m (c_j + Tl_j), \quad t \in [0,T].$$

Therefore

$$\|\Phi y\| \leq A\int_0^T p(s)\delta_H s + \sum_{j=1}^m (c_j + Tl_j). \tag{6.7}$$

3. *$\Phi : PC([0,T]) \to PC([0,T])$ maps bounded sets into equicontinuous sets. Let $y \in B$, $t^1, t^2 \in [0,T]$. Without loss of generality, assume that $t^1 < t^2$. Then*

$$D(\Phi y(t^1), \Phi y(t^2))$$

$$
\begin{aligned}
&= D\Big(\int_0^T G(t^1,s)\cdot f(s,y(s))\delta_H s + \sum_{j=1}^{m} W_j(t^1,y(t_j)), \\
&\qquad \int_0^T G(t^2,s)\cdot f(s,y(s))\delta_H s + \sum_{j=1}^{m} W_j(t^2,y(t_j))\Big) \\
&= D\bigg(\int_0^T (G(t^1,s)-G(t^2,s))\cdot f(s,y(s))\delta_H s \\
&\qquad + \sum_{j=1}^{m} (W_j(t^1,y(t_j)) \ominus_H W_j(t^2,y(t_j))), \widetilde{0}\bigg) \\
&\le D\bigg(\int_0^T (G(t^1,s)-G(t^2,s))\cdot f(s,y(s))\delta_H s, \widetilde{0}\bigg) \\
&\qquad + D\bigg(\sum_{j=1}^{m} (W_j(t^1,y(t_j)) \ominus_H W_j(t^2,y(t_j))), \widetilde{0}\bigg) \\
&\le D\bigg(\int_0^T (G(t^1,s)-G(t^2,s))\cdot f(s,y(s))\delta_H s, \widetilde{0}\bigg) \\
&\qquad + \sum_{j=1}^{m} D\bigg((W_j(t^1,y(t_j)) \ominus_H W_j(t^2,y(t_j))), \widetilde{0}\bigg) \\
&\le \int_0^T |G(t^1,s)-G(t^2,s)| D(f(s,y(s)),\widetilde{0})\Delta s \\
&\qquad + \sum_{j=1}^{m} D(W_j(t^1,y(t_j)), W_j(t^2,y(t_j)))) \\
&\le \int_0^T |G(t^1,s)-G(t^2,s)| p(s)\Delta s
\end{aligned}
$$

$$+\sum_{j=1}^{m} D(W_j(t^1,y(t_j)),W_j(t^2,y(t_j)))$$

$$\to 0, \quad as \quad t^1 \to t^2.$$

From here, using the Arzela-Ascoli theorem, it follows that $\{\Phi y\}_{y\in B}$ *is an equicontinuous family.*

4. *Define the set*

$$E(\Phi) = \{y \in PC([0,T]) : y = \lambda\Phi y, \quad 0 < \lambda < 1\}.$$

Let $y \in E(\Phi)$*. Then, using (6.6) and (6.7), we get*

$$\|y\| \le \lambda \left(A \int_0^T p(s)\delta_H s + \sum_{j=1}^{m} (c_j + T l_j) \right).$$

Therefore $E(\Phi)$ *is a bounded set.*

By Schaefer's fixed-point theorem, it follows that Φ *has at least one fixed point in* $PC([0,T])$*. This completes the proof.*

Exercise 6.1.5 *Prove that the BVP*

$$\begin{aligned}
\ominus_H \delta_H^2 y(t) &= \frac{t}{1+t^2}\cdot y(t), \quad t \in J\backslash\{t_j\}_{j=1}^m,\\
y(t_j^+)\ominus_H y(t_j^-) &= (1,1,1)+y(t_j^-),\\
\delta_H y(t_j^+)\ominus_H \delta_H y(t_j^-) &= (1,1,1)+4\cdot y(t_j^-), \quad j\in\{1,\ldots,m\},\\
y(0) &= y(T) = \widetilde{0},
\end{aligned}$$

has at least one solution in $PC([0,T])$.

6.2 TRIPLE POSITIVE SOLUTIONS OF BOUNDARY VALUE PROBLEMS FOR FUZZY *P*-LAPLACIAN IMPULSIVE DYNAMIC EQUATIONS

Consider the following BVP

$$\begin{aligned}
\delta_H(\phi_p(\delta_H y(t))) + a(t)\cdot f(y(t)) &= \widetilde{0}, \quad t\in[0,T]\backslash\{t_1\},\\
y(t_1^+)\ominus_H y(t_1^-) &= I(y(t_1)),\\
\delta_H y(t_1^+)\ominus_H \delta_H y(t_1^-) &= J(y(t_1)),\\
y(0) = \delta_H y(T) &= \widetilde{0},
\end{aligned} \tag{6.8}$$

where $\phi_p(s) = D(s,\widetilde{0})^{p-2} \cdot s$, $p > 1$, $\phi_p^{-1} = \phi_q$, $\dfrac{1}{p} + \dfrac{1}{q} = 1$,

(H1) $f \in \mathscr{C}(F(\mathbb{R}))$ is a non-negative function,

(H2) $a \in \mathscr{C}(\mathbb{T})$ is a non-negative function,

(H3) $I \in \mathscr{C}(F(\mathbb{R}))$ is a non-negative function, $J \in \mathscr{C}(F(\mathbb{R}))$ is a non-positive and non-increasing.

We will start our investigations with the following useful result.

Theorem 6.2.1 *The function y is a solution of the BVP* (6.8) *if and only if it satisfies the integral equation*

$$y(t) = \int_0^t \phi_q\left(\int_s^T a(\tau) \cdot f(y(\tau))\delta_H \tau\right) \delta_H s + W(t, y(t_1)),$$

where

$$W(t, y(t_1)) = \begin{cases} (-t) \cdot J(y(t_1)) & 0 \le t \le t_1 \\ I(y(t_1)) \ominus_H t_1 \cdot J(y(t_1)) & t_1 < t \le T. \end{cases}$$

Proof 6.2.2 *Note that the solution of the equation*

$$\delta_H\left(\phi_p(\delta_H y(t))\right) + a(t) \cdot f(y(t)) = \widetilde{0}, \quad t \in [0, T],$$

is given by the representation

$$y(t) = \int_0^t \phi_q\left(\int_s^T a(\tau) \cdot f(y(\tau))\delta_H \tau\right) \delta_H s + t \cdot c_1 + c_2,$$

where c_1 and c_2 are fuzzy real constants. Now, we integrate the first equation of (6.8) *from t to T and using that $\delta_H y(T) = \widetilde{0}$, we get*

$$\phi_p(\delta_H y(t)) = \int_t^T a(s) \cdot f(y(s))\delta_H s,$$

and

$$\delta_H y(t) = \phi_q\left(\int_t^T a(s)\cdot f(y(s))\delta_H s\right).$$

In particular,

$$\delta_H y(t_1^+) = \phi_q\left(\int_{t_1}^T a(s)\cdot f(y(s))\delta_H s\right),$$

and hence,

$$\begin{aligned}\delta_H y(t_1^-) &= \delta_H y(t_1^+)\ominus_H J(y(t_1))\\ &= \phi_q\left(\int_{t_1}^T a(s)f(y(s))\delta_H s\right)\ominus_H J(y(t_1)).\end{aligned}$$

1. *Let $t\in[0,t_1]$. Then*

$$\begin{aligned}y(0) &= c_2\\ &= \widetilde{0},\end{aligned}$$

and

$$\begin{aligned}y(t) &= \int_0^t \phi_q\left(\int_s^T a(\tau)\cdot f(y(\tau))\delta_H\tau\right)\delta_H s + t\cdot c_1,\\ \delta_H y(t) &= \phi_q\left(\int_t^T a(\tau)\cdot f(y(\tau))\delta_H\tau\right)+c_1,\\ \delta_H y(t_1^-) &= \phi_q\left(\int_{t_1}^T a(\tau)\cdot f(y(\tau))\delta_H\tau\right)+c_1\\ &= \phi_q\left(\int_{t_1}^T a(a)\cdot f(y(s))\delta s\right)\ominus_H J(y(t_1)).\end{aligned}$$

Then

$$c_1 = \ominus_H J(y(t_1)),$$

and

$$y(t) = \int_0^t \phi_q \left(\int_s^T a(\tau) \cdot f(y(\tau)) \delta_H \tau \right) \delta_H s \ominus_H t \cdot J(y(t_1)).$$

Hence,

$$y(t_1^-) = \int_0^{t_1} \phi_q \left(\int_s^T a(\tau) \ominus_H f(y(\tau)) \delta_H \tau \right) \delta_H s \ominus_H t_1 \cdot J(y(t_1)).$$

2. *Let* $t \in [t_1, T]$. *Then*

$$\begin{aligned} \delta_H y(t) &= \phi_q \left(\int_s^T a(\tau) \cdot f(y(\tau)) \delta_H \tau \right) + c_1, \\ \delta_H y(T) &= c_1 \\ &= \widetilde{0}, \end{aligned}$$

and

$$y(t) = \int_0^t \phi_q \left(\int_s^T a(\tau) \cdot f(y(\tau)) \delta_H \tau \right) \delta_H s + c_2.$$

From here,

$$\begin{aligned} y(t_1^+) &= \int_0^{t_1} \phi_q \left(\int_s^T a(\tau) \cdot f(y(\tau)) \delta_H \tau \right) \delta_H s + c_2, \\ y(t_1^+) \ominus_H y(t_1^-) &= c_2 + t_1 \cdot J(y(t_1)) \\ &= I(y(t_1)). \end{aligned}$$

Therefore

$$c_2 = I(y(t_1)) \ominus_H t_1 \cdot J(y(t_1)),$$

and

$$y(t)=\int_0^t \phi_q\left(\int_s^T a(\tau)\cdot f(y(\tau))\delta_H\tau\right)\delta_H s+I(y(t_1))\ominus_H t_1\cdot J(y(t_1)).$$

This completes the proof.

Let

$$E=\{y:[0,T]\to F(\mathbb{R}),\quad y\in\mathscr{C}([0,t_1]),\quad y\in\mathscr{C}((t_1,T]),\quad y(t_1^+)\in F(\mathbb{R})\},$$

equipped with the norm

$$\|y\|=\max\{\sup_{t\in[0,t_1]}|y(t)|,\quad \sup_{t\in(t_1,T]}|y(t)|\}.$$

Define the cone P as follows

$$P = \Big\{y\in E: y \quad \text{is concave, non-decreasing, and non-negative}$$

$$I(y(t_1))\geq\widetilde{0},\quad J(y(t_1))\leq\widetilde{0}\Big\}.$$

Note that for $y\in P$, we obtain

$$\|y\|=y(T).$$

Let

$$\eta=\inf\left[\frac{t_1+T}{2},T\right)\in\mathbb{T}.$$

Fix $0<\eta<r<T$. In addition, if $y\in P$, then

$$\begin{aligned} y(t) &\geq \frac{1}{2}\cdot\sup_{s\in[\frac{t_1}{2},t_1]} y(s)\\ &= \frac{1}{2}\cdot y(t_1),\quad t\in\left[\frac{t_1}{2},t_1\right], \end{aligned}$$

and

$$\begin{aligned} y(t) &\geq \frac{1}{2} \cdot \sup_{s\in[\eta,T]} y(s) \\ &= \frac{1}{2} \cdot y(T), \quad t \in [\eta, T]. \end{aligned}$$

Define, for $y \in P$,

$$\begin{aligned} \gamma(y) &= \max_{t\in[0,\eta]} y(t) \\ &= y(\eta), \\ \beta(y) &= \min_{t\in[\eta,r]} y(t) \\ &= y(\eta), \end{aligned}$$

and

$$\begin{aligned} \alpha(y) &= \max_{t\in[t_1,r]} y(t) \\ &= y(r). \end{aligned}$$

Thus, for each $y \in P$, we obtain

$$\begin{aligned} \gamma(y) &= \beta(y) \\ &\leq \alpha(y), \end{aligned}$$

and

$$\begin{aligned} \gamma(y) &= y(\eta) \\ &\geq \frac{1}{2} \cdot y(T) \\ &= \frac{1}{2}\|y\|. \end{aligned}$$

Hence,

$$\|y\| \leq 2 \cdot \gamma(y).$$

Denote

$$P(\gamma, d) = \{x \in P : \gamma(x) < d\}.$$

In our next result, we will use the following fixed-point theorem.

Theorem 6.2.3 *Let P be a cone in a real Banach space E and α, β, and γ be increasing, non-negative, continuous functionals on P such that for some $c > 0$ and $M > 0$*

$$\gamma(x) \le \beta(x) \le \alpha(x), \quad \|x\| \le M\gamma(x), \quad x \in \overline{P(\gamma, c)}.$$

Suppose that there exist positive numbers a and b with $a < b < c$ and

$$F : \overline{P(\gamma, c)} \to P,$$

is a completely continuous operator such that

1. $\gamma(Fx)c$, $x \in \partial P(\gamma, c)$,
2. $\beta(Fx) > b$, $x \in \partial P(\beta, b)$,
3. $P(\alpha, a) \neq \emptyset$ *and* $\alpha(Fx) < 0$, $x \in \partial P(\alpha, a)$.

Then F has at least three fixed points $x_1, x_2, x_3 \in \overline{P(\gamma, c)}$ such that

$$0 \le \alpha(x_1) < a < \alpha(x_2), \quad \beta(x_2) < b < \beta(x_3), \quad \gamma(x_3) < c.$$

For $y \in E$, define the operator

$$Fy(t) = \int_0^t \phi_q \left(\int_s^T a(\tau) \cdot f(y(\tau)) \delta_H \tau \right) \delta_H s + W(t, y(t_1)), \quad t \in [0, T].$$

Exercise 6.2.4 *Assume $(H1)$–$H(3)$. Prove that $F : E \to E$ is completely continuous.*

Define

$$M = \eta \cdot \phi_q \left(\int_0^T a(\tau) \cdot \Delta\tau \right),$$

$$N = \eta \cdot \phi_q \left(\int_\eta^T a(\tau) \Delta\tau \right),$$

$$m = r \cdot \phi_q \left(\int_0^T a(\tau) \Delta\tau \right).$$

Theorem 6.2.5 *Let* $(H1)$–$(H3)$ *hold,*

$$0 < a < b < \frac{M}{2N}c,$$

f, I, and J satisfy the following conditions

(C1)

$$f(w) \leq \phi_p\left(\frac{1}{2M}\cdot c\right), \quad \widetilde{0} \leq w \leq 2\cdot c,$$

(C2)

$$f(w) > \phi_p\left(\frac{1}{N}\cdot b\right), \quad b \leq w \leq 2\cdot b,$$

(C3)

$$f(w) < \phi_p\left(\frac{1}{2m}\cdot a\right), \quad \widetilde{0} \leq w \leq a,$$

(C4)

$$I(w(t_1)) \ominus_H t_1 \cdot J(w(t_1)) \leq \frac{1}{2}\cdot c, \quad w : \mathbb{T} \to \mathbb{R}.$$

Then the BVP (6.8) *has at least three fixed points* $y_1, y_2, y_3 \in \overline{P(\gamma, c)}$ *such that*

$$\begin{aligned} \widetilde{0} &\leq \alpha(y_1) < a < \alpha(y_2), \\ \beta(y_2) &< b < \beta(y_3), \\ \gamma(y_3) &< c. \end{aligned} \tag{6.9}$$

Proof 6.2.6 *Let* $y \in \overline{P(\gamma, c)}$. *By* $(H1)$ *and* $(H3)$, *it follows that* $Fy(t) \geq \widetilde{0}$, $t \in [0, T]$. *Next, by* $(H1)$ *and by the definition of the operator* F, *we get*

$$\delta_H^2(Fy)(t) \leq \widetilde{0}, \quad t \in [0, T].$$

Moreover, Fy *is concave on* $[0, t_1]$ *and* $(t_1, T]$. *Also,*

$$\begin{aligned} \delta_H(Fy)(t) &= \phi_q\left(\int_t^T a(\tau)\cdot f(y(\tau))\delta_H\tau\right) \ominus_H \chi_{[0,t_1]}(t)\cdot J(y(t_1)) \\ &\geq \widetilde{0}, \quad t \in [0, T], \end{aligned}$$

where $\chi_{[0,t_1]}$ *is the characteristic function of* $[0,t_1]$. *Thus,* Fy *is non-decreasing on* $[0,t_1]$ *and* $(t_1,T]$. *Since* $Fy(0)=\widetilde{0}$, *we conclude that*

$$Fy(t)\geq\widetilde{0},\quad t\in[0,t_1].$$

Also,

$$\begin{aligned}
I(F(y(t_1))) &= Fy(t_1^+)\ominus_H Fy(t_1^-)\\
&= W(t_1^+,y(t_1^+))\ominus_H W(t_1^-,y(t_1^-))\\
&= I(y(t_1^+))\ominus_H t_1\cdot J(y(t_1^+))\ominus_H t_1\cdot J(y(t_1))\\
&= I(y(t_1^+))\ominus_H t_1\cdot\left(J(y(t_1^+))+J(y(t_1))\right)\\
&\geq \widetilde{0},\\
J(F(y(t_1))) &= \delta_H(Fy)(t_1^+)\ominus_H \delta_H(Fy)(t_1^-)\\
&= J(y(t_1))\\
&\leq \widetilde{0}.
\end{aligned}$$

Thus,

$$\begin{aligned}
Fy(t_1^+) &\geq Fy(t_1^-)\\
&\geq \widetilde{0},
\end{aligned}$$

and

$$Fy(t)\geq\widetilde{0},\quad t\in(t_1,T].$$

Therefore, $Fy\in P$ *and in particular,*

$$F:\overline{P(\gamma,c)}\to P.$$

Let now, $y\in\partial P(\gamma,c)$. *Then*

$$\begin{aligned}
\gamma(y) &= \max_{t\in[0,\eta]} y(t)\\
&= y(\eta)\\
&= c.
\end{aligned}$$

Hence,

$$\begin{aligned}\|y\| &\leq 2\cdot\gamma(y)\\ &= 2\cdot c.\end{aligned}$$

Therefore,

$$\begin{aligned}\widetilde{0} &\leq y(t)\\ &\leq 2\cdot c, \quad t\in[0,T].\end{aligned}$$

Hence and $(C1)$, we obtain

$$f(y(s)) \leq \phi_p\left(\frac{1}{2M}\cdot c\right), \quad s\in[0,T].$$

Also,

$$\begin{aligned}\gamma(Fy) &= Fy(\eta)\\ &= \int_0^\eta \phi_q\left(\int_s^T a(\tau)\cdot f(y(\tau))\delta_H\tau\right)\delta_H s + W(\eta, y(t_1))\\ &= \int_0^\eta \phi_q\left(\int_s^T a(\tau)\cdot f(y(\tau))\delta_H\tau\right)\delta_H s + I(y(t_1)) \ominus_H t_1\cdot J(y(t_1))\\ &< \left(\eta\phi_q\left(\int_0^T a(\tau)\Delta\tau\right)\frac{1}{2M}\right)\cdot c + \frac{1}{2}\cdot a\\ &= \frac{1}{2}\cdot c + \frac{1}{2}\cdot c\\ &= c.\end{aligned}$$

Now, suppose that $y\in\partial P(\beta,b)$. Then

$$\begin{aligned}\beta(y) &= \min_{t\in[\eta,r]} y(t)\\ &= y(\eta)\\ &= b.\end{aligned}$$

Thus,

$$y(t) \geq b, \quad t \in [\eta, T],$$

and

$$\begin{aligned} \widetilde{0} &\leq y(t) \\ &\leq \|y\| \\ &= y(T), \quad t \in [\eta, T], \\ \|y\| &\leq 2 \cdot \gamma(y) \\ &= 2 \cdot \beta(y) \\ &= 2 \cdot b, \quad y \in P. \end{aligned}$$

Then

$$b \leq y(t) \leq 2 \cdot b, \quad t \in [\eta, T].$$

Hence and $(C2)$*, we arrive at*

$$f(y(s)) > \phi_p\left(\frac{1}{N} \cdot b\right), \quad s \in [\eta, T],$$

and

$$\begin{aligned} \beta(Fy) &= Fy(\eta) \\ &= \int_0^\eta \phi_q\left(\int_s^T a(\tau) \cdot f(y(\tau))\delta_H\tau\right)\delta_H s + W(\eta, y(t_1)) \\ &= \int_0^\eta \phi_q\left(\int_s^T a(\tau) \cdot f(y(\tau))\delta_H\tau\right)\delta_H s + I(y(t_1)) \ominus_H t_1 \cdot J(y(t_1)) \\ &\geq \left(\eta\phi_q\left(\int_\eta^T a(\tau)\Delta\tau\right)\frac{1}{N}\right) \cdot b \\ &= b. \end{aligned}$$

Note that $\frac{1}{2}\cdot a \in P(\alpha,a)$. *Then* $P(\alpha,a) \neq \emptyset$. *Let now,* $y \in \partial P(\alpha,a)$. *Then*

$$\begin{aligned}
\alpha(y) &= \max_{t\in[t_1,r]} y(t) \\
&= y(r) \\
&= a, \\
0 &\leq y(t) \\
&\leq a, \quad t \in [t_1,r].
\end{aligned}$$

Since y *is non-decreasing and* $y(t_1^+) \geq y(t_1)$, *we obtain*

$$\widetilde{0} \leq y(t) \leq a, \quad t \in [0,r].$$

By $(C3)$, *we obtain*

$$f(y(s)) \leq \phi_p\left(\frac{1}{2m}\cdot a\right), \quad s \in [0,r].$$

Then

$$\begin{aligned}
\alpha(Fy) &= Fy(r) \\
&= \int_0^r \phi_q\left(\int_s^T a(\tau)\cdot f(y(\tau))\delta_H\tau\right)\delta_H s + W(r, y(t_1)) \\
&= \int_0^r \phi_q\left(\int_s^T a(\tau)\cdot f(y(\tau))\delta_H\tau\right)\delta_H s + I(y(t_1)) \ominus_H t_1\cdot J(y(t_1)) \\
&< \left(r\phi_q\left(\int_0^T a(\tau)\Delta\tau\right)\frac{1}{2m}\right)\cdot a + \frac{1}{2}\cdot a \\
&= \frac{1}{2}\cdot a + \frac{1}{2}\cdot a \\
&= a.
\end{aligned}$$

Hence and with Theorem 6.2.3, we get that the problem (6.8) *has at least three positive solutions* $y_1, y_2, y_3 \in \overline{P(\gamma, c)}$ *which satisfy* (6.9). *This completes the proof.*

Exercise 6.2.7 *Find conditions for the function g so that the BVP*

$$\begin{aligned} \delta_H\left(\phi_p(\delta_H y(t))\right) + g(y(t)) &= \widetilde{0}, \quad t \in [0,T]\backslash\{t_1\}, \\ y(t_1^+) \ominus_H y(t_1^-) &= (4,4,4), \\ \delta_H y(t_1^+) \ominus_H \delta_H y(t_1^-) &= (-1,0,1), \\ y(0) = \delta_H y(T) &= \widetilde{0}, \end{aligned}$$

where $\phi_p(s) = |s|^{p-2}s$, $p > 1$, $\phi_p^{-1} = \phi_q$, $\dfrac{1}{p} + \dfrac{1}{q} = 1$, *has at least three different solutions.*

6.3 ADVANCED PRACTICAL PROBLEMS

Problem 6.3.1 *Find conditions for the function g so that the BVP*

$$\begin{aligned} \ominus_H \delta_H^2 y(t) &= \frac{t^2+t+1}{4+10t^8} \cdot g(y), \quad t \in J\backslash\{t_j\}_{j=1}^m, \\ y(t_j^+) \ominus_H y(t_j^-) &= \ominus_H \frac{3}{7} \cdot y(t_j^-), \\ \delta_H y(t_j^+) \ominus_H \delta_H y(t_j^-) &= \frac{3}{10} \cdot y(t_j^-), \quad j \in \{1,\ldots,m\}, \\ y(0) &= y(T) = \widetilde{0}, \end{aligned}$$

has at least one solution in $PC([0,T])$.

Problem 6.3.2 *Prove that the BVP*

$$\begin{aligned} \ominus_H \delta_H^2 y(t) &= \left(\frac{t^2+1}{1+t^4+3t^6+t^8}, t, 3\right), \quad t \in J\backslash\{t_j\}_{j=1}^m, \\ y(t_j^+) \ominus_H y(t_j^-) &= (1,-1,2), \\ \delta_H y(t_j^+) \ominus_H \delta_H y(t_j^-) &= (4,3,0), \quad j \in \{1,\ldots,m\}, \\ y(0) &= y(T) = \widetilde{0}, \end{aligned}$$

has at least one solution in $PC([0,T])$.

Problem 6.3.3 *Find conditions for the function g so that the BVP*

$$\begin{aligned}
-\delta_H^2 y(t) &= \frac{1}{8}\cdot g(y(t)), \quad t\in J\backslash\{t_j\}_{j=1}^m,\\
y(t_j^+)\ominus_H y(t_j^-) &= (1,1,1),\\
\delta_H y(t_j^+)\ominus_H \delta_H y(t_j^-) &= (-1,0,2), \quad j\in\{1,\ldots,m\},\\
y(0) &= y(T)=\widetilde{0},
\end{aligned}$$

has at least one solution in $PC([0,T])$.

Problem 6.3.4 *Find conditions for the function g in the BVP*

$$\begin{aligned}
\delta_H\left(\phi_p(\delta_H y(T))\right)+\frac{t}{1+e^t}\cdot g(y) &= \widetilde{0}, \quad t\in[0,T]\backslash\{t_1\},\\
y(t_1^+)\ominus_H y(t_1^-) &= (1,-1,2),\\
\delta_H y(t_1^+)\ominus_H \delta_H y(t_1^-) &= (-2,0,3),\\
y(0)=\delta_H y(T) &= \widetilde{0},
\end{aligned}$$

where $\phi_p(s)=|s|^{p-2}s$, $p>1$, $\phi_p^{-1}=\phi_q$, $\frac{1}{p}+\frac{1}{q}=1$, *has at least three different solutions.*

CHAPTER 7

Oscillations of Fuzzy Impulsive Dynamic Equations

7.1 OSCILLATIONS OF FIRST-ORDER FUZZY IMPULSIVE DYNAMIC EQUATIONS

Suppose that $\mathbb{T}$ is an unbounded above time scale and $0 \in \mathbb{T}$. Consider the problem

$$\begin{aligned} \delta_H y(t) &= f(t, y(t)), \quad t \geq 0, \\ y(t_k^+) &= I_k(y(t_k)), \quad k \in \mathbb{N}, \end{aligned} \tag{7.1}$$

where $f : [0, \infty) \times F(\mathbb{R}) \to F(\mathbb{R})$ is a given function, $I_k \in \mathscr{C}(F(\mathbb{R}))$, $t_k \in \mathbb{T}$, $k \in \mathbb{N}$,

$$0 = t_0 < t_1 < \ldots < t_m < t_{m+1} < \ldots < \infty,$$

$$\begin{aligned} y(t_k^-) &= \lim_{h \to 0+} y(t_k - h), \\ y(t_k^+) &= \lim_{h \to 0+} y(t_k + h), \quad k \in \mathbb{N}. \end{aligned}$$

Suppose that t_k, $k \in \mathbb{N}$, are right-dense. When t_k, $k \in \mathbb{N}$, is left-scattered, we have

$$y(t_k^-) = y(t_k).$$

DOI: 10.1201/9781003655367-7

Denote $J = [0,\infty)$. For a function $y \in \mathscr{C}(J)$, with y_k we denote its restriction to $J_k = [t_k, t_{k+1}]$, $k \in \mathbb{N}$. Define

$$\begin{aligned} PC &= \Big\{ y : [0,\infty) \to F(\mathbb{R}), \quad y_k \in \mathscr{C}(J_k), \quad k \in \mathbb{N}, \\ &\quad \exists y(t_k^-), \quad y(t_k^+), \quad k \in \mathbb{N}, \quad y(t_k^-) = y(t_k) \Big\}. \end{aligned}$$

Definition 7.1.1 *A function $y \in PC \bigcap \mathscr{C}^1((t_k, t_{k+1}))$, $k \in \mathbb{N}_0$, is said to be a solution of the equation (7.1) if y satisfies*

$$\delta_H y(t) = f(t, y(t)), \quad t \in J \backslash \{t_1, \ldots\},$$

and

$$y(t_k^+) = I_k(y(t_k^-)), \quad k \in \mathbb{N}.$$

Definition 7.1.2 *A function $\alpha \in PC \bigcap \mathscr{C}^1((t_k, t_{k+1}))$, $k \in \mathbb{N}_0$, is said to be a lower solution of the equation (7.1) if α satisfies*

$$\delta_H \alpha(t) \leq f(t, \alpha(t)), \quad t \in J \backslash \{t_1, \ldots\},$$

and

$$\alpha(t_k^+) \leq I_k(\alpha(t_k^-)), \quad k \in \mathbb{N}.$$

Definition 7.1.3 *A function $\beta \in PC \bigcap \mathscr{C}^1((t_k, t_{k+1}))$, $k \in \mathbb{N}_0$, is said to be an upper solution of the equation (7.1) if β satisfies*

$$\delta_H \beta(t) \geq f(t, \beta(t)), \quad t \in J \backslash \{t_1, \ldots\},$$

and

$$\beta(t_k^+) \geq I_k(\beta(t_k^-)), \quad k \in \mathbb{N}.$$

Theorem 7.1.4 *Let α and β be lower and upper solutions, respectively, of the equation (7.1) with $\alpha \leq \beta$ and*

$$\begin{aligned} \alpha(t_k^+) &\leq \min_{y \in (\alpha(t_k^-), \beta(t_k^-)]} I_k(y) \\ &\leq \max_{y \in (\alpha(t_k^-), \beta(t_k^-)]} I_k(y) \\ &\leq \beta(t_k^+), \quad k \in \mathbb{N}. \end{aligned} \tag{7.2}$$

Assume that α is eventually positive and non-decreasing, or β is eventually negative non-increasing. Then every solution y of (7.1) *such that $y \in [\alpha, \beta]$ is non-oscillatory.*

Proof 7.1.5 *1. Let α is eventually positive and non-decreasing. Then there exists a $T_0 \in J$ such that*

$$\alpha(t) > \widetilde{0}, \quad t \in [T_0, \infty).$$

Hence,

$$y(t) > \widetilde{0}, \quad t \in [T_0, \infty), \quad t \neq t_k, \quad k \in \mathbb{N}.$$

Note that there is a $k \in \mathbb{N}$ such that

$$\alpha(t_k) > \widetilde{0}.$$

Since

$$y(t_k^+) = I_k(y(t_k)),$$

using (7.2), *we get*

$$\alpha(t_k^+) \leq y(t_k^+).$$

Because α is non-decreasing, we have

$$\begin{aligned} \alpha(t_k + h) &\geq \alpha(t_k) \\ &> \widetilde{0}, \end{aligned}$$

for any $h > 0$. Therefore

$$\alpha(t_l^+) > \widetilde{0},$$

for any $t_l \geq t_k$ and then y is non-oscillatory.

2. *Suppose that β is eventually negative and non-increasing. Then there is a $T_1 \geq 0$ such that*

$$\beta(t) < \widetilde{0}, \quad t \in [T_1, \infty).$$

Hence,

$$y(t) < \widetilde{0}, \quad t \in J \backslash \{t_1, \ldots\}.$$

Note that there exists a $k \in \mathbb{N}$ such that

$$\beta(t_k) < \widetilde{0},$$

and

$$y(t_k^+) = I_k(y(t_k)).$$

By (7.2), it follows that

$$y(t_k^+) \leq \beta(t_k^+).$$

Since β is non-increasing, we have

$$\beta(t_k + h) \leq \beta(t_k)$$

for any $h > 0$. Thus,

$$\beta(t_l) < \widetilde{0}$$

and

$$y(t_l) < \widetilde{0},$$

for any $t_l \geq t_k$. Consequently, y is non-oscillatory. This completes the proof.

Theorem 7.1.6 *Let α and β be lower and upper solutions, respectively, of the equation (7.1). Assume that the sequences $\{\alpha(t_k)\}_{k\in\mathbb{N}}$ and $\{\beta(t_k)\}_{k\in\mathbb{N}}$ are oscillatory. Then every solution y of the equation (7.1) such that $y \in [\alpha, \beta]$ is oscillatory.*

Proof 7.1.7 *Assume that y is a solution of equation (7.1) such that $y \in [\alpha, \beta]$ and y are non-oscillatory.*

1. *Let y eventually be a positive solution of equation (7.1). Then there is a $T_0 \geq 0$ such that*

$$y(t) > \widetilde{0}, \quad t \in [T_0, \infty).$$

Hence, there is a $t_k \in [T_0, \infty)$ so that

$$\beta(t_l) > \widetilde{0},$$

for any $t_l \geq t_k$. This is a contradiction, because the sequence $\{\beta(t_k)\}_{k\in\mathbb{N}}$ is oscillatory.

2. *Let y eventually be a negative solution of equation (7.1). Then there is a $T_1 \geq 0$ such that*

$$y(t) < \widetilde{0}, \quad t \in [T_1, \infty).$$

From here, there is a $t_k \in [T_1, \infty)$ so that

$$\alpha(t_l) < \widetilde{0},$$

for any $t_l \geq t_k$. This is a contradiction, because the sequence $\{\alpha(t_k)\}_{k \in \mathbb{N}}$ is oscillatory. This completes the proof.

Example 7.1.8 *Suppose that there exist $g_1, g_2 \in \mathscr{C}(J)$ such that*

$$g_1(t) \leq f(t,y) \leq g_2(t), \quad t \in J, \quad y \in F(\mathbb{R}),$$

and

$$\begin{aligned} \int_0^t g_1(s)\delta_H s &\leq I_k\left(\int_0^t g_1(s)\delta_H s\right), \\ \int_0^t g_2(s)\delta_H s &\geq I_k\left(\int_0^t g_2(s)\delta_H s\right), \quad k \in \mathbb{N}, \quad t \geq 0. \end{aligned}$$

Let

$$\begin{aligned} \alpha(t) &= \int_0^t g_1(s)\delta_H s, \\ \beta(t) &= \int_0^t g_2(s)\delta_H s, \quad t \geq 0. \end{aligned}$$

Then

$$\begin{aligned} \delta_H \alpha(t) &= g_1(t) \\ &\leq f(t,y), \end{aligned}$$

$$\begin{aligned}\delta_H\beta(t) &= g_2(t)\\ &\geq f(t,y), \quad t\geq 0, \quad y\in\mathbb{R}.\end{aligned}$$

In particular, we have

$$\begin{aligned}\delta_H\alpha(t) &\leq f(t,\alpha(t)),\\ \delta_H\beta(t) &\geq f(t,\beta(t)), \quad t\in J\backslash\{t_1,\ldots\},\\ \alpha(t_k^+) &\leq I_k(\alpha(t_k^-)),\\ \beta(t_k^+) &\geq I_k(\beta(t_k)), \quad k\in\mathbb{N}.\end{aligned}$$

Therefore, α and β are lower and upper solutions, respectively, of the equation (7.1)*. Hence,*

1. *if*
$$g_1(t) > \widetilde{0}, \quad t\geq 0,$$
then α is positive and non-decreasing. Thus, y is non-oscillatory.

2. *if*
$$g_2(t) < \widetilde{0}, \quad t\geq 0,$$
then β is negative and non-increasing. Thus, y is non-oscillatory.

Therefore, if the sequences $\{\alpha(t_k)\}_{k\in\mathbb{N}}$ and $\{\beta(t_k)\}_{k\in\mathbb{N}}$ are both oscillatory, then y is oscillatory.

Exercise 7.1.9 *Let α and β are lower and upper solutions, respectively, of the equation* (7.1)*, and*

$$\begin{aligned}\alpha(t_k) &= \left((-1)^k(t_k)^2, t_k, t_k^3\right)\\ \beta(t_k) &= \left((-1)^k(t_k)^4, (-1)^k t_k^2, 1\right), \quad k\in\mathbb{N}.\end{aligned}$$

Prove that every solution y of the equation (7.1) *such that $y\in[\alpha,\beta]$ is oscillatory.*

7.2 OSCILLATIONS OF SECOND-ORDER FUZZY IMPULSIVE DYNAMIC EQUATIONS

Suppose that $\mathbb{T}$ is an unbounded above time scale and $0 \in \mathbb{T}$. Consider the equation

$$\begin{aligned}
\delta_H^2 y(t) + f(t, y^\sigma(t)) &= \widetilde{0}, \quad t \in J, \quad t \neq t_k, \quad k \in \mathbb{N}, \\
y(t_k^+) &= g_k(y(t_k^-)), \\
\delta_H y(t_k^+) &= h_k(\delta_H y(t_k^-)), \quad k \in \mathbb{N}, \\
y(t_0^+) &= y_0, \\
\delta_H y(t_0^+) &= y_0^\Delta,
\end{aligned} \tag{7.3}$$

where $J = [0, \infty)$, $y_0, y_0^\Delta \in F(\mathbb{R})$, $t_k \in \mathbb{T}$, $k \in \mathbb{N}$,

$$0 = t_0 < t_1 < t_2 < \ldots < t_k < \ldots,$$

$\lim\limits_{k \to \infty} t_k = \infty$, t_k, $k \in \mathbb{N}$, are right-dense,

(H8) $f \in \mathscr{C}(\mathbb{T} \times F(\mathbb{R}))$, x and $f(t,x)$ have the same sign for any $x \in F(\mathbb{R})$,

$$f(t,x) \geq p(t) \cdot \phi(x), \quad t \in J, \quad x \neq \widetilde{0},$$

$p \in \mathscr{C}_{rd}(\mathbb{T})$, $p \geq 0$ on $\mathbb{T}$, x and ϕ have the same sign for any $x \in F(\mathbb{R})$,

$$\delta_H \phi(x) \geq \widetilde{0}, \quad x \in F(\mathbb{R}),$$

(H9) $g_k, h_k \in \mathscr{C}(F(\mathbb{R}))$, $k \in \mathbb{N}$, and there are positive real constants a_k, a_k^*, b_k, b_k^*, $k \in \mathbb{N}$, such that

$$\begin{aligned}
a_k^* \cdot x &\leq g_k(x) \leq a_k \cdot x, \\
b_k^* \cdot x &\leq h_k(x) \leq b_k \cdot x, \quad k \in \mathbb{N}.
\end{aligned}$$

For $i \in \{1,2\}$, define

$$AC^i(J) = \Big\{y : J \to F(\mathbb{R}) \text{ is } i-\text{times } \delta_H\text{-differentiable}, \\ \delta_H^i y \text{ is absolutely continuous}\Big\},$$

$$PC = \{y : J \to F(\mathbb{R}) \text{ is rd-continuous except at } t_k, \\ \exists y(t_k^-),\ y(t_k^+),\ \delta_H y(t_k^-),\ \delta_H y(t_k^+), \\ y(t_k^-) = y(t_k),\ \delta_H y(t_k^-) = \delta_H y(t_k),\ k \in \mathbb{N}\}.$$

Definition 7.2.1 *A function $y \in PC \bigcap AC^i(J \backslash \{t_k\}_{k\in\mathbb{N}})$ is said to be a solution of the equation* (7.3), *if*

$$\delta_H^2 y(t) + f(t, y^\sigma(t)) = 0, \quad on \quad J \backslash \{t_k\}_{k\in\mathbb{N}},$$

and

$$\begin{aligned} y(t_k^+) &= g_k(y(t_k^-)), \\ \delta_H y(t_k^+) &= h_k(\delta_H y(t_k^-)), \quad k \in \mathbb{N}, \\ y(t_0^+) &= y_0, \\ \delta_H y(t_0^+) &= y_0^\Delta, \end{aligned}$$

for any $k \in \mathbb{N}$.

Theorem 7.2.2 *Suppose* $(H8)$, $(H9)$ *and* $y(t) > 0$, $t \geq t_0' \geq t_0$, *are a non-oscillatory solution of* (7.3). *If*

$$\int_{t_j}^{\infty} \prod_{t_j < t_k < s} \frac{b_k^*}{a_k} \Delta s = \infty, \tag{7.4}$$

for some $t_j \geq t_0$, then $\delta_H y(t_k^+) \geq \widetilde{0}$ and $\delta_H y(t) \geq \widetilde{0}$ for $t \in (t_k, t_{k+1}]$, where $t_k \geq t_0'$.

Proof 7.2.3 *Assume that there exists some $j \in \mathbb{N}$ such that $t_j \geq t_0'$ and $\delta_H y(t_j^-) < \widetilde{0}$. Hence,*

$$\begin{aligned} \delta_H y(t_j^+) &= h_j(\delta_H y(t_j^-)) \\ &\leq b_j^* \cdot \delta_H y(t_j^-) \\ &< \widetilde{0}. \end{aligned}$$

Let

$$\delta_H y(t_j^+) = \ominus_H \alpha,$$

where α is a positive fuzzy constant. For $t \in (t_{j+i-1}, t_{j+i}]$, $i \in \mathbb{N}$, we obtain

$$\begin{aligned} \delta_H^2 y(t) &= \ominus_H f(t, y^\sigma(t)) \\ &\leq (-p(t)) \cdot \phi(y^\sigma(t)) \\ &\leq \widetilde{0}. \end{aligned}$$

Therefore, $\delta_H y$ does not increase in $(t_{j+i-1}, t_{j+i}]$, $i \in \mathbb{N}$. Then

$$\begin{aligned} \delta_H y(t_{j+1}^-) &\leq \delta_H y(t_j^+) \\ &= \ominus_H \alpha \\ &< \widetilde{0}, \\ \delta_H y(t_{j+2}^-) &\leq \delta_H y(t_{j+1}^+) \\ &= h_{j+1} \cdot \left(\delta_H y(t_{j+1}^-)\right) \\ &\leq b_{j+1}^* \cdot \delta_H y(t_{j+1}) \\ &\leq \ominus_H b_{j+1}^* \cdot \alpha \\ &< \widetilde{0}. \end{aligned}$$

By induction, we obtain the following.

$$
\begin{aligned}
\delta_H y(t) &\leq \left(- \prod_{t_j<t_k<t} b_k^* \right) \cdot \alpha \\
&< \widetilde{0}, \quad t \in (t_{j+n}, t_{j+n+1}].
\end{aligned}
$$

From (H9), we have

$$
\begin{aligned}
y(t_k^+) &= g_k(y(t_k^-)) \\
&\leq a_k \cdot y(t_k^-).
\end{aligned}
$$

Hence and with Theorem 4.6.1, we arrive at

$$
\begin{aligned}
y(t) &\leq \left(\prod_{t_j<t_k<t} a_k \right) \cdot y(t_j^+) \ominus_H \left(\int_{t_j}^{t} \prod_{s<t_k<t} a_k \prod_{t_j<t_k<s} b_k^* \Delta s \right) \cdot \alpha \\
&= \left(\prod_{t_j<t_k<t} a_k \right) \cdot \left(y(t_j^+) \ominus_H \left(\int_{t_j}^{t} \prod_{t_j<t_k<s} \frac{b_k^*}{a_k} \Delta s \right) \cdot \alpha \right) \\
&\to -\infty, \quad as \quad t \to \infty,
\end{aligned}
$$

because $y(t_j^+) > \widetilde{0}$*. This is a contradiction. Consequently*

$$
\delta_H y(t_k^-) \geq \widetilde{0}, \quad t_k \geq t_0'.
$$

Applying (H9), we get

$$
\begin{aligned}
\delta_H y(t_k^+) &= h_k \cdot (\delta_H y(t_k^-)) \\
&\geq b_k \cdot \delta_H y(t_k^-) \\
&\geq \widetilde{0}.
\end{aligned}
$$

Since $\delta_H y$ *decreases in* $(t_k, t_{k+1}]$, $t_k \geq t_0'$, *we have* $\delta_H y(t) \geq \widetilde{0}$, $t \in (t_k, t_{k+1}]$, $t_k \geq t_0'$. *This completes the proof.*

Exercise 7.2.4 *Suppose (H8), (H9) hold and $y(t) < \widetilde{0}$, $t \geq t_0' \geq t_0$, is a non-oscillatory solution of (7.3). If (7.4) holds, prove that $\delta_H y(t_k^+) \leq \widetilde{0}$ and $\delta_H y(t) \leq \widetilde{0}$ for $t \in (t_k, t_{k+1}]$, where $t_k \geq t_0'$.*

Theorem 7.2.5 *Suppose (H8), (H9), (7.4) hold and there exists a positive integer k_0 such that $a_k^* \geq 1$, $k \geq k_0$. If*

$$\int_{t_0}^{\infty} \prod_{t_0 < t_k < t} \frac{1}{b_k} p(t) \Delta t = \infty, \tag{7.5}$$

then the equation (7.3) is oscillatory.

Proof 7.2.6 *Suppose that (7.3) has a non-oscillatory solution y. Without loss of generality, assume that y is an eventually positive solution of equation (7.3) and $k_0 = 1$. Then there exists a $t_0' \geq t_0$ such that $y(t) > \widetilde{0}$, $t \geq t_0'$. By Theorem 4.6.1, it follows that*

$$\delta_H y(t) \geq \widetilde{0}, \quad t \in (t_k, t_{k+1}], \quad t_k \geq t_0'.$$

Let

$$w(t) \cdot \phi(y(t)) = \delta_H y(t), \quad t \geq t_0'. \tag{7.6}$$

Then $w(t_k^+) \geq 0$, $t_k \geq t_0'$, and $w(t) > 0$, $t \in (t_k, t_{k+1}]$, $t_k \geq t_0'$. We have

$$\begin{aligned} w^{\Delta}(t) \cdot \phi(y^{\sigma}(t)) &\leq w^{\Delta}(t) \cdot \phi(y^{\sigma}(t)) + w(t) \cdot \delta_H \phi(y(t)) \\ &= \delta_H^2 y(t) \\ &= \ominus_H f(t, y^{\sigma}(t)) \\ &\leq (-p(t)) \cdot \phi(y^{\sigma}(t), \quad t \geq t_0', \quad t \neq t_k, \quad k \in \mathbb{N}. \end{aligned}$$

Therefore

$$w^{\Delta}(t) \leq -p(t), \quad t \geq t_0', \quad t \neq t_k, \quad k \in \mathbb{N}.$$

Next,

$$w(t_k^+) \cdot \phi(y(t_k^+)) = \delta_H y(t_k^+)$$

$$
\begin{aligned}
&= h_k(\delta_H(y(t_k^-))) \\
&\leq b_k \cdot \delta_H(y(t_k^-)) \\
&= (b_k w(t_k^-)) \cdot \phi(y(t_k^-)), \quad k \in \mathbb{N}.
\end{aligned}
$$

On the other hand, we obtain the following.

$$
\begin{aligned}
w(t_k^+) \cdot \phi(y(t_k^+)) &= w(t_k^+) \cdot \phi(g_k(y(t_k^-))) \\
&\geq w(t_k^+) \cdot \phi(a_k^* \cdot y(t_k^-)) \\
&\geq w(t_k^+) \cdot \phi(y(t_k^-)), \quad k \in \mathbb{N}.
\end{aligned}
$$

Consequently

$$
w(t_k^+) \cdot \phi(y(t_k^-)) \leq (b_k w(t_k^-)) \cdot \phi(y(t_k^-)), \quad k \in \mathbb{N},
$$

or

$$
w(t_k^+) \leq b_k w(t_k^-), \quad k \in \mathbb{N}.
$$

Hence and with Theorem 4.6.1, we arrive at

$$
\begin{aligned}
w(t) &\leq w(t_0) \prod_{t_0<t_k<t} b_k - \int_{t_0}^{t} \prod_{s<t_k<t} b_k p(s)\Delta s \\
&= \prod_{t_0<t_k<t} b_k \left(w(t_0) - \int_{t_0}^{t} \prod_{t_0<t_k<s} \frac{1}{b_k} p(s)\Delta s \right) \\
&\to -\infty, \quad as \quad t \to \infty.
\end{aligned}
$$

This is a contradiction. This completes the proof.

Theorem 7.2.7 *Suppose* $(H8)$, $(H9)$, (7.4) *hold and*

$$
\phi(a \cdot b) \geq \psi(a) \cdot \phi(b) \quad for \quad any \quad a \in \mathbb{R}, \quad b \in F(\mathbb{R}), \quad a \cdot b > \widetilde{0},
$$

where $\psi \in \mathscr{C}(\mathbb{R})$ is a non-negative function. If

$$\int_{t_0}^{\infty} \prod_{t_0<t_k<t} \frac{\psi(a_k^*)}{b_k} p(t)\Delta t = \infty,$$

then the equation (7.3) *is oscillatory.*

Proof 7.2.8 *Suppose that* (7.3) *has a non-oscillatory solution y. Without loss of generality, assume that y is an eventually positive solution of the equation* (7.3)*. Then there exists a $t_0' \geq t_0$ such that $y(t) > 0$, $t \geq t_0'$. Define w by* (7.6)*. We have $w(t_k^+) \geq 0$, $t_k \geq t_0'$, and using the proof of the previous theorem, we get*

$$\begin{aligned} (w(t_k^+)\psi(a_k^*)) \cdot \phi(y(t_k^-)) &\leq w(t_k^+) \cdot \phi(a_k^* \cdot y(t_k^-)) \\ &\leq (b_k w(t_k^-)) \cdot \phi(y(t_k^-)), \quad k \in \mathbb{N}, \end{aligned}$$

whereupon

$$w(t_k^+)\psi(a_k^*) \leq b_k w(t_k^-), \quad k \in \mathbb{N},$$

or

$$w(t_k^+) \leq \frac{b_k}{\psi(a_k^*)} w(t_k^-), \quad k \in \mathbb{N}.$$

Hence, applying Theorem 4.6.1, we arrive at

$$\begin{aligned} w(t) &\leq w(t_0) \prod_{t_0<t_k<t} \frac{b_k}{\psi(a_k^*)} - \int_{t_0}^{t} \prod_{s<t_k<t} \frac{b_k}{\psi(a_k^*)} p(s)\Delta s \\ &= \prod_{t_0<t_k<t} \frac{b_k}{\psi(a_k^*)} \left(w(t_0) - \int_{t_0}^{t} \prod_{t_0<t_k<s} \frac{\psi(a_k^*)}{b_k} p(s)\Delta s \right) \\ &\to -\infty, \quad as \quad t \to \infty. \end{aligned}$$

This is a contradiction. This completes the proof.

Corollary 7.2.9 *Suppose* (*H*8), (*H*9), (7.4) *hold and there exists a positive integer* k_0 *such that*

$$a_k^* \geq 1, \quad b_k \leq 1, \quad k \geq k_0.$$

If

$$\int_{t_0}^{\infty} p(t)\Delta t = \infty,$$

then the equation (7.3) *is oscillatory.*

Proof 7.2.10 *Without loss of generality, suppose that* $k_0 = 1$. *Then*

$$\begin{aligned} \int_{t_0}^{\infty} \prod_{t_0 < t_k < t} \frac{1}{b_k} p(t)\Delta t &\geq \int_{t_0}^{\infty} p(t)\Delta t \\ &= \infty. \end{aligned}$$

Hence and with Theorem 7.2.5, it follows that equation (7.3) *is oscillatory. This completes the proof.*

Corollary 7.2.11 *Suppose* (*H*8), (*H*9), (7.4) *hold, and there exists a positive integer* k_0 *and a constant* $\alpha > 0$ *such that*

$$a_k^* \geq 1, \quad \frac{1}{b_k} \geq \left(\frac{t_{k+1}}{t_k}\right)^{\alpha}, \quad k \geq k_0.$$

If

$$\int_{t_0}^{\infty} t^{\alpha} p(t)\Delta t = \infty,$$

then the equation (7.3) *is oscillatory.*

Proof 7.2.12 *Without loss of generality, suppose that* $k_0 = 1$. *Then*

$$\int_{t_0}^{t} \prod_{t_0 < t_k < s} \frac{1}{b_k} p(s)\Delta s$$

$$
\begin{aligned}
&= \int_{t_0}^{t_1} p(s)\Delta s + \frac{1}{b_1}\int_{t_1}^{t_2} p(s)\Delta s + \cdots + \frac{1}{b_1 b_2 \ldots b_n}\int_{t_0}^{t} p(s)\Delta s \\
&\geq \frac{1}{t_1^{\alpha}}\left(\int_{t_1}^{t_2} t_2^{\alpha} p(s)\Delta s + \int_{t_2}^{t_3} t_3^{\alpha} p(s)\delta_H s + \cdots + \int_{t_n}^{t} t_{n+1}^{\alpha} p(s)\Delta s\right) \\
&\geq \frac{1}{t_1^{\alpha}}\left(\int_{t_1}^{t_2} s^{\alpha} p(s)\Delta s + \int_{t_2}^{t_3} s^{\alpha} p(s)\delta_H s + \cdots + \int_{t_n}^{t} s^{\alpha} p(s)\Delta s\right) \\
&= \frac{1}{t_1^{\alpha}}\int_{t_1}^{t} s^{\alpha} p(s)\Delta s \\
&\to \infty, \quad as \quad t \to \infty.
\end{aligned}
$$

Hence and with Theorem 7.2.5, it follows that the equation (7.3) is oscillatory. This completes the proof.

Exercise 7.2.13 *Let $(H8)$, $(H9)$, (7.4) hold and*

$$\phi(a \cdot b) \geq \psi(a) \cdot \phi(b) \quad for \quad any \quad a \in \mathbb{R}, \quad b \in F(\mathbb{R}), \quad a \cdot b > \widetilde{0},$$

where $\psi \in \mathscr{C}(\mathbb{R})$ is a non-negative function. Let also, there exists a positive integer k_0 and a constant $\alpha > 0$ such that

$$\frac{\psi(a_k^*)}{b_k} \geq \left(\frac{t_{k+1}}{t_k}\right)^{\alpha}, \quad k \geq k_0.$$

If

$$\int_{t_0}^{\infty} t^{\alpha} p(t)\Delta t = \infty,$$

prove that the equation (7.3) is oscillatory.

7.3 OSCILLATIONS OF HIGHER-ORDER LINEAR FUZZY IMPULSIVE DYNAMIC EQUATIONS

Let $\mathbb{T}$ is an unbounded above time scale and $0 \in \mathbb{T}$. Denote $J = [0, \infty)$. In this section we investigate the following linear higher-order fuzzy impulsive dynamic equation

$$
\begin{aligned}
\delta_H^n x(t) + p(t)\cdot x^{\sigma}(t) &= \widetilde{0}, \quad t \in J\backslash\{t_k\}_{k\in\mathbb{N}},\\
\delta_H^j x(t_k^+) &= a_k^j \cdot \delta_H^j x(t_k), \quad j \in \{0,\ldots,n-1\}, \quad k \in \mathbb{N},\\
x(t_0^+) &= x_0,\\
\delta_H^j x(t_0^+) &= x_0^{\Delta^j}, \quad j \in \{1,\ldots,n-1\},
\end{aligned} \tag{7.7}
$$

where n is even, $x_0, x_0^{\Delta^j} \in F(\mathbb{R})$, $j \in \{1,\ldots,n-1\}$,

$$
0 = t_0 < t_1 < \ldots < t_m < \ldots,
$$

$\lim\limits_{k\to\infty} t_k = \infty$, t_k, $k \in \mathbb{N}$, are right-dense, $a_k^j > 0$, $j \in \{0,\ldots,n-1\}$, $k \in \mathbb{N}$,

(H10) $\displaystyle\int_{t_0}^{\infty} \prod_{t_0<t_k<s} \frac{a_k^j}{a_k^{j-1}} \Delta s = \infty,$

$p \in \mathscr{C}_{rd}(\mathbb{T})$, $p > 0$ on $\mathbb{T}$. For $i \in \{1,\ldots,n\}$, denote

$$
\begin{aligned}
AC^i(J) &= \Big\{ y : J \to F(\mathbb{R}) \quad \text{is} \quad i-\text{times} \quad \delta_H-\text{differentiable},\\
&\qquad \delta_H^i y \quad \text{is} \quad \text{absolutely} \quad \text{continuous} \Big\},\\
PC &= \{ y : J \to F(\mathbb{R}) \quad \text{is} \quad \text{rd-continuous} \quad \text{except} \quad \text{at} \quad t_k,\\
&\qquad \exists y(t_k^-), \quad y(t_k^+), \quad \delta_H^j y(t_k^-), \quad \delta_H^j y(t_k^+),\\
&\qquad y(t_k^-) = y(t_k), \quad \delta_H^j y(t_k^-) = \delta_H^j y(t_k),\\
&\qquad j \in \{1,\ldots,n-1\}, \quad k \in \mathbb{N}\}.
\end{aligned}
$$

Definition 7.3.1 *A function $x \in PC\bigcap AC^n(J\backslash\{t_k\}_{k\in\mathbb{N}})$ is said to be a solution of the equation (7.7) if*

$$\delta_H^n x(t) + p(t)\cdot x^{\sigma}(t) = \widetilde{0}, \quad a.e. \quad on \quad J\backslash\{t_k\}_{k\in\mathbb{N}},$$

and

$$\begin{aligned}
\delta_H^j x(t_k^+) &= a_k^j \cdot \delta_H^j x(t_k), \quad j\in\{0,\ldots,n-1\}, \quad k\in\mathbb{N},\\
x(t_0^+) &= x_0,\\
\delta_H^j x(t_0^+) &= x_0^{\Delta^j}, \quad j\in\{1,\ldots,n-1\}.
\end{aligned}$$

Theorem 7.3.2 *Let x be a solution of the equation (7.7) and $(H10)$ holds. Suppose that there exist a $j\in\{1,\ldots,n-1\}$ and a $T\geq t_0$ such that*

$$\delta_H^{j+1}x(t) > \widetilde{0}, \quad \delta_H^j x(t) > \widetilde{0}, \quad t\geq T.$$

Then there exists a $T_1 \geq T$ so that

$$\delta_H^{j-1}x(t) > \widetilde{0}, \quad t\geq T_1.$$

Proof 7.3.3 *Without loss of generality, let $T = t_0$. Since*

$$\delta_H^{j+1}x(t) \geq \widetilde{0}, \quad \delta_H^j x(t) > \widetilde{0}, \quad t\in(t_k,t_{k+1}],$$

for any $k\in\mathbb{N}$, we conclude that $\delta_H^j x$ is monotonically non-decreasing on $(t_k,t_{k+1}]$ for any $k\in\mathbb{N}$. For $t\in(t_1,t_2]$, we get

$$\delta_H^j x(t) \geq \delta_H^j x(t_1^+),$$

and hence,

$$\delta_H^{j-1}x(t_2) \geq \delta_H^{j-1}x(t_1^+) + (t_2-t_1)\cdot\delta_H^j x(t_1^+).$$

As above, we obtain

$$\delta_H^{j-1}x(t_3) \geq \delta_H^{j-1}x(t_2^+) + (t_3-t_2)\cdot\delta_H^j x(t_2^+).$$

Because

$$\delta_H^j x(t_2) \geq \delta_H^j x(t_1^+),$$

we get

$$\begin{aligned}
\delta_H^{j-1} x(t_3) &\geq \delta_H^{j-1} x(t_2^+) + (t_3 - t_2) \cdot \delta_H^j x(t_2^+) \\
&= a_2^{j-1} \cdot \delta_H^{j-1} x(t_2) + (a_2^j (t_3 - t_2)) \cdot \delta_H^j x(t_2) \\
&\geq a_2^{j-1} \cdot \left(\delta_H^{j-1} x(t_1^+) + (t_2 - t_1) \cdot \delta_H^j x(t_1^+) \right) \\
&\quad + (a_2^j (t_2 - t_1)) \cdot \delta_H^j x(t_2) \\
&= a_2^{j-1} \cdot \Bigg(\delta_H^{j-1} x(t_1^+) + (t_2 - t_1) \cdot \delta_H^j x(t_1^+) \\
&\quad + \left(\frac{a_2^j}{a_2^{j-1}} (t_3 - t_2) \right) \cdot \delta_H^j x(t_1^+) \Bigg).
\end{aligned}$$

Applying induction, we get

$$\begin{aligned}
\delta_H^{j-1} x(t_m) &\geq \left(a_{m-1}^{j-1} \ldots a_3^{j-1} a_2^{j-1} \right) \cdot \Bigg(\delta_H^{j-1} x(t_1^+) \\
&+ \Bigg(\int_{t_1}^{t_2} \Delta t + \left(\frac{a_2^j}{a_2^{j-1}} \right) \cdot \int_{t_2}^{t_3} \Delta t \\
&+ \cdots + \left(\frac{a_2^j a_3^j \ldots a_{m-1}^j}{a_2^{j-1} a_3^{j-1} \ldots a_m^{j-1}} \right) \cdot \int_{t_{m-1}}^{t_m} \Delta t \Bigg) \cdot \delta_H^j x(t_1^+) \Bigg) \\
&\geq \left(a_{m-1}^{j-1} \ldots a_3^{j-1} a_2^{j-1} \right) \cdot \left(\delta_H^{j-1} x(t_1^+) + \int_{t_1}^{t_m} \prod_{t_1 < t_k < t_m} \frac{a_k^j}{a_k^{j-1}} \Delta t \cdot \delta_H^j x(t_1^+) \right).
\end{aligned}$$

Hence, and $(H10)$*, it follows that there exists a sufficiently large* $N > 0$ *such that when* $m \geq N$*,* $m \in \mathbb{N}$*, we have*

$$\delta_H^{j-1}x(t_m) > \widetilde{0}.$$

Now, using that

$$\delta_H^{j}x(t) > \widetilde{0} \quad on \quad (t_m, t_{m+1}],$$

we conclude that

$$\begin{aligned}\delta_H^{j-1}x(t) &> \delta_H^{j-1}x(t_m)\\ &> \widetilde{0}, \quad t \in (t_m, t_{m+1}].\end{aligned}$$

Therefore, there exists a $T_1 \geq T$ *such that*

$$\delta_H^{j-1}x(t) > \widetilde{0}, \quad t \geq T_1.$$

This completes the proof.

Exercise 7.3.4 *Let* x *be a solution of the equation* (7.7) *and* $(H10)$ *holds. Suppose that there exist a* $j \in \{1, \ldots, n-1\}$ *and a* $T \geq t_0$ *such that*

$$\delta_H^{j+1}x(t) < \widetilde{0}, \quad \delta_H^{j}x(t) < \widetilde{0}, \quad t \geq T.$$

Prove that there exists a $T_1 \geq T$ *so that*

$$\delta_H^{j-1}x(t) < \widetilde{0}, \quad t \geq T_1.$$

Theorem 7.3.5 *Let* x *be a solution of the equation* (7.7) *and* $(H1)$ *holds. Suppose that there exist a* $j \in \{1, \ldots, n\}$ *and a* $T \geq t_0$ *such that*

$$x(t) > \widetilde{0}, \quad \delta_H^{j}x(t) \leq \widetilde{0}, \quad t \geq T, \quad \delta_H^{j}x \not\equiv 0 \quad on \quad [T, \infty).$$

Then $\delta_H^{j-1}x(t) > \widetilde{0}$ *for all sufficiently large* t*.*

Proof 7.3.6 *Without loss of generality, suppose that* $T = t_0$*. Assume that there exists a* $t_m \geq T$ *such that*

$$\delta_H^{j-1}x(t_m) \leq \widetilde{0}.$$

Since

$$\delta_H^j x(t) \leq \widetilde{0}, \quad t \in [T, \infty),$$

We have that $\delta_H^{j-1} x$ *is monotonically non-increasing on* $(t_k, t_{k+1}]$ *for any* $k \geq m$. *Because* $\delta_H^j x \not\equiv \widetilde{0}$ *on* $[T, \infty)$, *there is a* $t_l \geq t_m$ *so that* $\delta_H^j x \not\equiv \widetilde{0}$ *on* $(t_l, t_{l+1}]$. *Without loss of generality, assume that* $l = m$. *Then*

$$\begin{aligned} \delta_H^{j-1} x(t_{m+1}) &< \delta_H^{j-1} x(t_m^+) \\ &= a_m^{j-1} \delta_H^{j-1} x(t_m) \\ &\leq \widetilde{0}. \end{aligned}$$

For $t \in (t_{m+1}, t_{m+2}]$, *we have*

$$\begin{aligned} \delta_H^{j-1} x(t_{m+2}) &< \delta_H^{j-1} x(t_{m+1}^+) \\ &= a_{m+1}^{j-1} \delta_H^{j-1} x(t_{m+1}) \\ &< \widetilde{0}. \end{aligned}$$

By induction, we get

$$\delta_H^{j-1} x(t) < \widetilde{0}, \quad t \in (t_{m+k}, t_{m+k+1}]$$

for any $k \in \mathbb{N}$. *Thus,*

$$\delta_H^j x(t) < \widetilde{0}, \quad \delta_H^{j-1} x(t) < \widetilde{0}, \quad t \geq t_m.$$

By Exercise 7.3.4, we get that there is a $t_r \geq t_m$ *so that*

$$\delta_H^{j-2} x(t) < \widetilde{0}, \quad t \geq t_r.$$

And so on, we get that $x(t) < \widetilde{0}$ *for all sufficiently large* t. *This is a contradiction. Consequently,* $\delta_H^{j-1} x(t) > \widetilde{0}$ *for all sufficiently large* t. *This completes the proof.*

Theorem 7.3.7 *Let x be a solution of the equation (7.7) and $(H10)$ holds. Suppose that there exists a $T \geq t_0$ so that $x(t) > \widetilde{0}$, $t \geq T$. Then there exists a $T' \geq T$ and $l \in \{1,3,\ldots,n-1\}$ such that*

$$\begin{aligned} \delta_H^j x(t) &> \widetilde{0}, \quad j \in \{0,1,\ldots,l\}, \\ (-1)^{j-1} \cdot \delta_H^j x(t) &> \widetilde{0}, \quad j \in \{l+1,\ldots,n-1\}, \qquad (7.8) \\ \delta_H^n x(t) &\leq \widetilde{0}, \quad t \in [T',\infty). \end{aligned}$$

Proof 7.3.8 *Without loss of generality, suppose that $T = t_0$. By (7.7), we get*

$$\begin{aligned} \delta_H^n x(t) &= (-p(t)) \cdot x^{\sigma}(t) \\ &\leq \widetilde{0}, \quad t \geq t_0. \end{aligned}$$

Hence and with Theorem 7.3.5, it follows that $\delta_H^{n-1} x(t) > \widetilde{0}$ for all sufficiently large t. Without loss of generality, suppose that $\delta_H^{n-1} x(t) > \widetilde{0}$ for $t \geq t_0$. Hence, there exists a $T_1 \geq T$ so that one of the following two possibilities holds.

(a1) $\delta_H^{n-1} x(t) > \widetilde{0}, \delta_H^{n-2} x(t) > \widetilde{0}, t \in [T_1,\infty)$,

(a2) $\delta_H^{n-1} x(t) > \widetilde{0}, \delta_H^{n-2} x(t) < \widetilde{0}, t \in [T_1,\infty)$.

1. *Let $(a1)$ holds. Then, by Theorem 7.3.2, we get $\delta_H^{n-3} x(t) > \widetilde{0}$ for all sufficiently large t. And so on, we obtain*

   $$\delta_H^{n-1} x(t) > \widetilde{0}, \quad \delta_H^{n-2} x(t) > \widetilde{0}, \quad \ldots, \delta_H x(t) > \widetilde{0}, \quad x(t) > \widetilde{0},$$

   for all sufficiently large t.

2. *Let $(a2)$ hold. Then, by Theorem 7.3.5, we get $\delta_H^{n-3} x(t) > \widetilde{0}$ for all sufficiently large t. Then there exists a $T_2 \geq T_1$ so that one of the following possibilities holds.*

   (a3) $\delta_H^{n-3} x(t) > \widetilde{0}$, $\delta_H^{n-4} x(t) > \widetilde{0}$, $t \in [T_2,\infty)$,

(a4) $\delta_H^{n-3}x(t) > \widetilde{0}$, $\delta_H^{n-4}x(t) < \widetilde{0}$, $t \in [T_2, \infty)$.

And so on, we get (7.8). This completes the proof.

Theorem 7.3.9 *Suppose that $(H10)$ holds and*

$$\int_{t_0}^{\infty} \prod_{t_0 < t_k < t} \frac{1}{a_k^{n-1}} p(t)\Delta t = \infty.$$

Then every solution of the equation (7.7) is oscillatory.

Proof 7.3.10 *Let x be a non-oscillatory solution f the equation (7.7). Without loss of generality, suppose that x is an eventually positive solution of the equation (7.7). Then there exists a $t_1 \in [t_0, \infty)$ so that*

$$x(t) > \widetilde{0}, \quad t \in [t_1, \infty).$$

Hence and with Theorem 7.3.7, it follows that there exists a $t_2 \in [t_1, \infty)$ so that

$$\delta_H^n x(t) \leq \widetilde{0}, \quad \delta_H^{n-1}x(t) > \widetilde{0}, \quad \delta_H x(t) > \widetilde{0}, \quad x(t) > \widetilde{0}, \quad t \in [t_2, \infty).$$

By (7.7), we get

$$\begin{aligned} \delta_H^{n-1}x(t_1) &= \delta_H^{n-1}x(t_0^+) \ominus_H \int_{t_0}^{t_1} p(t) \cdot x^{\sigma}(t)\Delta t \\ &\leq \delta_H^{n-1}x(t_0^+) \ominus_H x^{\sigma}(t_0^+) \cdot \int_{t_0}^{t_1} p(t)\Delta t. \end{aligned}$$

Similarly,

$$\begin{aligned} \delta_H^{n-1}x(t_2) &\leq \delta_H^{n-1}x(t_1^+) \ominus_H x^{\sigma}(t_1^+) \cdot \int_{t_1}^{t_2} p(t)\Delta t \end{aligned}$$

$$\leq a_1^{n-1} \cdot \delta_H^{n-1}x(t_1) \ominus_H x^{\sigma}(t_0^+) \cdot \int_{t_1}^{t_2} p(t)\Delta t$$

$$\leq a_1^{n-1} \cdot \left(x^{n-1}(t_0^+) \ominus_H x^\sigma(t_0^+) \cdot \int_{t_0}^{t_1} p(t)\Delta t \right)$$

$$\ominus_H x^\sigma(t_0^+) \cdot \int_{t_1}^{t_2} p(t)\Delta t$$

$$= a_1^{n-1} \cdot \left(x^{n-1}(t_0^+) \ominus_H x^\sigma(t_0^+) \cdot \int_{t_0}^{t_1} p(t)\Delta t \ominus_H \frac{1}{a_1^{n-1}} \int_{t_1}^{t_2} p(t)\Delta t \cdot x^\sigma(t_0^+) \right).$$

By induction, we arrive at

$$\delta_H^{n-1} x(t_{m+1}) \leq \left(a_1^{n-1} \dots a_m^{n-1} \right) \cdot \left(\delta_H^{n-1} x(t_0^+) \ominus_H x^\sigma(t_0^+) \cdot \left(\int_{t_0}^{t_1} p(t)\Delta t \right.\right.$$

$$\left.\left. + \frac{1}{a_1^{n-1}} \cdot \int_{t_1}^{t_2} p(t)\Delta t + \dots + \frac{1}{a_1^{n-1} \dots a_m^{n-1}} \cdot \int_{t_m}^{t_{m+1}} p(t)\Delta t \right)\right)$$

$$\leq \left(a_1^{n-1} \dots a_m^{n-1} \right) \cdot \left(\delta_H^{n-1} x(t_0^+) \right.$$

$$\left. \ominus_H x^\sigma(t_0^+) \cdot \left(\int_{t_0}^{t_{m+1}} \prod_{t_0 < t_k < t} \frac{1}{a_k^{n-1}} p(t)\Delta t \right)\right)$$

$$\to -\infty, \quad as \quad t_{m+1} \to \infty.$$

This is a contradiction. This completes the proof.

Corollary 7.3.11 *Assume that* $(H10)$ *holds and there exists a positive integer* k_0 *such that*

$$a_k^{n-1} \leq 1, \quad k \geq k_0.$$

If

$$\int_{t_0}^{\infty} p(t)\Delta t = \infty,$$

then every solution of the equation (7.7) *is oscillatory.*

Proof 7.3.12 *Without loss of generality, suppose that* $k_0 = 1$. *Then*

$$\int_{t_0}^{t} \prod_{t_0<t_k<s} \frac{1}{a_k^{n-1}} p(s)\Delta s \geq \int_{t_0}^{t} p(s)\Delta s$$

$$\to \infty, \quad as \quad t \to \infty.$$

Hence and with Theorem 7.3.9, it follows that every solution of the equation (7.7) *is oscillatory. This completes the proof.*

Corollary 7.3.13 *Assume that* $(H10)$ *holds and there exists a positive integer* k_0 *and a positive integer* α *such that*

$$\frac{1}{a_k^{n-1}} \geq \left(\frac{t_{k+1}}{t_k}\right)^{\alpha}, \quad k \geq k_0.$$

Let also,

$$\int_{t_0}^{\infty} t^{\alpha} p(t)\Delta t = \infty.$$

Then every solution of equation (7.7) *is oscillatory.*

Proof 7.3.14 *Without loss of generality, suppose that* $k_0 = 1$. *Then*

$$\int_{t_0}^{t_1} p(t)\Delta t + \frac{1}{a_1^{n-1}} \int_{t_1}^{t_2} p(t)\Delta t + \cdots + \frac{1}{a_1^{n-1} \dots a_m^{n-1}} \int_{t_m}^{t_{m+1}} p(t)\Delta t$$

$$\geq \frac{1}{a_1^{n-1}} \int_{t_1}^{t_2} p(t)\Delta t + \frac{1}{a_1^{n-1} a_2^{n-1}} \int_{t_2}^{t_3} p(t)\Delta t$$

$$+\cdots+\frac{1}{a_1^{n-1}\dots a_m^{n-1}}\int_{t_m}^{t_{m+1}} p(t)\Delta t$$

$$\geq \left(\frac{t_2}{t_1}\right)^{\alpha}\int_{t_1}^{t_2} p(t)\Delta t+\left(\frac{t_3}{t_1}\right)^{\alpha}\int_{t_2}^{t_3} p(t)\Delta t+\cdots+\left(\frac{t_{m+1}}{t_1}\right)^{\alpha}\int_{t_m}^{t_{m+1}} p(t)\Delta t$$

$$=\frac{1}{t_1^{\alpha}}\left(\int_{t_1}^{t_2} t_2^{\alpha}p(t)\Delta t+\int_{t_2}^{t_3} t_3^{\alpha}p(t)\Delta t+\cdots+\int_{t_m}^{t_{m+1}} t_{m+1}^{\alpha}p(t)\Delta t\right)$$

$$\geq\frac{1}{t_1^{\alpha}}\left(\int_{t_1}^{t_2} t^{\alpha}p(t)\Delta t+\int_{t_2}^{t_3} t^{\alpha}p(t)\Delta t+\cdots+\int_{t_m}^{t_{m+1}} t^{\alpha}p(t)\Delta t\right)$$

$$=\frac{1}{t_1^{\alpha}}\int_{t_1}^{t_{m+1}} t^{\alpha}p(t)\Delta t$$

$$\to\infty,\quad as\quad m\to\infty.$$

Hence and with Theorem 7.3.9, it follows that every solution of the equation (7.7) is oscillatory. This completes the proof.

Exercise 7.3.15 *Prove that every solution of the equation*

$$\begin{aligned}
\delta_H^n x(t)+\frac{1}{4t}\cdot x^{\sigma}(t) &= \widetilde{0},\quad t\geq\frac{1}{2},\quad t\neq k,\quad k\in\mathbb{N},\\
\delta_H^j x(k^+) &= \frac{k}{k+1}\cdot\delta_H^j x(k),\quad j\in\{0,1,\dots,n-1\},\\
x\left(\frac{1}{2}\right) &= \widetilde{0},
\end{aligned}$$

$$\delta_H^j x\left(\frac{1}{2}\right) = (4+j,1,0), \quad j \in \{0,\ldots,n-1\},$$

$$x(t) = (t^2,t^3,t^4), \quad t \in \left[\frac{1}{2},1\right],$$

where n is even, is oscillatory.

7.4 ADVANCED PRACTICAL PROBLEMS

Problem 7.4.1 *Let $f(t,y(t)) = f(y(t))$. Find conditions for a and f such that the zero solution of equation* (3.18) *is stable.*

Problem 7.4.2 *Let $f(t,y(t)) = g(t)+h(t)\cdot l(y)$. Find conditions for a, f, and l so that the zero solution of the equation* (3.18) *is stable.*

Problem 7.4.3 *Let α and β are lower and upper solutions, respectively, of the equation* (7.1), *and*

$$\alpha(t_k) = \left((-1)^k \frac{(t_k)^6}{1+(t_k)^2}, \frac{1}{1+t^2}, 1\right),$$

$$\beta(t_k) = \left((-1)^k (t_k)^8, t, 2\right), \quad k \in \mathbb{N}.$$

Prove that every solution y of the equation (7.1) *such that $\alpha \le y \le \beta$ is oscillatory.*

Problem 7.4.4 *Let $y_0, y_0^\Delta \in F(\mathbb{R})$, $\gamma \ge 3$, $\mu(t) \le ct$, $t \in \mathbb{T}$, $c > 0$. Prove that the equation*

$$\delta_H^2 y(t) + \frac{1}{t(\sigma(t))^2}\cdot(y(\sigma(t)))^\gamma = \widetilde{0}, \quad t \ge 1, \quad t \ne k, \quad k \in \mathbb{N},$$

$$y(k^+) = \frac{k+1}{k}\cdot y(k^-),$$

$$\delta_H y(k^+) = \delta_H y(k^-),$$

$$
\begin{aligned}
y(1) &= y_0, \\
\delta_H y(1) &= y_0^{\Delta},
\end{aligned}
$$

is oscillatory.

Problem 7.4.5 *Prove that every solution of the equation*

$$
\begin{aligned}
\delta_H^n x(t) + \frac{t^2+t+1}{4t^4+t^2+12} \cdot x^{\sigma}(t) &= \widetilde{0}, \quad t \geq \frac{1}{8}, \quad t \neq k, \quad k \in \mathbb{N}, \\
\delta_H^j x(k^+) &= \frac{k^2+k+1}{k^2+1} \cdot \delta_H^j x(k), \quad j \in \{0,1,\ldots,n-1\}, \\
x\left(\frac{1}{8}\right) &= (1,1,1), \\
\delta_H^j x\left(\frac{1}{8}\right) &= (-1+j, j, 1), \quad j \in \{0,\ldots,n-1\}, \\
x(t) &= \left(\frac{t+1}{t^2+5}, 1, 2\right), \quad t \in \left[\frac{1}{8}, 1\right],
\end{aligned}
$$

where n is even, is oscillatory.

CHAPTER 8

Linear Fuzzy Impulsive Dynamic Systems

Suppose that $\mathbb{T}$ is an unbounded above time scale with forward jump operator and delta differentiation operator σ and Δ, respectively. Let also, $0 \in \mathbb{T}$, $\{t_k\}_{k\in\mathbb{N}} \subset \mathbb{T}$ be such that

$$0 \leq t_0 < t_1 < \ldots < t_k < \ldots,$$

$\lim\limits_{k\to\infty} t_k = \infty$, and t_k, $k \in \mathbb{N}$, are right-dense. Denote $J = [t_0, \infty)$.

8.1 HOMOGENEOUS LINEAR FUZZY IMPULSIVE DYNAMIC SYSTEMS

Consider the following IVP

$$\begin{aligned} \delta_H x &= A(t)\cdot x, \quad t \in J\backslash\{t_k\}_{k\in\mathbb{N}}, \\ x(t_k^+) &= x(t_k^-) + B_k \cdot x(t_k), \quad k \in \mathbb{N}, \\ x(t_0) &= \eta, \end{aligned} \tag{8.1}$$

where $x : \mathbb{T} \to F(\mathbb{R})^n$, $\eta \in F(\mathbb{R})^n$, A is an $n \times n$ rd-continuous matrix-valued function, B_k, $k \in \mathbb{N}$, are constants $n \times n$ matrices. Define

DOI: 10.1201/9781003655367-8

$$\Omega = \Big\{x : J \to F(\mathbb{R})^n, \quad x \in \mathscr{C}((t_k, t_{k+1})), \quad k \in \mathbb{N}_0,$$
$$\exists x(t_k^+), x(t_k^-), \quad x(t_k^-) = x(t_k), \quad k \in \mathbb{N}\Big\},$$
$$\Omega^{(1)} = \{x \in \Omega : x \in \mathscr{C}^1((t_k, t_{k+1})), \quad k \in \mathbb{N}_0\}.$$

Definition 8.1.1 *A function $x \in \Omega^{(1)}$ is said to be a solution of the IVP* (8.1) *if*

$$\delta_H x = A(t) \cdot x, \quad on \quad J \backslash \{t_k\}_{k \in \mathbb{N}},$$

and

$$x(t_k^+) = x(t_k^-) + B_k \cdot x(t_k), \quad k \in \mathbb{N},$$
$$x(t_0) = \eta.$$

Theorem 8.1.2 *Any solution x of the IVP* (8.1) *is also a solution of the impulsive integral equation*

$$x(t) = \eta + \int_{t_0}^{t} A(s) \cdot x(s) \delta_H s + \sum_{t_0 < t_j \leq t} B_j \cdot x(t_j), \quad t \geq t_0, \tag{8.2}$$

and conversely.

Proof 8.1.3 *1. Let x be a solution of the IVP* (8.1). *For $t \in [t_0, t_1]$, we have that x satisfies the IVP*

$$\delta_H x = A(t) \cdot x, \quad t \in [t_0, t_1],$$
$$x(t_0) = \eta.$$

Then

$$x(t) = \eta + \int_{t_0}^{t} A(s) \cdot x(s) \delta_H s, \quad t \in [t_0, t_1].$$

Let $t \in [t_1, t_2]$. Then x satisfies the IVP

$$\begin{aligned} \delta_H x &= A(t) \cdot x, \quad t \in (t_1, t_2], \\ x(t_1^+) &= x(t_1) + B_1 \cdot x(t_1). \end{aligned}$$

For its solution we have

$$\begin{aligned} x(t) &= x(t_1) + B_1 \cdot x(t_1) + \int_{t_1}^{t} A(s) \cdot x(s) \delta_H s \\ &= \eta + \int_{t_0}^{t_1} A(s) \cdot x(s) \delta_H s + B_1 \cdot x(t_1) \\ &\quad + \int_{t_1}^{t} A(s) \cdot x(s) \delta_H s \\ &= \eta + \int_{t_0}^{t} A(s) \cdot x(s) \delta_H s + B_1 \cdot x(t_1). \end{aligned}$$

By induction, we get (8.2).

2. *The converse statement we leave to the reader as an exercise. This completes the proof.*

Theorem 8.1.4 *Let $y, p \in \mathscr{C}_{frd}(\mathbb{T})$, $y, p \geq \widetilde{0}$ on $\mathbb{T}$, $c \in F(\mathbb{R})$, $c \geq \widetilde{0}$, $b_k \in [0, \infty)$, $k \in \mathbb{N}$, and*

$$y(t) \leq c + \int_{t_0}^{t} p(s) \cdot y(s) \delta_H s + \sum_{t_0 < t_k < t} b_k \cdot y(t_k), \quad t \geq t_0.$$

Then

$$y(t) \leq \left(\prod_{t_0 < t_k < t} (1 + b_k) e_p(t, t_0) \right) \cdot c, \quad t \geq t_0.$$

Proof 8.1.5 *Let*

$$v(t) = c + \int_{t_0}^{t} p(s) \cdot y(s) \delta_H s + \sum_{t_0 < t_k < t} b_k \cdot y(t_k), \quad t \geq t_0.$$

Then

$$\begin{aligned} y(t) &\leq v(t), \\ \delta_H v(t) &= p(t) \cdot y(t) \\ &\leq p(t) \cdot v(t), \quad t \geq t_0, \\ v(t_k^+) &= v(t_k) + b_k \cdot y(t_k) \\ &\leq v(t_k) + b_k \cdot v(t_k) \\ &= (1 + b_k) \cdot v(t_k), \\ v(t_0) &= c. \end{aligned}$$

Hence and with Theorem 4.6.1, it follows that

$$\begin{aligned} v(t) &\leq \left(\prod_{t_0 < t_k < t} (1 + b_k) e_p(t, t_0) \right) \cdot v(t_0) \\ &= \left(\prod_{t_0 < t_k < t} (1 + b_k) e_p(t, t_0) \right) \cdot c, \quad t \geq t_0. \end{aligned}$$

Therefore

$$y(t) \leq \left(\prod_{t_0 < t_k < t} (1 + b_k) e_p(t, t_0) \right) \cdot c, \quad t \geq t_0.$$

This completes the proof.

Theorem 8.1.6 *The solution x of the IVP* (8.1) *satisfies the following estimate.*

$$\|x(t)\| \le \|\eta\| \prod_{t_0<t_k\le t} (1+\|B_k\|) e^{\int_{t_0}^{t} \|A(s)\| \delta_H s}, \quad t \ge t_0.$$

Proof 8.1.7 *By* (8.2), *we get*

$$\begin{aligned}
\|x(t)\| &= \left\| \eta + \int_{t_0}^{t} A(s)\cdot x(s)\delta_H s + \sum_{t_0<t_j\le t} B_j \cdot x(t_j) \right\| \\
&\le \|\eta\| + \left\| \int_{t_0}^{t} A(s)\cdot x(s)\delta_H s \right\| + \left\| \sum_{t_0<t_j\le t} B_j \cdot x(t_j) \right\| \\
&\le \|\eta\| + \int_{t_0}^{t} \|A(s)\cdot x(s)\|\Delta s + \sum_{t_0<t_j\le t} \|B_j \cdot x(t_j)\| \\
&\le \|\eta\| + \int_{t_0}^{t} \|A(s)\|\|x(s)\|\Delta s + \sum_{t_0<t_j\le t} \|B_j\|\|x(t_j)\|, \quad t \ge t_0.
\end{aligned}$$

From the last inequality and from Theorem 8.1.4, it follows that

$$\begin{aligned}
\|x(t)\| &\le \|\eta\| \prod_{t_0<t_j\le t} (1+\|B_j\|) e_{\|A(\cdot)\|}(t,t_0) \\
&\le \|\eta\| \prod_{t_0<t_k\le t} (1+\|B_k\|) e^{\int_{t_0}^{t} \|A(s)\| \delta_H s}, \quad t \ge t_0.
\end{aligned}$$

This completes the proof.

With $\Phi_A(t,s)$, $0 \le s \le t$, we denote the transition matrix of the dynamic system

$$\delta_H x = A(t)x.$$

Definition 8.1.8 *Define the impulsive transition matrix* $S_A(t,s)$*,* $0 \leq s \leq t$*, as follows.*

$$S_A(t,s) = \begin{cases} \Phi_A(t,s) \quad \text{if} \quad t_{k-1} \leq s \leq t \leq t_k \\ \Phi_A(t,t_k^+)(I+B_k)\Phi_A(t,s) \quad \text{if} \quad t_{k-1} \leq s < t_k < t < t_{k+1} \\ \Phi_A(t,t_k^+)\left(\prod_{s<t_j\leq t}(I+B_j)\Phi_A(t_j,t_{j-1}^+)\right)(I+B_i)\Phi_A(t_i,s) \\ \text{if} \quad t_{i-1} \leq s < t_i < \ldots < t_k < t < t_{k+1}. \end{cases}$$

Note that

$$S_A(t_k,s)=\Phi_A(t_k,t_{k-1}^+)\left(\prod_{s<t_j\leq t_{k-1}}(I+B_j)\Phi_A(t_j,t_{j-1}^+)\right)(I+B_i)\Phi_A(t_i,s)$$

for $t_{i-1} \leq s < t_i < \ldots < t_k$, and hence,

$$\begin{aligned} S_A(t,s) &= \Phi_A(t,t_k^+)(I+B_k)\Phi_A(t_k,t_{k-1}^+) \\ &\quad \times \left(\prod_{s<t_j\leq t_{k-1}}(I+B_j)\Phi_A(t_j,t_{j-1}^+)\right)(I+B_i)\Phi_A(t_i,s) \\ &= \Phi_A(t,t_{k-1})(I+B_k)S_A(t_k,s) \end{aligned}$$

if $t_{i-1} \leq s < t_i < \ldots < t_k < t < t_{k+1}$.

Theorem 8.1.9 *The IVP* (8.1) *has a unique solution given by*

$$x(t) = S_A(t,t_0)\cdot\eta, \quad t \geq t_0. \tag{8.3}$$

Proof 8.1.10 *For* $t \in [t_0,t_1)$*, the unique solution of the IVP* (8.1) *is given by*

$$\begin{aligned} x(t) &= \Phi_A(t,t_0)\cdot\eta \\ &= S_A(t,t_0)\cdot\eta. \end{aligned}$$

Let $t \in [t_1, t_2)$. Consider the IVP

$$\begin{aligned} \delta_H x &= A(t)\cdot x, \quad t \in (t_1, t_2), \\ x(t_1^+) &= x(t_1) + B_1 \cdot x(t_1). \end{aligned}$$

The unique solution of the last IVP is given by

$$\begin{aligned} x(t) &= \Phi_A(t, t_1^+)\cdot x(t_1^+) \\ &= (\Phi_A(t, t_1^+)(I + B_1))\cdot x(t_1) \\ &= (\Phi_A(t, t_1^+)(I + B_1)\Phi_A(t_1, t_0))\cdot \eta \\ &= S_A(t, t_0)\cdot \eta. \end{aligned}$$

By induction, we arrive at (8.3). *This completes the proof.*

Next, we have.

1.

$$\begin{aligned} S_A(t_k^+, s) &= \Phi_A(t_k^+, t_k^+)(I + B_k)S_A(t_k, s) \\ &= (I + B_k)S_A(t_k, s), \quad t_k \geq s, \quad k \in \mathbb{N}. \end{aligned}$$

2.

$$\begin{aligned} S_A(t, t_k^+) &= \Phi_A(t, t_k^+), \\ S_A(t, t_k) &= \Phi_A(t, t_k) \\ &= \Phi_A(t, t_k^+)(I + B_k)\Phi_A(t_k, t_k) \\ &= S_A(t, t_k^+)(I + B_k), \end{aligned}$$

whereupon

$$S_A(t,t_k^+) = S_A(t,t_k)(I+B_k)^{-1}, \quad t_k \le t, \quad k \in \mathbb{N}.$$

3.

$$\begin{aligned} S_A(t,s) &= \Phi_A(t,t_k^+)(I+B_k)S_A(t_k,s) \\ &= S_A(t,t_k^+)(I+B_k)S_A(t_k,s) \\ &= S_A(t,t_k^+)S_A(t_k^+,s), \quad t_0 \le s \le t_k \le t, \quad k \in \mathbb{N}. \end{aligned}$$

4. Using Theorem 8.1.6, we get

$$\begin{aligned} \|x(t)\| &= \|S_A(t,t_0)\cdot\eta\| \\ &\le \|\eta\| \prod_{t_0<t_k\le t}(1+\|B_k\|)e^{\int_{t_0}^{t}\|A(s)\|\delta_H s}, \quad t \ge t_0, \end{aligned}$$

for any $\eta \in \mathbb{R}^n$. Therefore

$$\|S_A(t,t_0)\| \le \prod_{t_0<t_k\le t}(1+\|B_k\|)e^{\int_{t_0}^{t}\|A(s)\|\delta_H s}, \quad t \ge t_0.$$

5. $S_A(\sigma(t),s) = (I+\mu(t)A(t))S_A(t,s), 0 \le s \le t.$

Proof 8.1.11 *We have*

$$\begin{aligned} S_A(\sigma(t),s) &= S_A(t,s)+\mu(t)S_A^{\Delta}(t,s) \\ &= S_A(t,s)+\mu(t)A(t)S_A(t,s) \\ &= (I+\mu(t)A(t))S_A(t,s), \quad 0 \le s \le t. \end{aligned}$$

This completes the proof.

8.2 NON-HOMOGENEOUS LINEAR FUZZY IMPULSIVE DYNAMIC SYSTEMS

Let $l^\infty(F(\mathbb{R})^n)$ be space of all sequences $\{c_k\}_{k\in\mathbb{N}}$, $c_k \in F(\mathbb{R})^n$, $k \in \mathbb{N}$, such that

$$\sup_{k\in\mathbb{N}} \|c_k\| < \infty.$$

Then $l^\infty(F(\mathbb{R}^n))$ is a Banach space with the norm

$$\|c\| = \sup_{k\geq 1} \|c_k\|.$$

Consider the IVP

$$\begin{aligned} \delta_H x &= A(t)\cdot x + h(t), \quad t \in J\backslash\{t_k\}_{k\in\mathbb{N}}, \\ x(t_k^+) &= x(t_k) + B_k \cdot x(t_k) + c_k, \quad k \in \mathbb{N}, \\ x(t_0^+) &= \eta, \end{aligned} \tag{8.4}$$

where η, A, B_k, $k \in \mathbb{N}$, are as in the previous section, $\{c_k\}_{k\in\mathbb{N}} \in l^\infty(F(\mathbb{R})^n)$, $h : J \to F(\mathbb{R})^n$ is a given function.

Theorem 8.2.1 *The IVP* (8.4) *has a unique solution given by*

$$x(t) = S_A(t,t_0)\cdot\eta + \int_{t_0}^{t} S_A(t,\sigma(s))\cdot h(s)\delta_H s + \sum_{t_0<t_j<t} S_A(t,t_j^+)\cdot c_j, \tag{8.5}$$

$t \geq t_0$.

Proof 8.2.2 *For $t \in [t_0,t_1]$, we have that*

$$\begin{aligned} x(t) &= \Phi_A(t,t_0)\cdot\eta + \int_{t_0}^{t} \Phi_A(t,\sigma(s))\cdot h(s)\delta_H s \\ &= S_A(t,t_0)\cdot\eta + \int_{t_0}^{t} S_A(t,\sigma(s))\cdot h(s)\delta_H s. \end{aligned}$$

Let $t \in (t_1, t_2]$. *Consider the IVP*

$$\begin{aligned}\delta_H x &= A(t)\cdot x + h(t), \quad t \in (t_1, t_2],\\ x(t_1^+) &= x(t_1) + B_1 \cdot x(t_1) + c_1.\end{aligned}$$

Then

$$\begin{aligned}x(t) &= \Phi_A(t, t_1^+)\cdot x(t_1^+) + \int_{t_1^+}^{t} \Phi_A(t, \sigma(s))\cdot h(s)\delta_H s\\ &= (\Phi_A(t, t_1^+)(I + B_1))\cdot x(t_1) + c_1 + \int_{t_1^+}^{t} \Phi_A(t, \sigma(s))\cdot h(s)\delta_H s\\ &= (S_A(t, t_1^+)(I + B_1))\\ &\quad \cdot\left(S_A(t_1, t_0)\cdot \eta + \int_{t_0}^{t_1} S_A(t_1, \sigma(s))\cdot h(s)\delta_H s\right)\\ &\quad + S_A(t, t_1^+)\cdot c_1 + \int_{t_1^+}^{t} S_A(t, \sigma(s))\cdot h(s)\delta_H s\\ &= (S_A(t, t_1^+)(I + B_1)S_A(t_1, t_0))\cdot \eta\\ &\quad + \int_{t_0}^{t_1} (S_A(t, t_1^+)(I + B_1)S_A(t_1, \sigma(s)))\cdot h(s)\delta_H s\\ &\quad + S_A(t, t_1^+)\cdot c_1 + \int_{t_1^+}^{t} S_A(t, \sigma(s))\cdot h(s)\delta_H s\\ &= S_A(t, t_0)\cdot \eta + S_A(t, t_1^+)\cdot c_1\end{aligned}$$

$$+\int_{t_0}^{t_1} S_A(t,\sigma(s))\cdot h(s)\delta_H s+\int_{t_1^+}^{t} S_A(t,\sigma(s))\cdot h(s)\delta_H s$$

$$= S_A(t,t_0)\cdot\eta+S_A(t,t_1^+)\cdot c_1+\int_{t_0}^{t} S_A(t,\sigma(s))\cdot h(s)\delta_H s,$$

$t\in(t_1,t_2]$. *By induction, we get* (8.5). *This completes the proof.*

Exercise 8.2.3 *Prove that the IVP*

$$\begin{aligned}\delta_H y &= (-A^T(t))\cdot y^\sigma+h(t), \quad t\in J\backslash\{t_k\}_{k\in\mathbb{N}},\\ y(t_k^+) &= (I+C_k)\cdot y(t_k)+c_k, \quad k\in\mathbb{N},\\ y(t_0^+) &= \eta,\end{aligned}$$

where

$$C_k=-B_k^T(I+B_k^T)^{-1}, \quad k\in\mathbb{N},$$

has a unique solution given by

$$y(t)=S_{\ominus A^T}(t,t_0)\cdot\eta+\int_{t_0}^{t} S_{\ominus A^T}(t,s)\cdot h(s)\delta_H s+\sum_{t_0<t_j<t} S_{\ominus A^T}(t,t_j^+)\cdot c_j,$$

$t\geq t_0$.

8.3 BOUNDEDNESS AND STABILITY OF LINEAR FUZZY IMPULSIVE DYNAMIC SYSTEMS

Definition 8.3.1 *The fuzzy impulsive dynamic system* (8.1) *is said to be exponentially stable if there exists a positive constant* λ *with* $-\lambda\in\mathscr{R}^+$ *such that for any* $t_0\geq 0$, *there exists an* $N=N(t_0)\geq 1$ *such that the solution of* (8.1) *through* $(t_0,x(t_0))$ *satisfies*

$$\|x(t)\|\leq N\|x(t_0)\|e_{-\lambda}(t,t_0), \quad t\geq t_0.$$

Definition 8.3.2 *The fuzzy dynamic system* (8.1) *is said to be uniformly exponentially stable if it is exponentially stable and the constant N is chosen independently of $t_0 \geq 0$.*

Theorem 8.3.3 *Suppose that there exists a positive constant θ such that*

$$t_{k+1} - t_k < \theta, \quad k \in \mathbb{N}.$$

If the solution of the following IVP

$$\begin{aligned} \delta_H x &= A(t) \cdot x, \quad t \in J \backslash \{t_k\}_{k \in \mathbb{N}}, \\ x(t_k^+) &= x(t_k) + B_k \cdot x(t_k) + c_k, \quad k \in \mathbb{N}, \\ x(t_0^+) &= \widetilde{0}, \end{aligned} \tag{8.6}$$

is bounded for any $c = \{c_k\}_{k \in \mathbb{N}} \in l^\infty(F(\mathbb{R})^n)$, then there exist positive constants $N = N(t_0) \geq 1$, λ with $-\lambda \in \mathscr{R}^+$ such that

$$\|S_A(t, t_0)\| \leq N e_{-\lambda}(t, t_0), \quad t \geq t_0,$$

and it is exponentially stable.

Proof 8.3.4 *By Theorem 8.2.1, it follows that the IVP* (8.6) *has a unique solution x given by*

$$x(t) = \sum_{t_0 < t_j < t} S_A(t, t_j^+) \cdot c_j, \quad t \geq t_0.$$

For fixed $t \geq t_0$, define the operator $U_t : l^\infty(F(\mathbb{R})^n) \to F(\mathbb{R})^n$ as follows

$$U_t(c) = x(t), \quad c \in l^\infty(F(\mathbb{R})^n).$$

Note that the operator U_t is bounded and

$$\begin{aligned} \|U_t(c)\| &= \|x(t)\| \\ &= \| \sum_{t_0 < t_j < t} S_A(t, t_j^+) \cdot c_j \| \end{aligned}$$

$$\leq \sum_{t_0<t_j<t} \|S_A(t,t_j^+)\| \|c\|_{l^\infty}$$

$$< \infty.$$

Since the solution x of the IVP (8.6) is bounded, by the uniform boundedness principle, it follows that there exists a constant $K>0$ such that

$$\|x(t)\| \leq K\|c\|_{l^\infty},$$

for any $c \in l^\infty(F(\mathbb{R})^n)$ and for any $t_0 \geq 0$. Therefore

$$\| \sum_{t_0<t_j<t} S_A(t,t_j^+) \cdot c_j\| \leq K\|c\|_{l^\infty(\mathbb{R}^n)}, \tag{8.7}$$

for any $c \in l^\infty(F(\mathbb{R})^n)$ and $t_0 \geq 0$. Fix $t_0 \geq 0$. Then there exists $i \in \mathbb{N}$ such that $t_0 \in [t_{i-1},t_i)$. Define the sequences $\{\beta_j\}_{j\in\mathbb{N}}$ and $\{c_j\}_{j\in\mathbb{N}} \in l^\infty(F(\mathbb{R})^n)$ as follows

$$\beta_j = \begin{cases} 0 \quad if \quad j<i \\ \\ \|S_A(t_j^+,t_0)\| \quad if \quad j \geq i, \end{cases}$$

and

$$c_j = \begin{cases} \tilde{0} \quad if \quad j<i \\ \\ \left(\frac{1}{\beta_j} S_A(t_j^+,t_0)\right) \cdot y \quad if \quad j \geq i, \end{cases}$$

respectively. Here, $y \in F(\mathbb{R})^n \backslash \{\tilde{0}\}$ is arbitrarily chosen. Note that

$$\|c\|_{l^\infty} = 1.$$

Consequently

$$\|S_A(t,t_0) \cdot y\| \sum_{t_0<t_j<t} \frac{1}{\beta_j} \leq K\|y\|,$$

for any $t \geq t_0$. Since $y \in F(\mathbb{R})^n \backslash \{\tilde{0}\}$ was arbitrarily chosen, by the last inequality we conclude that

$$\|S_A(t,t_0)\| \sum_{t_0<t_j<t} \frac{1}{\beta_j} \leq K, \tag{8.8}$$

for any $t \geq t_0$. *If* $t_k > t_i$ *and* $t \in [t_0, t_{k+1})$, *then*

$$\|S_A(t_k^+, t_0)\| \sum_{t_0 < t_j < t} \frac{1}{\beta_j} \leq K,$$

for any $t \in [t_0, t_{k+1})$. *Therefore*

$$\beta_k \sum_{t_0 < t_j < t} \frac{1}{\beta_j} \leq K, \tag{8.9}$$

for any $t \in [t_0, t_{k+1})$. *If* $t \in [t_0, t_{i+1})$, *using* (8.8), *we obtain*

$$\|S_A(t, t_0)\| \leq K\beta_i$$

for any $t \in (t_i, t_{i+1})$. *Without loss of generality, suppose that* $K > 1$. *Now, if* $t \in [t_0, t_{i+2})$, *using* (8.9), *we obtain*

$$\beta_{i+1}\left(\frac{1}{\beta_i} + \frac{1}{\beta_{i+1}}\right) \leq K,$$

whereupon

$$\beta_{i+1} \leq (K-1)\beta_i.$$

If $t \in [t_0, t_{i+3})$, *using* (8.9), *we obtain*

$$\beta_{i+2}\left(\frac{1}{\beta_i} + \frac{1}{\beta_{i+1}} + \frac{1}{\beta_{i+2}}\right) \leq K.$$

Hence,

$$\begin{aligned} 1 + \frac{\beta_{i+2}}{(K-1)\beta_i} + \frac{\beta_{i+2}}{\beta_i} &\leq 1 + \frac{\beta_{i+2}}{\beta_{i+1}} + \frac{\beta_{i+2}}{\beta_i} \\ &\leq K, \end{aligned}$$

that is

$$\frac{\beta_{i+2}}{\beta_i}\left(\frac{1}{K-1} + 1\right) \leq K - 1.$$

It follows that

$$\beta_{i+2} \leq \frac{(K-1)^2}{K}\beta_i.$$

By induction, if $t \in (t_0, t_{i+j})$, *we get*

$$\beta_{i+j} \leq \frac{(K-1)^j}{K^{j-1}} \beta_i, \quad j \geq 1. \tag{8.10}$$

Assume that (8.10) *is true for any* $j \geq l-1$. *If* $t \in [t_0, t_{i+l})$, *using* (8.9), *we have that*

$$\beta_{i+l}\left(\frac{1}{\beta_i} + \frac{1}{\beta_{i+1}} + \cdots + \frac{1}{\beta_{i+l}}\right) \leq K.$$

It follows that

$$\begin{aligned} 1+\beta_{i+l}\left(\frac{1}{\beta_i} + \frac{1}{(K-1)\beta_i} + \frac{K}{(K-1)^2\beta_i} + \cdots + \frac{K^{l-2}}{(K-1)^{l-1}\beta_i}\right) & \\ \leq \; 1+\beta_{i+l}\left(\frac{1}{\beta_i} + \frac{1}{\beta_{i+1}} + \cdots + \frac{1}{\beta_{i+l-1}}\right) & \\ \leq \; K, & \end{aligned}$$

whereupon

$$\frac{\beta_{i+l}}{\beta_i}\left(1 + \frac{1}{K-1} + \frac{K}{(K-1)^2} + \cdots + \frac{K^{l-2}}{(K-1)^{l-1}}\right) \leq K-1,$$

and so,

$$\beta_{i+l} \leq \frac{(K-1)^l}{K^{l-1}} \beta_i.$$

Let $j \geq 1$ *be fixed and let* $\{c_l\}_{l \in \mathbb{N}} \in l^\infty(F(\mathbb{R})^n)$ *be a sequence defined as follows*

$$c_l = \begin{cases} \widetilde{0} & \text{if} \quad l \neq i+j \\ \\ S_A(t_l^+, t_0) \cdot y & \text{if} \quad l = i+j, \end{cases} \tag{8.11}$$

where $y \in F(\mathbb{R})^n \backslash \{0\}$ *is arbitrarily chosen. Then, from* (8.7) *and* (8.10), *we obtain that*

$$\|S_A(t,t_0)\| \;\leq\; K\|S_A(t_{i+j}^+, t_0)\|$$

$$\begin{aligned} &\leq K\beta_{i+j} \\ &\leq \frac{(K-1)^j}{K^{j-2}}\beta_i, \end{aligned}$$

whereupon

$$\|S_A(t,t_0)\| \leq \frac{(K-1)^j}{K^{j-2}}\beta_i, \quad t \in [t_0, t_{i+j+1}).$$

Since

$$t_{k+1} - t_k \leq \theta, \quad k \in \mathbb{N},$$

it follows that

$$t - t_0 \leq t_{i+j+1} - t_i < (j+2)\theta \quad \textit{for} \quad t \in [t_{i+j}, t_{i+j+1}),$$

that is,

$$j > \frac{1}{\theta}(t - t_0) - 2.$$

Therefore

$$\|S_A(t,t_0)\| \leq \frac{K^4}{(K-1)^2}\beta_i \left(1 - \frac{1}{K}\right)^{\frac{1}{\theta}(t-t_0)} \qquad \textit{for} \quad t \in [t_{i+j}, t_{i+j+1}).$$

Now, define the positive function λ with $-\lambda \in \mathscr{R}^+$ as the solution of the inequality

$$e_{-\lambda}(t,t_0) \geq \left(1 - \frac{1}{K}\right)^{\frac{1}{\theta}(t-t_0)}, \quad t \in [t_{i+j}, t_{i+j+1}).$$

Let

$$N = \max\left\{\frac{K^4}{(K-1)^2}\beta_i, \sup_{t_0 \leq t \leq t_i} \frac{S_A(t,t_0)\|}{e_{-\lambda}(t,t_0)}\right\}.$$

Then

$$\|S_A(t,t_0)\| \leq N e_{-\lambda}(t,t_0), \quad t \geq t_0 \geq 0.$$

This completes the proof.

Exercise 8.3.5 *Suppose that there exist positive constants* θ, b, *and* M *such that*

$$t_{k+1} - t_k < \theta, \quad k \in \mathbb{N},$$

$$\sup_{k\in\mathbb{N}} \|B_k\| \leq b,$$

$$\int_{t_k}^{t_{k+1}} \|A(s)\|\Delta s \leq M, \quad k \in \mathbb{N}.$$

If the solution of the IVP (8.6) *is bounded for any* $c \in l^\infty(\mathbb{R}^n)$, *prove that it is uniformly exponentially stable.*

8.4 ADVANCED PRACTICAL PROBLEMS

Problem 8.4.1 *Prove that the IVP*

$$\delta_H y = (-A^T(t)) \cdot y^\sigma, \quad t \geq t_0, \quad t \neq t_k, \quad k \in \mathbb{N},$$

$$y(t_k^+) = (I + C_k) \cdot y(t_k), \quad k \in \mathbb{N},$$

$$y(t_0^+) = \eta,$$

where

$$C_k = -B_k^T(I + B_k^T), \quad k \in \mathbb{N},$$

has a unique solution given by

$$y(t) = y(t_0) \ominus_H \int_{t_0}^{t} A^T(s) \cdot y^\sigma(s) \delta_H s + \sum_{t_0 < t_j < t} B_j^T \cdot y(t_j), \quad t \geq t_0.$$

Problem 8.4.2 *Let there exists a constant* $\theta > 0$ *such that*

$$t_{k+1} - t_k \leq \theta, \quad k \in \mathbb{N}.$$

Prove that for each $\lambda > 0$ *such that* $\dfrac{1}{\lambda} > \theta$ *we have* $-\lambda \in \mathscr{R}^+$.

Problem 8.4.3 *Prove that for any* $\lambda > 0$ *so that* $-\lambda \in \mathscr{R}^+$ *we have*

$$e_{-\lambda}(t,t_0) \le e^{-\lambda(t-t_0)}, \quad t \ge t_0.$$

Problem 8.4.4 *Let there exist positive constants* γ, θ, b *and* M *such that*

$$\gamma \;<\; t_{k+1} - t_k < \theta,$$

$$\sup_{k\in\mathbb{N}} \|B_k\| \;\le\; b,$$

$$\int_{t_k}^{t_{k+1}} \|A(s)\|\Delta s \;\le\; M, \quad k \in \mathbb{N}.$$

Prove that if the solution of the IVP (8.6) *is bounded, then the solution of the IVP* (8.4) *is bounded.*

Problem 8.4.5 *Let*

$$A(t) \;=\; \begin{pmatrix} -1 & 0 \\ 0 & -2 \end{pmatrix},$$

$$I + B_k \;=\; \begin{pmatrix} \frac{1}{1+k} & 0 \\ 0 & \frac{1}{1+2k} \end{pmatrix}, \quad k \in \mathbb{N}.$$

Prove that the system (8.1) *is uniformly exponentially stable.*

APPENDIX A

Appendix: Fuzzy Sets

A.1 BASIC DEFINITIONS

Definition A.1.1 *If X is a collection of objects denoted generically by x, then a fuzzy set $\widetilde{A}$ in X is a set of ordered pairs*

$$\widetilde{A} = \{(x, u_{\widetilde{A}}(x)) : x \in X\},$$

where $u_{\widetilde{A}}(x)$ is called the membership function or grade of membership, or degree of compatibility, or degree of truth, of x in $\widetilde{A}$ to the membership space M. Here M is a subset of non-negative real numbers whose supremum is finite.

A fuzzy set is obviously a generalization of a classical set, and the membership function is a generalization of the characteristic function. Note that the membership function is not limited to values between 0 and 1.

Definition A.1.2 *If*

$$\sup_{x \in X} u_{\widetilde{A}}(x) = 1,$$

then the fuzzy set $\widetilde{A}$ is called normal.

A non-empty fuzzy set $\widetilde{A}$ can always be normalized by dividing $u_{\widetilde{A}}(x)$ by $\sup_{x \in X} u_{\widetilde{A}}(x)$. For convenience, we will assume that the fuzzy sets are normalized.

DOI: 10.1201/9781003655367-A

Example A.1.3 *Let* $X = \mathbb{R}$,

$$u_{\widetilde{A}}(x) = \begin{cases} 0 & \text{if} \quad x \leq 10 \\ \dfrac{1}{1+(x-10)^2} & \text{if} \quad x > 10. \end{cases}$$

Then

$$\widetilde{A} = \{(x, u_{\widetilde{A}}(x)) : x \in \mathbb{R}\}$$

is a fuzzy set.

Example A.1.4 *Let*

$$X = \{1,2,3,4,5,6,7,8,9,10\}$$

and

$$\begin{aligned} \widetilde{A} &= \{(1,0.2),(2,0.5),(3,0.8),(4,1),(5,0.7), \\ &\quad (6,0.3),(7,0),(8,0),(9,0),(10,0)\}. \end{aligned}$$

Here

$$\begin{aligned} u_{\widetilde{A}}(1) &= 0.2, \\ u_{\widetilde{A}}(2) &= 0.5, \\ u_{\widetilde{A}}(3) &= 0.8, \\ u_{\widetilde{A}}(4) &= 1, \\ u_{\widetilde{A}}(5) &= 0.7, \\ u_{\widetilde{A}}(6) &= 0.3, \end{aligned}$$

$$\begin{aligned} u_{\widetilde{A}}(7) &= 0, \\ u_{\widetilde{A}}(8) &= 0, \\ u_{\widetilde{A}}(9) &= 0, \\ u_{\widetilde{A}}(10) &= 0. \end{aligned}$$

Definition A.1.5 *The support $S(\widetilde{A})$ of a fuzzy set $\widetilde{A}$ is the set*

$$S(\widetilde{A}) = \{x \in X : u_{\widetilde{A}}(x) > 0\}.$$

Example A.1.6 *Consider Example A.1.4. Then*

$$S(\widetilde{A}) = \{1,2,3,4,5,6\}.$$

Definition A.1.7 *Let $\alpha \in \mathbb{R}_+$. Then we define the α-level set or α-cut as follows*

$$A_\alpha = \{x \in X : u_{\widetilde{A}}(x) \geq \alpha\},$$

and strong α-level set

$$A'_\alpha = \{x \in X : u_{\widetilde{A}}(x) > \alpha\}.$$

Sometimes the α-cut of $\widetilde{A}$ is denoted by $\left[u_{\widetilde{A}}\right]^\alpha$.

Example A.1.8 *Consider Example A.1.4. Then*

$$\begin{aligned} A_{0.2} &= \{1,2,3,4,5,6\}, \\ A_{0.5} &= \{2,3,4,5\}, \\ A_{0.8} &= \{3,4\}, \\ A_1 &= \{4\}, \\ A'_{0.8} &= \{4\}. \end{aligned}$$

Definition A.1.9 *A fuzzy set $\widetilde{A}$ is said to be convex, if*

$$u_{\widetilde{A}}(\lambda x_1 + (1-\lambda)x_2) \geq \min\{u_{\widetilde{A}}(x_1), u_{\widetilde{A}}(x_2)\}$$

for any $x_1, x_2 \in X$ and for any $\lambda \in [0,1]$.

Definition A.1.10 *For a finite fuzzy set $\widetilde{A}$ the cardinality $|\widetilde{A}|$ is defined as*

$$|\widetilde{A}| = \sum_{x \in X} u_{\widetilde{A}}(x).$$

The number

$$\|\widetilde{A}\| = \frac{|\widetilde{A}|}{|X|}$$

is called the relative cardinality of $\widetilde{A}$.

Example A.1.11 *Consider Example A.1.4. The cardinality of $\widetilde{A}$ is*

$$\begin{aligned} |\widetilde{A}| &= u_{\widetilde{A}}(1) + \cdots + u_{\widetilde{A}}(10) \\ &= 0.2 + 0.5 + 0.8 + 1 + 0.7 + 0.3 \\ &= 3.5 \end{aligned}$$

and the relative cardinality of $\widetilde{A}$ is

$$\begin{aligned} \|\widetilde{A}\| &= \frac{|\widetilde{A}|}{|X|} \\ &= \frac{3.5}{10} \\ &= 0.35. \end{aligned}$$

A.2 BASIC OPERATIONS WITH FUZZY SETS

Definition A.2.1 *The membership function $u_{\widetilde{C}}$ of the intersection $\widetilde{C} = \widetilde{A} \bigcap \widetilde{B}$ of two fuzzy sets $\widetilde{A}$ and $\widetilde{B}$ is point-wise defined by*

$$u_{\widetilde{C}}(x) = \min\{u_{\widetilde{A}}(x), u_{\widetilde{B}}(x)\}, \quad x \in X.$$

Definition A.2.2 *The membership function $u_{\widetilde{D}}$ of the union $\widetilde{D} = \widetilde{A} \bigcup \widetilde{B}$ of two fuzzy sets $\widetilde{A}$ and $\widetilde{B}$ is point-wise defined by*

$$u_{\widetilde{D}}(x) = \max\{u_{\widetilde{A}}(x), u_{\widetilde{B}}(x)\}, \quad x \in X.$$

Definition A.2.3 *The membership function of the complement of a fuzzy set $\widetilde{A}$ is defined by*

$$u_{C\widetilde{A}}(x) = 1 - u_{\widetilde{A}}(x), \quad x \in X.$$

Example A.2.4 *Let*

$$\begin{aligned} X &= \{1,2,3,4,5,6,7,8,9,10\}, \\ \widetilde{A} &= \{(1,0.2),(2,0.5),(3,0.8),(4,1),(5,0.7),(6,0.3)\}, \\ \widetilde{B} &= \{(3,0.2),(4,0.4),(5,0.6),(6,0.8),(7,1),(8,1)\}. \end{aligned}$$

Then

$$\begin{aligned} \widetilde{A} \bigcap \widetilde{B} &= \{(3,0.2),(4,0.4),(5,0.6),(6,0.3)\}, \\ \widetilde{A} \bigcup \widetilde{B} &= \{(1,0.2),(2,0.5),(3,0.8),(4,1),(5,0.7), \\ &\quad (6,0.8),(7,1),(8,1)\}, \\ C\widetilde{B} &= \{(1,1),(2,1),(3,0.8),(4,0.6),(5,0.4), \\ &\quad (6,0.2),(9,1),(10,1)\}. \end{aligned}$$

Definition A.2.5 *Let $\widetilde{A}_1,\ldots,\widetilde{A}_n$ be fuzzy sets in $X_1,\ldots,X_n$. The cartesian product is then a fuzzy set in $X_1\times\ldots\times X_n$ with the membership function defined by*

$$u_{\widetilde{A}_1\times\ldots\times\widetilde{A}_n}(x)=\min_{l\in\{1,\ldots,n\}}\{u_{\widetilde{A}_l}(x_l)\},\quad x=(x_1,\ldots,x_n)\in X_1\times\ldots\times X_n.$$

Example A.2.6 *Let*

$$\widetilde{A} = \{(1,0.1),(2,0.2),(3,0.4)\},$$

$$\widetilde{B} = \{(4,0.1),(5,0.3),(6,0.2)\}.$$

Then

$$\widetilde{A}\times\widetilde{B} = \{((1,4),0.1),((1,5),0.1),((1,6),0.1),$$

$$((2,4),0.1),((2,5),0.2),((2,6),0.2)$$

$$((3,4),0.1),((3,5),0.3),((3,6),0.2)\}.$$

Definition A.2.7 *The mth power of a fuzzy set $\widetilde{A}$ is a fuzzy set with the membership function*

$$u_{\widetilde{A}^m}(x)=\left(u_{\widetilde{A}}(x)\right)^m,\quad x\in X.$$

Example A.2.8 *Let*

$$\widetilde{A}=\{(1,0.1),(2,0.3),(3,1)\}.$$

Then

$$\widetilde{A}^3=\{(1,0.001),(2,0.027),(3,1)\}.$$

Definition A.2.9 *The algebraic sum $\widetilde{C}=\widetilde{A}+\widetilde{B}$ of two fuzzy sets $\widetilde{A}$ and $\widetilde{B}$ is a fuzzy set defined by*

$$\widetilde{C}=\{(x,u_{\widetilde{A}+\widetilde{B}}(x)):x\in X\},$$

where

$$u_{\widetilde{A}+\widetilde{B}}(x)=u_{\widetilde{A}}(x)+u_{\widetilde{B}}(x)-u_{\widetilde{A}}(x)u_{\widetilde{B}}(x),\quad x\in X.$$

Example A.2.10 *Let*

$$\begin{aligned} X &= \{3,5,7\}, \\ \widetilde{A} &= \{(3,0.5),(5,1),(7,0.6)\}, \\ \widetilde{B} &= \{(3,1),(5,0.6)\}. \end{aligned}$$

Then

$$\widetilde{A}+\widetilde{B}=\{(3,0.5),(5,1),(7,0.6)\}.$$

Definition A.2.11 *The bounded sum $\widetilde{C}=\widetilde{A}\oplus\widetilde{B}$ of two fuzzy sets $\widetilde{A}$ and $\widetilde{B}$ is defined as*

$$\widetilde{C}=\{(x,u_{\widetilde{A}\oplus\widetilde{B}}(x)):x\in X\},$$

where

$$u_{\widetilde{A}\oplus\widetilde{B}}(x)=\min\{1,u_{\widetilde{A}}(x)+u_{\widetilde{B}}(x)\},\quad x\in X.$$

Example A.2.12 *Let*

$$\begin{aligned} X &= \{1,2,3\}, \\ \widetilde{A} &= \{(1,0.2),(2,0.9)\}, \\ \widetilde{B} &= \{(1,0.9),(3,0.8)\}. \end{aligned}$$

Then

$$\widetilde{A}\oplus\widetilde{B}=\{(1,1),(2,0.9),(3,0.8)\}.$$

Definition A.2.13 *The bounded difference $\widetilde{C}=\widetilde{A}\ominus\widetilde{B}$ of two fuzzy sets $\widetilde{A}$ and $\widetilde{B}$ is defined as*

$$\widetilde{C}=\{(x,u_{\widetilde{A}\ominus\widetilde{B}}(x)):x\in X\},$$

where

$$u_{\widetilde{A}\ominus\widetilde{B}}(x)=\max\{0,u_{\widetilde{A}}(x)+u_{\widetilde{B}}(x)-1\},\quad x\in X.$$

Example A.2.14 *Let*

$$\begin{aligned} X &= \{1,3,4\}, \\ \widetilde{A} &= \{(1,0.1),(3,0.7),(4,0.9)\}, \\ \widetilde{B} &= \{(3,0.2),(4,0.9)\}. \end{aligned}$$

Then

$$\widetilde{A} \ominus \widetilde{B} = \{(4,0.8)\}.$$

Definition A.2.15 *The algebraic product $\widetilde{C} = \widetilde{A} \cdot \widetilde{B}$ of two fuzzy sets $\widetilde{A}$ and $\widetilde{B}$ is defined as*

$$\widetilde{A} \cdot \widetilde{B} = \{(x, u_{\widetilde{A}}(x) u_{\widetilde{B}}(x)) : x \in X\}.$$

Example A.2.16 *Let*

$$\begin{aligned} X &= \{1,2,3\}, \\ \widetilde{A} &= \{(1,0.2),(2,0.1)\}, \\ \widetilde{B} &= \{(1,0.4),(2,0.3),(3,0.4)\}. \end{aligned}$$

Then

$$\widetilde{A} \cdot \widetilde{B} = \{(1,0.08),(2,0.03)\}.$$

A.3 FUZZY NUMBERS

With $F(\mathbb{R})$ we denote the class of fuzzy subsets u of the real axis, satisfying the following properties.

1. u is normal, i.e., there exists $x_0 \in \mathbb{R}$ such that $u(x_0) = 1$,
2. u is fuzzy-convex, i.e.,

$$u(tx + (1-t)y) \geq \min\{u(x), u(y)\}, \quad \forall t \in [0,1], \quad x, y \in \mathbb{R},$$

3. u is upper semi-continuous on $\mathbb{R}$,

4. $\mathrm{cl}\{x \in \mathbb{R} : u(x) > 0\}$ is compact, where cl denotes the closure of a set

Definition A.3.1 *$F(\mathbb{R})$ is called the space of fuzzy numbers.*

Note that $\mathbb{R}$ is understood as $\mathbb{R} = \{\kappa_{\{x\}} : x \in \mathbb{R}\}$, where $\kappa_A = 1$ if $x \in A$ and $\kappa_A = 0$ if $x \notin A$. We have $\mathbb{R} \subset F(\mathbb{R})$. As a matter of convenience, we will denote a real fuzzy number $\widetilde{M}$ with membership function $u_{\widetilde{M}}(x)$ with u and for $0 < \alpha \leq 1$, denote

$$[u]^\alpha = \{x \in \mathbb{R} : u(x) \geq \alpha\},$$

and

$$[u]^0 = \mathrm{cl}\{x \in \mathbb{R} : u(x) > 0\}.$$

We refer $\underline{u}$ and $\overline{u}$ as the lower and upper branches on u, respectively. Note that $[u]^\alpha = [\underline{u}^\alpha, \overline{u}^\alpha]$ and $[u]^\alpha$ is a bounded closed interval.

Definition A.3.2 *For $u, v \in F(\mathbb{R})$ and $\lambda \in \mathbb{R}$, the sum $u + v$ and the product $\lambda \cdot u$ are defined by*

$$[u+v]^\alpha \;=\; [u]^\alpha + [v]^\alpha$$

and

$$[\lambda \cdot u]^\alpha \;=\; \lambda [u]^\alpha,$$

respectively, where

$$[u]^\alpha + [v]^\alpha = \{x + y : x \in [u]^\alpha, y \in [v]^\alpha\}$$

and

$$\lambda [u]^\alpha = \{\lambda x : x \in [u]^\alpha\}.$$

Denote $\widetilde{0} = \kappa_{\{0\}}$.

Theorem A.3.3 *For any $u \in F(\mathbb{R})$, we have*

$$\widetilde{0} + u = u + \widetilde{0} = u.$$

Proof A.3.4 *Let $u \in F(\mathbb{R})$. Then*

$$\begin{aligned}
[\widetilde{0}]^{\alpha} + [u]^{\alpha} &= \{x + y : \kappa_{\{0\}}(x) \geq \alpha, \quad y \in [u]^{\alpha}\} \\
&= \{y + x : y \in [u]^{\alpha}, \quad \kappa_{\{0\}}(x) \geq \alpha\} \\
&= [u]^{\alpha} + [\widetilde{0}]^{\alpha} \\
&= \{y : y \in [u]^{\alpha}\} \\
&= [u]^{\alpha}.
\end{aligned}$$

This completes the proof.

Remark A.3.5 *By Theorem A.3.3, it follows that $\widetilde{0}$ is the zero of $F(\mathbb{R})$.*

Theorem A.3.6 *For any $a, b \in \mathbb{R}$ with $a, b \geq 0$ or $a, b \leq 0$ and for any $u \in F(\mathbb{R})$, we have*

$$(a + b) \cdot u = a \cdot u + b \cdot u.$$

Proof A.3.7 *Let $a, b \in \mathbb{R}$, $a, b \geq 0$. Then*

$$\begin{aligned}
[(a + b) \cdot u]^{\alpha} &= \{(a + b)x : x \in [u]^{\alpha}\} \\
&= \{ax + bx : x \in [u]^{\alpha}\} \\
&= [a \cdot u]^{\alpha} + [b \cdot u]^{\alpha} \\
&= a[u]^{\alpha} + b[u]^{\alpha}.
\end{aligned}$$

The case $a, b \in \mathbb{R}$, $a, b \leq 0$, we leave to the reader as an exercise. This completes the proof.

Theorem A.3.8 *For any $\lambda \in \mathbb{R}$ and any $u, v \in F(\mathbb{R})$, we have*

$$\lambda \cdot (u+v) = \lambda \cdot u + \lambda \cdot v.$$

Proof A.3.9 *We have*

$$\begin{aligned} [\lambda \cdot (u+v)]^\alpha &= \{\lambda x + \lambda y : x \in [u]^\alpha, \quad y \in [v]^\alpha\} \\ &= [\lambda \cdot u]^\alpha + [\lambda \cdot v]^\alpha. \end{aligned}$$

This completes the proof.

Theorem A.3.10 *For any $\lambda, \mu \in \mathbb{R}$ and any $u \in F(\mathbb{R})$, we have*

$$\lambda \cdot (\mu \cdot u) = (\lambda \mu) \cdot u.$$

Proof A.3.11 *We have*

$$\begin{aligned} [\lambda \cdot (\mu \cdot u)]^\alpha &= \lambda [\mu \cdot u]^\alpha \\ &= \lambda (\mu [u]^\alpha) \\ &= (\lambda \mu)[u]^\alpha \\ &= [(\lambda \mu) \cdot u]^\alpha. \end{aligned}$$

This completes the proof.

Definition A.3.12 *A fuzzy number $\widetilde{A} = (a, b, c)$ is said to be a triangular fuzzy number if its membership is given by*

$$u_{\widetilde{A}}(x) = \begin{cases} 0 \quad if \quad x < a \\ \dfrac{x-a}{b-a} \quad if \quad a \le x \le b \\ \dfrac{c-x}{c-b} \quad if \quad b \le x \le c \\ 0 \quad if \quad x > c. \end{cases}$$

The α-cut of a triangular fuzzy number $\widetilde{A} = (a_1, a_2, a_3)$ is defined by

$$[u]^\alpha = [(a_2 - a_1)\alpha + a_1, -(a_3 - a_2)\alpha + a_3].$$

A triangular fuzzy number $\widetilde{A} = (a_1, a_2, a_3)$ is positive if $a_i > 0$, $i \in \{1,2,3\}$, and negative if $a_i < 0$, $i \in \{1,2,3\}$.

Definition A.3.13 *Two triangular fuzzy numbers*

$$\widetilde{A} = (a_1, a_2, a_3) \quad \text{and} \quad \widetilde{B} = (b_1, b_2, b_3)$$

are said to be equal if

$$a_i = b_i, \quad i \in \{1,2,3\}.$$

Definition A.3.14 *Let $\widetilde{A} = (a_1, a_2, a_3)$ and $\widetilde{B} = (b_1, b_2, b_3)$ be two triangular fuzzy numbers. Define*

1. *Addition*
$$\widetilde{A} + \widetilde{B} = (a_1 + b_1, a_2 + b_2, a_3 + b_3).$$

2. *Subtraction*
$$\widetilde{A} - \widetilde{B} = (a_1 - b_3, a_2 - b_2, a_3 - b_1).$$

3. *Multiplication*
$$\widetilde{A} \times \widetilde{B} = \Big(\min\{a_1b_1, a_1b_3, a_3b_1, a_3b_3\}, a_2b_2 \\ \max\{a_1b_1, a_1b_3, a_3b_1, a_3b_3\} \Big).$$

Example A.3.15 *Let*

$$\widetilde{A} = (1,2,3),$$

$$\widetilde{B} = (-3,-1,2).$$

Here

$$\begin{aligned} a_1 &= 1, \\ a_2 &= 2, \\ a_3 &= 3, \\ b_1 &= -3, \\ b_2 &= -1, \\ b_3 &= 2. \end{aligned}$$

Then

$$\begin{aligned} a_1 b_1 &= -3, \\ a_1 b_3 &= 2, \\ a_3 b_1 &= -9, \\ a_3 b_3 &= 6, \\ a_2 b_2 &= -2, \\ \min\{a_1 b_1, a_1 b_3, a_3 b_1, a_3 b_3\} &= -9, \\ \max\{a_1 b_1, a_1 b_3, a_3 b_1, a_3 b_3\} &= 6. \end{aligned}$$

Hence,

$$\widetilde{A} + \widetilde{B} = (-2, 1, 5),$$

$$\widetilde{A} - \widetilde{B} = (-1, 3, 6),$$

$$\widetilde{A} \times \widetilde{B} = (-9, -2, 6).$$

Suppose that X is a metric vector space with the induced topology, and in particular, $X = \mathbb{R}^n$, $n \geq 1$, of real vectors equipped with the standard addition and scalar multiplication operations. With $K(X)$ we will denote the space of non-empty compact sets of X, and with $K_c(X)$ we will denote the space of compact convex sets in X.

Definition A.3.16 *Let $A, B \in K(X)$. We define the H-difference of A and B as the set $C \in K(X)$ such that*

$$A \ominus_H B = C \iff \begin{cases} (i) & A = B + C \quad or \\ (ii) & B = A + (-1) \cdot C. \end{cases}$$

Theorem A.3.17 *If the H-difference $A \ominus_H B$ exists, then it is unique.*

Proof A.3.18

1. *Let $C = A \ominus_H B$ exists in the case (i). Then $A = B + C$ and C is unique.*

2. *Let $A \ominus_H B$ exists in the case (ii). Suppose that (ii) is satisfied for C and D, i.e., suppose that*

   $$B = A + (-1) \cdot C \quad and \quad B = A + (-1) \cdot D.$$

   Hence,

   $$A + (-1) \cdot C = A + (-1) \cdot D,$$

   whereupon

   $$(-1) \cdot C = (-1) \cdot D$$

   and $C = D$.

3. *If (i) is satisfied for C and (ii) is satisfied for D, i.e.,*

   $$A = B + C \quad and \quad B = A + (-1) \cdot D,$$

then

$$\begin{aligned} B &= A+(-1)\cdot D \\ &= B+C+(-1)\cdot D. \end{aligned}$$

Hence,

$$\{0\} = C+(-1)\cdot D,$$

which is possible if and only if

$$C = D = \{c\},$$

is singleton. This completes the proof.

Remark A.3.19 *A necessary condition for $A \ominus_H B$ to exist is that A contains a translation of B or B contains a translation of A. In fact, for any given $c \in C$, we get*

$$B+\{c\} = A \quad for \quad (i),$$

and

$$A+\{-c\} \subseteq B \quad for \quad (ii).$$

Remark A.3.20 *If $A \ominus_H B$ exists, then $B \ominus_H A$ exists and*

$$B \ominus_H A = (-1)\cdot(A \ominus_H B).$$

Remark A.3.21 *Note that*

$$A \ominus_H A = \{0\}.$$

Theorem A.3.22 *Let $A \ominus_H B$ exist. Then*

$$(A+B) \ominus_H B = A.$$

Proof A.3.23 *Let*

$$C = (A+B) \ominus_H B.$$

1. *Suppose that*

$$A+B=B+C.$$

Then

$$A=C.$$

2. *Suppose that*

$$B=(A+B)+(-1)\cdot C.$$

Then

$$B=B+(A+(-1)\cdot C),$$

whereupon

$$A+(-1)\cdot C=\{0\}.$$

Therefore, A and C are singleton sets and $A=C$. This completes the proof.

Theorem A.3.24 *Let $A\ominus_H B$ exist. Then*

$$((-1)\cdot B)\ominus_H((-1)\cdot A)$$

exists and

$$A\ominus_H B=((-1)\cdot B)\ominus_H((-1)\cdot A).$$

Proof A.3.25 *Let $C=A\ominus_H B$. Then*

$$A=B+C \quad \text{or} \quad B=A+(-1)\cdot C.$$

Hence,

$$(-1)\cdot A=(-1)\cdot B+(-1)\cdot C,$$

or

$$(-1)\cdot B=(-1)\cdot A+(-1)\cdot((-1)\cdot C).$$

Therefore

$$((-1)\cdot B)\ominus_H((-1)\cdot A),$$

exists and

$$\begin{aligned}((-1)\cdot B)\ominus_H((-1)\cdot A) &= C\\ &= A\ominus_H B.\end{aligned}$$

This completes the proof.

Theorem A.3.26 *Let $A \ominus_H B$ exist. Then*

$$(A+(-1)\cdot B)+B=C, \tag{A.1}$$

if and only if

$$A+(-1)\cdot B=C\ominus_H B. \tag{A.2}$$

Proof A.3.27 1. *Suppose* (A.1). *Then, by the definition for H-difference, we get* (A.2).

2. *Suppose* (A.2). *Then*

   (a) $(A+(-1)\cdot B)+B=C$ *or*

   (b)

$$\begin{aligned} B &= C+(-1)\cdot(A+(-1)\cdot B) \\ &= C+(-1)\cdot A+B, \end{aligned}$$

   whereupon

$$C+(-1)\cdot A=\{0\}.$$

   Therefore, C and A are singleton sets and $A=C$. Because $A\ominus_H B$ exists, we obtain that B is a singleton set and $A=B$. Thus,

$$A=B=C,$$

   and (A.1) *holds. This completes the proof.*

Theorem A.3.28 *Let $A\ominus_H B$ exist. Then*

$$A\ominus_H B=B\ominus_H A=C,$$

if and only if $C=\{0\}$ and $A=B$.

Proof A.3.29 1. *Let*

$$A\ominus_H B=B\ominus_H A=C.$$

We have the following cases.

(a) *$A = B + C$ and $B = A + C$. Hence,*

$$\begin{aligned} A &= B + C \\ &= A + C + C. \end{aligned}$$

Then $C = \{0\}$ and $A = B$.

(b) *$A = B + C$ and $A = B + (-1) \cdot C$. Then*

$$B + C = B + (-1) \cdot C,$$

and $C = (-1) \cdot C$. Therefore, C is a singleton set, and $C = \{0\}$. From here, we conclude that $A = B$.

(c) *$B = A + (-1) \cdot C$ and $B = A + C$. Then*

$$A + (-1) \cdot C = A + C,$$

and

$$(-1) \cdot C = C.$$

Hence, $C = \{0\}$ and $A = B$.

(d) *$B = A + (-1) \cdot C$ and $A = B + (-1) \cdot C$. Then*

$$\begin{aligned} B &= A + (-1) \cdot C \\ &= B + (-1) \cdot C + (-1) \cdot C. \end{aligned}$$

Hence, $C = \{0\}$ and $A = B$.

2. *If $C = \{0\}$ and $A = B$, then*

$$A \ominus_H B = B \ominus_H A.$$

This completes the proof.

Remark A.3.30 *In general,*

$$A + (-1) \cdot B = B + (-1) \cdot A,$$

does not imply $A = B$. *Really, let*

$$A = [\underline{a}, \overline{a}], \quad B = [\underline{b}, \overline{b}].$$

Then

$$\begin{aligned} A + (-1) \cdot B &= [\underline{a} - \overline{b}, \overline{a} - \underline{b}] \\ &= [\underline{b} - \overline{a}, \overline{b} - \underline{a}], \end{aligned}$$

whereupon

$$\underline{a} + \overline{a} = \underline{b} + \overline{b},$$

which does not require $A = B$.

Definition A.3.31 *Let* X *and* Y *be two non-empty subsets of a metric space* (M, d). *We define their Hausdorff distance* $D(X, Y)$ *as follows.*

$$D(X,Y) = \max \left\{ \sup_{x \in X} \inf_{y \in Y} d(x,y), \sup_{y \in Y} \inf_{x \in X} d(x,y) \right\}.$$

Example A.3.32 *Consider the metric space* $\mathbb{R}$ *of real numbers with the usual metric* d *induced by the absolute value,*

$$d(x,y) = |x - y|.$$

Take

$$X = [0,1], \quad Y = [-1,0).$$

Then

$$D(X,Y) = 1.$$

Definition A.3.33 *If* $u, v \in F(\mathbb{R})$ *and* $u \ominus_H v$ *exists, then its* α*-cuts are given by*

$$\begin{aligned} [u \ominus_H v]^\alpha &= \Big[\min\{\underline{u}^\alpha - \underline{v}^\alpha, \overline{u}^\alpha - \overline{v}^\alpha\}, \\ &\qquad \max\{\underline{u}^\alpha - \underline{v}^\alpha, \overline{u}^\alpha - \overline{v}^\alpha\} \Big]. \end{aligned}$$

Definition A.3.34 *Define the Hausdorff distance between fuzzy numbers as follows,* $D : F(\mathbb{R}) \times F(\mathbb{R}) \to [0,\infty)$,

$$D(u,v) = \sup_{\alpha\in[0,1]} \max\{|\underline{u}^{\alpha} - \underline{v}^{\alpha}|, |\overline{u}^{\alpha} - \overline{v}^{\alpha}|\}, \quad u,v \in F(\mathbb{R}),$$

where

$$[u]^{\alpha} = [\underline{u}^{\alpha}, \overline{u}^{\alpha}], \quad [v]^{\alpha} = [\underline{v}^{\alpha}, \overline{v}^{\alpha}].$$

The Hausdorff distance between fuzzy numbers satisfies the following properties.

1. $D(u+w, v+w) = D(u,v)$, $\forall u,v,w \in F(\mathbb{R})$,
2. $D(k\cdot u, k\cdot v) = |k|D(u,v)$, $\forall k \in \mathbb{R}$, $\forall u,v \in F(\mathbb{R})$,
3. $D(u+v, w+e) \le D(u,w) + D(v,e)$, $\forall u,v,w,e \in F(\mathbb{R})$,
4. $D(u \ominus_H v, e \ominus_H w) \le D(u,e) + D(v,w)$, $\forall u,v,w,e \in F(\mathbb{R})$.

Thus, $(F(\mathbb{R}), D)$ is a complete metric space. Denote

$$\|u\|_F = D(u, \widetilde{0}).$$

For $I \subseteq \mathbb{R}$, with $\mathscr{C}_f(I)$ denote the space of all continuous fuzzy functions on I.

Definition A.3.35 *For each* $x,y \in \mathscr{C}_f(I)$, *define*

$$D(x,y) = \sup_{t\in I} D(x(t), y(t)).$$

A.4 MULTIPLICATION OF COMPACT INTERVALS

Let $u = [u^-, u^+]$ and $v = [v^-, v^+]$ be two compact intervals.

Definition A.4.1 *Define multiplication* $u \star v$ *of the compact intervals* u *and* v *as follows.*

$$u \star v = \Big[\min\{u^-v^-, u^-v^+, u^+v^-, u^+v^+\}, \max\{u^-v^-, u^-v^+, u^+v^-, u^+v^+\}\Big].$$

Example A.4.2 *Let* $u = [-1, 2]$, $v = [1, 4]$. *Then*

$$\begin{aligned} u \star v &= \Big[\min\{-1, -4, 2, 8\}, \max\{-1, -4, 2, 8\}\Big] \\ &= [-4, 8]. \end{aligned}$$

Below, we will give other kinds of multiplications of compact intervals. Now, we introduce the following notations.

$$\begin{aligned} I^1_{u,v} &= \begin{vmatrix} u^- & u^+ \\ v^- & v^+ \end{vmatrix}, \\ I^2_{u,v} &= \begin{vmatrix} u^+ & u^- \\ v^- & v^+ \end{vmatrix}, \\ I^3_{u,v} &= \begin{vmatrix} u^- & u^- \\ v^- & v^+ \end{vmatrix}, \\ I^4_{u,v} &= \begin{vmatrix} u^+ & u^+ \\ v^- & v^+ \end{vmatrix}, \\ I^5_{u,v} &= \begin{vmatrix} u^- & u^+ \\ v^- & v^- \end{vmatrix}, \\ I^6_{u,v} &= \begin{vmatrix} u^- & u^+ \\ v^+ & v^+ \end{vmatrix}. \end{aligned}$$

Suppose that $[a^-, a^+] \subseteq [u^-, u^+]$, $[b^-, b^+] \subseteq [v^-, v^+]$.

Definition A.4.3 *Define type 1 multiplication of compact intervals* $[a^-, a^+]$ *and* $[b^-, b^+]$ *as follows.*

$$[a^-, a^+] \circ [b^-, b^+] = \{a * b : a \in [a^-, a^+], \quad b \in [b^-, b^+]\},$$

where if $I^1_{u,v} \leq 0$, *then*

$$a * b = \begin{cases} ab & \text{if} \quad ab \in [u^- v^+, u^+ v^-] \\ u^- v^+ & \text{if} \quad ab < u^- v^+ \\ u^+ v^- & \text{if} \quad ab > u^+ v^-, \end{cases}$$

if $I^1_{u,v} \geq 0$, *then*

$$a * b = \begin{cases} ab & if \quad ab \in [u^+v^-, u^-v^+] \\ u^+v^- & if \quad ab < u^+v^- \\ u^-v^+ & if \quad ab > u^-v^+. \end{cases}$$

Example A.4.4 *Let* $\mathbb{T} = \mathbb{Z}$,

$$\begin{aligned} [u^-, u^+]_{\mathbb{T}} &= [-4, 3]_{\mathbb{T}}, \\ [v^-, v^+]_{\mathbb{T}} &= [-5, 1]_{\mathbb{T}}, \\ [a^-, a^+]_{\mathbb{T}} &= [-2, 1]_{\mathbb{T}}, \\ [b^-, b^+]_{\mathbb{T}} &= [-1, 1]_{\mathbb{T}}. \end{aligned}$$

We have

$$\begin{aligned} [a^-, a^+]_{\mathbb{T}} &= \{-2, -1, 0, 1\}, \\ [b^-, b^+]_{\mathbb{T}} &= \{-1, 0, 1\}, \\ I^1_{u,v} &= \begin{vmatrix} u^- & u^+ \\ v^- & v^+ \end{vmatrix} \\ &= \begin{vmatrix} -4 & 3 \\ -5 & 1 \end{vmatrix} \\ &= -4 + 15 \\ &= 11. \end{aligned}$$

Then

$$[u^+v^-, u^-v^+] = [-15, -4],$$

$$\begin{aligned}
(-2)*(-1) &= u^-v^+ \\
&= -4, \\
(-2)*0 &= u^-v^+ \\
&= -4, \\
(-2)*1 &= u^-v^+ \\
&= -4, \\
(-1)*(-1) &= u^-v^+ \\
&= -4, \\
(-1)*0 &= u^-v^+ \\
&= -4, \\
(-1)*1 &= u^-v^+ \\
&= -4, \\
0*(-1) &= u^-v^+ \\
&= -4, \\
0*0 &= u^-v^+
\end{aligned}$$

$$\begin{aligned}
&= -4, \\
0 * 1 &= u^- v^+ \\
&= -4, \\
1 * (-1) &= u^- v^+ \\
&= -4, \\
1 * 0 &= u^- v^+ \\
&= -4, \\
1 * 1 &= u^- v^+ \\
&= -4.
\end{aligned}$$

Thus,

$$[-2,1]_{\mathbb{T}} \circ [-1,1]_{\mathbb{T}} = \{-4\}.$$

Example A.4.5 *Let* $\mathbb{T} = 2^{\mathbb{N}_0}$ *and*

$$\begin{aligned}
[u^-, u^+]_{\mathbb{T}} &= [1,8]_{\mathbb{T}} \\
&= \{1,2,4,8\}, \\
[v^-, v^+]_{\mathbb{T}} &= [1,4]_{\mathbb{T}} \\
&= \{1,2,4\},
\end{aligned}$$

$$\begin{aligned} [a^-, a^+]_{\mathbb{T}} &= [1,4]_{\mathbb{T}} \\ &= \{1,2,4\}, \\ [b^-, b^+]_{\mathbb{T}} &= [1,2]_{\mathbb{T}} \\ &= \{1,2\}. \end{aligned}$$

We have

$$\begin{aligned} [u^-v^+, u^+v^-] &= [4,8]_{\mathbb{T}} \\ &= \{4,8\}, \\ I^1_{u,v} &= \begin{vmatrix} u^- & u^+ \\ v^- & v^+ \end{vmatrix} \\ &= \begin{vmatrix} 1 & 8 \\ 1 & 4 \end{vmatrix} \\ &= 4-8 \\ &= -4, \\ 1*1 &= u^-v^+ \\ &= 4, \\ 1*2 &= u^-v^+ \\ &= 4, \end{aligned}$$

$$\begin{aligned} 2*1 &= u^- v^+ \\ &= 4, \\ 2*2 &= u^- v^+ \\ &= 4, \\ 4*1 &= u^- v^+ \\ &= 4, \\ 4*2 &= u^+ v^- \\ &= 8. \end{aligned}$$

Thus,

$$[1,4]_{\mathbb{T}} \circ [1,2]_{\mathbb{T}} = \{4,8\}.$$

Exercise A.4.6 *Let* $\mathbb{T} = \mathbb{Z}$ *and*

$$\begin{aligned} [u^-, u^+]_{\mathbb{T}} &= [-1,10]_{\mathbb{T}}, \\ [v^-, v^+]_{\mathbb{T}} &= [-3,7]_{\mathbb{T}}, \\ [a^-, a^+]_{\mathbb{T}} &= [1,5]_{\mathbb{T}}, \\ [b^-, b^+]_{\mathbb{T}} &= [-1,6]_{\mathbb{T}}. \end{aligned}$$

Find

$$[a^-, a^+]_{\mathbb{T}} \circ [b^-, b^+]_{\mathbb{T}}.$$

Definition A.4.7 *Define type 2 multiplication of the compact intervals* $[a^-,a^+]$ *and* $[b^-,b^+]$ *as follows.*

$$[a^-,a^+] \circledcirc [b^-,b^+] = \{a*b : a \in [a^-,a^+], \quad b \in [b^-,b^+]\},$$

where if $I^2_{u,v} \le 0$, *then*

$$a*b = \begin{cases} ab & \text{if} \quad ab \in [u^+v^+, u^-v^-] \\ u^+v^+ & \text{if} \quad ab < u^+v^+ \\ u^-v^- & \text{if} \quad ab > u^-v^-, \end{cases}$$

if $I^2_{u,v} \ge 0$, *then*

$$a*b = \begin{cases} ab & \text{if} \quad ab \in [u^-v^-, u^+v^+] \\ u^-v^- & \text{if} \quad ab < u^-v^- \\ u^+v^+ & \text{if} \quad ab > u^+v^+. \end{cases}$$

Example A.4.8 *Let* $\mathbb{T} = \mathbb{Z}$ *and*

$$\begin{aligned} [u^-,u^+]_{\mathbb{T}} &= [-1,3]_{\mathbb{T}} \\ &= \{-1,0,1,3\}, \\ [v^-,v^+]_{\mathbb{T}} &= [0,2]_{\mathbb{T}} \\ &= \{0,1,2\}, \\ [a^-,a^+]_{\mathbb{T}} &= [-1,0]_{\mathbb{T}} \\ &= \{-1,0\}, \end{aligned}$$

$$
\begin{aligned}
[b^-, b^+]_{\mathbb{T}} &= [0,1]_{\mathbb{T}} \\
&= \{0,1\}.
\end{aligned}
$$

We have

$$
\begin{aligned}
I^2_{u,v} &= \begin{vmatrix} u^+ & u^- \\ v^- & v^+ \end{vmatrix} \\
&= \begin{vmatrix} 3 & -1 \\ 0 & 2 \end{vmatrix} \\
&= 6, \\
[u^-v^-, u^+v^+]_{\mathbb{T}} &= [0,6]_{\mathbb{T}} \\
&= \{0,1,2,3,4,5,6\}, \\
(-1)*0 &= u^-v^- \\
&= 0, \\
(-1)*1 &= u^-v^- \\
&= 0, \\
0*0 &= u^-v^- \\
&= 0, \\
0*1 &= u^-v^- \\
&= 0.
\end{aligned}
$$

Thus,

$$
[-1,0]_{\mathbb{T}} \circledcirc [0,1]_{\mathbb{T}} = \{0\}.
$$

Example A.4.9 *Consider Example A.4.5. We have*

$$\begin{aligned}
I^2_{u,v} &= \begin{vmatrix} u^+ & u^- \\ v^- & v^+ \end{vmatrix} \\
&= u^+v^+ - u^-v^- \\
&= 32 - 1 \\
&= 31, \\
[u^-v^-, u^+v^+] &= [1,32]_{\mathbb{T}} \\
&= \{1,2,4,8,16,32\}, \\
1*1 &= u^-v^- \\
&= 1, \\
1*2 &= 2, \\
2*1 &= 2, \\
2*2 &= 4, \\
4*1 &= 4, \\
4*2 &= 8.
\end{aligned}$$

Thus,

$$[1,4]_{\mathbb{T}} \circledcirc [1,2]_{\mathbb{T}} = \{1,2,4,8\}.$$

Exercise A.4.10 *Let* $\mathbb{T} = \mathbb{Z}$ *and*

$$\begin{aligned} [u^-, u^+]_{\mathbb{T}} &= [-17, 7]_{\mathbb{T}}, \\ [v^-, v^+]_{\mathbb{T}} &= [-8, 7]_{\mathbb{T}}, \\ [a^-, a^+]_{\mathbb{T}} &= [-10, 5]_{\mathbb{T}}, \\ [b^-, b^+]_{\mathbb{T}} &= [-1, 6]_{\mathbb{T}}. \end{aligned}$$

Find

$$[a^-, a^+]_{\mathbb{T}} \circledcirc [b^-, b^+]_{\mathbb{T}}.$$

Definition A.4.11 *Define type 3 multiplication of the compact intervals* $[a^-, a^+]$ *and* $[b^-, b^+]$ *as follows.*

$$[a^-, a^+] \boxtimes [b^-, b^+] = \{a * b : a \in [a^-, a^+], \quad b \in [b^-, b^+]\},$$

where if $I^3_{u,v} \leq 0$, *then*

$$a * b = \begin{cases} ab & \text{if} \quad ab \in [u^- v^+, u^- v^-] \\ u^- v^+ & \text{if} \quad ab < u^- v^+ \\ u^- v^- & \text{if} \quad ab > u^- v^-, \end{cases}$$

if $I^3_{u,v} \geq 0$, *then*

$$a * b = \begin{cases} ab & \text{if} \quad ab \in [u^- v^-, u^- v^+] \\ u^- v^- & \text{if} \quad ab < u^- v^- \\ u^- v^+ & \text{if} \quad ab > u^- v^+. \end{cases}$$

Example A.4.12 *Consider Example A.4.5. We have*

$$
\begin{aligned}
I^3_{u,v} &= \begin{vmatrix} u^- & u^- \\ v^- & v^+ \end{vmatrix} \\
&= \begin{vmatrix} 1 & 1 \\ 1 & 4 \end{vmatrix} \\
&= 4-1 \\
&= 3, \\
[u^-v^-, u^-v^+]_{\mathbb{T}} &= [1,4]_{\mathbb{T}}, \\
1*1 &= u^-v^- \\
&= 1, \\
1*2 &= 2, \\
2*1 &= 2, \\
2*2 &= u^-v^+ \\
&= 4, \\
4*1 &= u^-v^+ \\
&= 4, \\
4*2 &= u^-v^+ \\
&= 4.
\end{aligned}
$$

Thus,

$$\begin{aligned}[1,4]_{\mathbb{T}} \boxtimes [1,2]_{\mathbb{T}} &= [1,4]_{\mathbb{T}} \\ &= \{1,2,4\}.\end{aligned}$$

Example A.4.13 *Consider Example A.4.8. We have*

$$\begin{aligned}
I^3_{u,v} &= \begin{vmatrix} u^- & u^- \\ v^- & v^+ \end{vmatrix} \\
&= \begin{vmatrix} -1 & -1 \\ 0 & 2 \end{vmatrix} \\
&= -2-0 \\
&= -2, \\
[u^-v^+, u^-v^-]_{\mathbb{T}} &= [-2,0]_{\mathbb{T}} \\
&= \{-2,-1,0\}, \\
(-1)*0 &= 0, \\
(-1)*1 &= -1, \\
0*0 &= u^-v^- \\
&= 0, \\
0*1 &= u^-v^- \\
&= 0.
\end{aligned}$$

Thus,

$$\begin{aligned}[-1,0]_{\mathbb{T}} \boxtimes [0,1]_{\mathbb{T}} &= [-1,0]_{\mathbb{T}} \\ &= \{-1,0\}.\end{aligned}$$

Exercise A.4.14 *Let* $\mathbb{T} = 3\mathbb{Z}$ *and*

$$\begin{aligned}[u^-,u^+]_{\mathbb{T}} &= [-15,3]_{\mathbb{T}}, \\ [v^-,v^+]_{\mathbb{T}} &= [0,30]_{\mathbb{T}}, \\ [a^-,a^+]_{\mathbb{T}} &= [0,3]_{\mathbb{T}}, \\ [b^-,b^+]_{\mathbb{T}} &= [-3,0]_{\mathbb{T}}.\end{aligned}$$

Find

$$[a^-,a^+]_{\mathbb{T}} \boxtimes [b^-,b^+]_{\mathbb{T}}.$$

Definition A.4.15 *Define type 4 multiplication of the compact intervals* $[a^-,a^+]$ *and* $[b^-,b^+]$ *as follows.*

$$[a^-,a^+] \boxdot [b^-,b^+] = \{a*b : a \in [a^-,a^+], \quad b \in [b^-,b^+]\},$$

where if $I^4_{u,v} \leq 0$, *then*

$$a*b = \begin{cases} ab & \text{if} \quad ab \in [u^+v^+, u^+v^-] \\ u^+v^+ & \text{if} \quad ab < u^+v^+ \\ u^+v^- & \text{if} \quad ab > u^+v^-, \end{cases}$$

if $I^4_{u,v} \geq 0$, *then*

$$a*b = \begin{cases} ab & \text{if} \quad ab \in [u^+v^-, u^+v^+] \\ u^+v^- & \text{if} \quad ab < u^+v^- \\ u^+v^+ & \text{if} \quad ab > u^+v^+. \end{cases}$$

Example A.4.16 *Consider Example A.4.5. We have*

$$\begin{aligned}
I^4_{u,v} &= \begin{vmatrix} u^+ & u^+ \\ v^- & v^+ \end{vmatrix} \\
&= \begin{vmatrix} 8 & 8 \\ 1 & 4 \end{vmatrix} \\
&= 32 - 8 \\
&= 24, \\
[u^+v^-, u^+v^+]_{\mathbb{T}} &= [8, 32]_{\mathbb{T}} \\
&= \{8, 16, 32\}, \\
1 * 1 &= u^+v^- \\
&= 8, \\
1 * 2 &= u^+v^- \\
&= 8, \\
2 * 1 &= u^+v^- \\
&= 8, \\
2 * 2 &= u^+v^- \\
&= 8,
\end{aligned}$$

$$
\begin{aligned}
4 * 1 &= u^+ v^- \\
&= 8, \\
4 * 2 &= u^+ v^- \\
&= 8.
\end{aligned}
$$

Thus,

$$[1,4]_{\mathbb{T}} \boxdot [1,2]_{\mathbb{T}} = \{8\}.$$

Example A.4.17 *Consider Example A.4.8. We have*

$$
\begin{aligned}
I^4_{u,v} &= \begin{vmatrix} u^+ & u^+ \\ v^- & v^+ \end{vmatrix} \\
&= \begin{vmatrix} 3 & 3 \\ 0 & 2 \end{vmatrix} \\
&= 6, \\
[u^+ v^-, u^+ v^+]_{\mathbb{T}} &= [0,6]_{\mathbb{T}} \\
&= \{0,1,2,3,4,5,6\}, \\
(-1) * 0 &= u^+ v^- \\
&= 0, \\
(-1) * 1 &= u^+ v^- \\
&= 0,
\end{aligned}
$$

$$
\begin{aligned}
0 * 1 &= u^+ v^- \\
&= 0, \\
1 * 1 &= 1.
\end{aligned}
$$

Thus,

$$
\begin{aligned}
[-1,0]_{\mathbb{T}} \boxdot [0,1]_{\mathbb{T}} &= [0,1]_{\mathbb{T}} \\
&= \{0,1\}.
\end{aligned}
$$

Exercise A.4.18 *Let* $\mathbb{T} = \mathbb{Z}$ *and*

$$
\begin{aligned}
[u^-, u^+]_{\mathbb{T}} &= [-10,10]_{\mathbb{T}}, \\
[v^-, v^+]_{\mathbb{T}} &= [0,7]_{\mathbb{T}}, \\
[a^-, a^+]_{\mathbb{T}} &= [1,3]_{\mathbb{T}}, \\
[b^-, b^+]_{\mathbb{T}} &= [0,4]_{\mathbb{T}}.
\end{aligned}
$$

Find

$$
[a^-, a^+]_{\mathbb{T}} \boxdot [b^-, b^+]_{\mathbb{T}}.
$$

Definition A.4.19 *Define type 5 multiplication of the compact intervals* $[a^-, a^+]$ *and* $[b^-, b^+]$ *as follows.*

$$
[a^-, a^+] \otimes [b^-, b^+] = \{a * b : a \in [a^-, a^+], \quad b \in [b^-, b^+]\},
$$

where if $I^5_{u,v} \leq 0$, *then*

$$
a * b = \begin{cases} ab & \text{if} \quad ab \in [u^- v^-, u^+ v^-] \\ u^- v^- & \text{if} \quad ab < u^- v^- \\ u^+ v^- & \text{if} \quad ab > u^+ v^-, \end{cases}
$$

if $I^5_{u,v} \geq 0$, *then*

$$a * b = \begin{cases} ab & \text{if} \quad ab \in [u^+v^-, u^-v^-] \\ u^+v^- & \text{if} \quad ab < u^+v^- \\ u^-v^- & \text{if} \quad ab > u^-v^-. \end{cases}$$

Example A.4.20 *Consider Example A.4.5. Then*

$$\begin{aligned} I^5_{u,v} &= \begin{vmatrix} u^- & u^+ \\ v^- & v^- \end{vmatrix} \\ &= \begin{vmatrix} 1 & 8 \\ 1 & 1 \end{vmatrix} \\ &= 1 - 8 \\ &= -7, \\ [u^-v^-, u^+v^-]_{\mathbb{T}} &= [1, 8]_{\mathbb{T}} \\ &= \{1, 2, 4, 8\}, \\ 1 * 1 &= 1, \\ 1 * 2 &= 2, \\ 2 * 1 &= 2, \\ 2 * 2 &= 4, \\ 4 * 1 &= 4, \\ 4 * 2 &= 8. \end{aligned}$$

Thus,

$$\begin{aligned}[1,4]_{\mathbb{T}} \otimes [1,2]_{\mathbb{T}} &= [1,8]_{\mathbb{T}} \\ &= \{1,2,4,8\}.\end{aligned}$$

Example A.4.21 *Consider Example A.4.8. We have*

$$\begin{aligned}I^5_{u,v} &= \begin{vmatrix} u^- & u^+ \\ v^- & v^- \end{vmatrix} \\ &= \begin{vmatrix} -1 & 3 \\ 0 & 0 \end{vmatrix} \\ &= 0, \\ [u^-v^-, u^+v^-]_{\mathbb{T}} &= [0,0]_{\mathbb{T}} \\ &= \{0\}.\end{aligned}$$

Thus,

$$\begin{aligned}[-1,0]_{\mathbb{T}} \otimes [0,1]_{\mathbb{T}} &= [0,0]_{\mathbb{T}} \\ &= \{0\}.\end{aligned}$$

Exercise A.4.22 *Let* $\mathbb{T} = 3^{\mathbb{N}_0}$ *and*

$$\begin{aligned}[u^-, u^+]_{\mathbb{T}} &= [1,27]_{\mathbb{T}}, \\ [v^-, v^+]_{\mathbb{T}} &= [3,27]_{\mathbb{T}}, \\ [a^-, a^+]_{\mathbb{T}} &= [1,9]_{\mathbb{T}}, \\ [b^-, b^+]_{\mathbb{T}} &= [3,9]_{\mathbb{T}}.\end{aligned}$$

Find

$$[a^-,a^+]_{\mathbb{T}} \otimes [b^-,b^+]_{\mathbb{T}}.$$

Definition A.4.23 *Type 6 multiplication of the compact intervals $[a^-,a^+]$ and $[b^-,b^+]$ is defined as follows.*

$$[a^-,a^+] \odot [b^-,b^+] = \{a*b : a \in [a^-,a^+], \quad b \in [b^-,b^+]\},$$

where if $I^6_{u,v} \le 0$, then

$$a*b = \begin{cases} ab & \text{if} \quad ab \in [u^-v^+, u^+v^+] \\ u^-v^+ & \text{if} \quad ab < u^-v^+ \\ u^+v^+ & \text{if} \quad ab > u^+v^+, \end{cases}$$

if $I^6_{u,v} \ge 0$, then

$$a*b = \begin{cases} ab & \text{if} \quad ab \in [u^+v^+, u^-v^+] \\ u^+v^+ & \text{if} \quad ab < u^+v^+ \\ u^-v^+ & \text{if} \quad ab > u^-v^+. \end{cases}$$

Example A.4.24 *Consider Example A.4.5. We have*

$$\begin{aligned} I^6_{u,v} &= \begin{vmatrix} u^- & u^+ \\ v^+ & v^+ \end{vmatrix} \\ &= \begin{vmatrix} 1 & 8 \\ 4 & 4 \end{vmatrix} \\ &= 4 - 32 \\ &= -28, \end{aligned}$$

$$
\begin{aligned}
[u^- v^+, u^+ v^+]_{\mathbb{T}} &= [4,32]_{\mathbb{T}} \\
&= \{4,8,16,32\}, \\
1*1 &= u^- v^+ \\
&= 4, \\
1*2 &= u^- v^+ \\
&= 4, \\
2*1 &= u^- v^+ \\
&= 4, \\
2*2 &= u^- v^+ \\
&= 4, \\
4*1 &= u^- v^+ \\
&= 4, \\
4*2 &= u^- v^+ \\
&= 8.
\end{aligned}
$$

Thus,

$$
\begin{aligned}
[1,4]_{\mathbb{T}} \odot [1,2]_{\mathbb{T}} &= \{4,8\} \\
&= [4,8]_{\mathbb{T}}.
\end{aligned}
$$

Example A.4.25 *Consider Example A.4.8. We have*

$$\begin{aligned} I^6_{u,v} &= \begin{vmatrix} u^- & u^+ \\ v^+ & v^+ \end{vmatrix} \\ &= \begin{vmatrix} -1 & 3 \\ 2 & 2 \end{vmatrix} \\ &= -2-6 \\ &= -8, \\ [u^-v^+, u^+v^+]_{\mathbb{T}} &= [-2,6]_{\mathbb{T}} \\ &= \{-2,-1,0,1,2,3,4,5,6\}, \\ (-1)*0 &= 0, \\ (-1)*1 &= -1, \\ 0*0 &= 0, \\ 0*1 &= 0. \end{aligned}$$

Thus,

$$\begin{aligned} [-1,0]_{\mathbb{T}} \odot [0,1]_{\mathbb{T}} &= \{-1,0\} \\ &= [-1,0]_{\mathbb{T}}. \end{aligned}$$

Exercise A.4.26 *Let* $\mathbb{T} = 2\mathbb{Z}$ *and*

$$\begin{aligned} [u^-,u^+]_{\mathbb{T}} &= [-20,20]_{\mathbb{T}}, \\ [v^-,v^+]_{\mathbb{T}} &= [-40,40]_{\mathbb{T}}, \\ [a^-,a^=]_{\mathbb{T}} &= [-2,4]_{\mathbb{T}}, \\ [b^-,b^+]_{\mathbb{T}} &= [-4,8]_{\mathbb{T}}. \end{aligned}$$

Find

$$[a^-,a^+]_{\mathbb{T}} \odot [b^-,b^+]_{\mathbb{T}}.$$

Bibliography

[1] Anthony, J. M., and H. Sherwood. Fuzzy groups redefined, J. Math. Anal. Appl., 69, 1979, 124–130.

[2] Aubin, J.-P., and H. Frankowska. Set-Valued Analysis. Basel: Birkhäuser, 1990.

[3] Bede, B., Rudas, I. J., and Bencsik, A. L. First order linear fuzzy differential equations under generalized differentiability, Information Sciences, 177, (2007), 1648–1662.

[4] Belohlavek, R., and V. Vychodil. Fuzzy Equational Logic. Berlin: Springer, 2005.

[5] Bobylev, V. N. Support function of a fuzzy set and its characteristic properties. Matematicheskie Zametki 37 (4), 1985, 507–513.

[6] Buckley, J. J., and E. Eslami. Fuzzy plane geometry I: Points and lines; II: Circles and polygons, Fuzzy Sets Syst., 8687, 1997, 179–187.

[7] Chang, C. L. Fuzzy Sets and Pattern Recognition. PhD diss., Dept. of Electrical Engineering, Univ. of California, Berkeley, 1967.

[8] Chang, C. L. Fuzzy topological spaces. J. Math. Anal. Appl., 24 (1), 1968, 182–190.

[9] Das, P. S. Fuzzy groups and level sub-groups. J. Math. Anal. Appl., 84, 1981, 264–269.

[10] Demirci, M. Fuzzy functions and their fundamental properties, Fuzzy Sets Syst., 106, 1999, 239–246.

[11] Diamond, P., and P. Kloeden. Metric Spaces of Fuzzy Sets: Theory and Applications. Singapore: World Scientific, 1994.

[12] Formato, F., Gerla, G., and L. Scarpati. Fuzzy sub-groups and similarities, Soft Computing, 3, 1999, 1–6.

[13] Steinlage, and R. H. Warren. Compactness in fuzzy topological spaces. J. Math. Anal. Appl., 62, 1978, 547–562.

[14] Georgiev, S. Functional dynamic equations on time scales, Springer, 2019.

[15] Georgiev, S., and Zennir, K. Boundary value problems on time scales, Vol. I, CRC, 2021.

[16] Georgiev, S. and Zennir, K. Boundary value problems on time scales, Vol. II, CRC, 2021.

[17] Georgiev, S. Dynamic geometry on time scales, CRC, 2021.

[18] Georgiev, S. Fuzzy dynamic equations, dynamic inclusions and optimal control problems on time scales, Springer, 2021.

[19] Gerla, G. Sharpness relation and decidable fuzzy sets. IEEE Trans. Automat. Contr., 27 (5), 1982, 11–13.

[20] Goguen, J. A. L-fuzzy sets. J. Math. Anal. Appl., 18, 1967, 145–174.

[21] Goguen, J. A. The fuzzy Tychonoff theorem. J. Math. Anal. Appl., 43, 1973, 734–742.

[22] Hilger, S. Ein Makettenakalkul mit Anwendung auf Zentrumsmannigfaltigkeiten, Ph.D. Thesis, Universität Würzburg, 1988.

[23] Hutton, B. Normality in fuzzy topological spaces. J. Math. Anal. Appl., 50, 1975, 74–79.

[24] Katsaras, A. K., and D. B. Liu. Fuzzy vector spaces and fuzzy topological vector spaces, J. Math. Anal. Appl., 58, 1977, 135–146.

[25] Klement, E. P. Fuzzy σ-algebras and fuzzy measurable functions, Fuzzy Sets Syst., 4, 1980, 83–93.

[26] Klement, E. P., M. L. Puri, and D. A. Ralescu. Limit theorems for fuzzy random variables, Proc. Royal Society of London A 407, 1986, 171–182.

[27] Kosko, B. Fuzzy Thinking: The New Science of Fuzzy Logic, Hyperion, 1993.

[28] Lakshmikantham, V., and Vatsala, A. S. Basic Theory of Fuzzy Difference Equations, J. Fuzzy Diff. Equ., 8 (2002), 957–968.

[29] Lowen, R. Fuzzy topological spaces and fuzzy compactness. J. Math. Anal. Appl., 56, 1976, 621–633.

[30] Lowen, R. fuzzy topological spaces induced by Lipschitz functions, Mathematische Nachrichten 120, 1985, 249–265.

[31] Lupulescu, V. Hukuhara differentiability of interval-valued functions and interval differential equations on time scales, Information Sciences, 248, 2013, 50–67.

[32] Mamdani, E.H., and Assilian, S. An experiment in linguistic synthesis with a fuzzy logic controller, Inter. J. Man-Machine Stud., 7 (1), 1975, 1–13.

[33] Najariyan, M., Mazandarani, M., and Balas, V. E. Solving first order linear fuzzy differential equations system, In: Balas V., Jain L., Balas M. (eds) Soft Computing Applications. SOFA 2016. Advances in Intelligent Systems and Computing, vol 634. Springer, Cham, 2018.

[34] Puri, M. L., and D. A. Ralescu. Differentials of fuzzy functions, J. Math. Anal. Appl., 91, 1983, 552–558.

[35] Rosenfeld, A. Fuzzy groups. J. Math. Anal. Appl., 35, 1971, 512–517.

[36] Ruspini, E. H. Recent developments in fuzzy clustering. In Fuzzy Sets and Possibility Theory: Recent Developments, ed. R. R. Yager, 133–147. New York: Pergamon Press, 1982.

[37] Sanchez, E. Solution of fuzzy equations with extended operations. Fuzzy Sets Syst., 12, 1984, 237–248.

[38] Sugeno, M. Industrial applications of fuzzy control, Elsevier Science Pub. Co.1985.

[39] Skolem, T., Bemerkungen zum Komprehensionsaxiom [Remarks on the axiom of comprehension]. Z. Math. Logik. Grundlagen. Math., 3, 1957, 1–17.

[40] Vasavi, C., Kumar, G. S., and Murty, M. S., Generalized differentiability and integrability for fuzzy set-valued functions on time scales, Soft Computing, 20, 2016, 1093–1104.

[41] Vasavi, C., Kumar, G. S., and Murty, M. S., Fuzzy dynamic equations on time scales under second type Hukuhara delta derivative, Inter. Journal of Chem. Sci., 14, 2016, 49–66.

[42] Vasai, C., Kumar, G. S., and Murty, M. S., Fuzzy Hukuhara delta differential and applications to fuzzy dynamic equations on time scales, J. Uncertain Sys., 10, 2016, 163–180.

[43] Wang, L.-X., Adaptive fuzzy systems and control: design and stability analysis, Prentice Hall, 1994.

[44] Zadeh, L. A. Fuzzy Sets, Information & Control, (1965) 338–353.

[45] Zadeh, L. A., Fuzzy sets and systems. In Proc. Symp. on System Theory, 29–37, 1965. Brooklyn, NY: Polytechnic Press.

[46] Zadeh, L. A. Similarity relations and fuzzy orderings. Inf. Sci., 3 (2), 1971, 177–200.

[47] Zadeh, L. A., Probability measures and fuzzy events, J. Math. Anal. Appl., 23 (2), 1968, 421–427.

Index